AF576081

Mesoporous Materials for Drug Delivery and Theranostics

Mesoporous Materials for Drug Delivery and Theranostics

Editors

Valentina Cauda
Giancarlo Canavese

MDPI • Basel • Beijing • Wuhan • Barcelona • Belgrade • Manchester • Tokyo • Cluj • Tianjin

Editors
Valentina Cauda
Politecnico di Torino
Italy

Giancarlo Canavese
Politecnico di Torino
Italy

Editorial Office
MDPI
St. Alban-Anlage 66
4052 Basel, Switzerland

This is a reprint of articles from the Special Issue published online in the open access journal *Pharmaceutics* (ISSN 1999-4923) (available at: https://www.mdpi.com/journal/pharmaceutics/special_issues/mesoporous_material).

For citation purposes, cite each article independently as indicated on the article page online and as indicated below:

LastName, A.A.; LastName, B.B.; LastName, C.C. Article Title. *Journal Name* **Year**, *Article Number*, Page Range.

ISBN 978-3-03943-939-3 (Hbk)
ISBN 978-3-03943-940-9 (PDF)

Cover image courtesy of Valentina Cauda.

Contents

About the Editors

Valentina Cauda is an Associate Professor at the Department of Applied Science and Technology (DISAT), Politecnico di Torino, head of the TrojaNanoHorse Lab (in brief TNHLab), and co-founder of the Interdepartmental laboratory PolitoBIOMed Lab.

Thanks to her ERC Starting Grant project (TrojaNanoHorse, GA 678151), which started in March 2016, she now leads a multidisciplinary research group of 18 people, including chemists, biologists, physicists, engineers, and nanotechnologists. Her main research topic concerns theranostic nanomaterials: the research team develops metal oxide nanomaterials from wet synthesis, chemical functionalization, and physical–chemical characterization up to their coating by lipidic bilayer from both artificial and natural origins, aimed for drug delivery, tumor cell targeting, bio-imaging. Metal oxide nanomaterials, like zinc oxide, mesoporous silica, titania, and metal (gold, silver) nanostructures, as well as liposomes and cell-derived extracellular vesicles, are investigated.

Valentina Cauda graduated in Chemical Engineering in 2004 at Politecnico di Torino and then received her Ph.D. in Material Science and Technology in 2008. After a short period at the University of Madrid, she worked as a Post Doc at the University of Munich, Germany, on nanoparticles for drug delivery and tumor cell targeting. From 2010 to 2015, she worked as a Senior Post Doc at the Istituto Italiano di Tecnologia in Torino, and then she moved, as Associate Professor, to Politecnico di Torino. In 2010, she received the prize for young researchers at the Chemistry Department of the University of Munich, in 2013, she received the Giovedì Scienza award, in 2015, the Zonta Prize for Chemistry, and, in 2017, the USERN Prize for Biological Sciences. She has 113 scientific publications and a Hirsch Factor of 35 (updated on November 2020). She holds 4 international patents concerning the use of metal oxide nanoparticles in nanomedicine. Prof. Cauda is the principal investigator of several industrial, national, and international projects raising more than € 5 million funds in all. The most relevant are the recently granted ERC Proof-of-Concept *XtraUS* N. 957563, the FET Open RIA *MIMIC-KEY*, the Marie-Slodowska Curie Action *MINT* N. 842964 (where she acts as supervisor of an incoming Post-Doc from abroad), and the ERC Starting Grant *Trojananohorse*.

More details available at https://areeweb.polito.it/TNHlab/.

Giancarlo Canavese is an Assistant Professor at the Department of Applied Science and Technology (DISAT), Politecnico di Torino and a member of the Interdepartmental laboratory PolitoBIOMed Lab and the Materials and Processes for Micro & Nano Technologies (Chilab Laboratory) at DISAT.

His main research topics concern sonodynamic technique enhanced by engineered nanostructures for theranostic applications, interactions of oxide nanoparticles under ultrasound activation with biomaterial structure, such as cells and extracellular vesicles, and acoustophoretic labs-on-chips for drug delivery, tumor biomarkers detection, and bio-imaging.

Giancarlo Canavese received his ME degree in Mechanical Engineering in 2004 and his Ph.D. degree in Biomedical Engineering in 2008 from Politecnico di Torino, Italy. From 2010 to 2015, he worked as a Senior Post-Doc at the Istituto Italiano di Tecnologia, Italy, where he studied nanostructured piezoelectric materials for biosensors and energy devices. From 2013–2014, he spent six months as a visiting scientist at the Houston Methodist Research Institute, Texas US, where he worked on a microfluidic lab-on-chip device to study drug delivery in microgravity conditions on the International Space Station.

He has 90 scientific publications and a Hirsch Factor of 27 (updated on November 2020). He holds 8 international patents about micro and nanotechnologies for biomedical applications.

Editorial

Mesoporous Materials for Drug Delivery and Theranostics

Valentina Cauda * and Giancarlo Canavese *

Department of Applied Science & Technology, Politecnico di Torino, Corso Duca degli Abruzzi 24, 10129 Turin, Italy
* Correspondence: valentina.cauda@polito.it (V.C.); giancarlo.canavese@polito.it (G.C.)

Received: 6 November 2020; Accepted: 17 November 2020; Published: 18 November 2020

Mesoporous materials, especially those made of silica or silicon, are capturing great interest in the field of nanomedicine. Thanks to their exceptional porous structure and surface area, their homogeneous and tunable pore size, the ease of surface functionalization, the capability to establish host–guest interactions with other molecules protecting them from the external environment, and finally their biocompatibility, mesoporous materials enable a broad series of biologically relevant interventions and interactions with cells. The deep investigation on mesoporous nanoparticles has contributed to develop smart and stimuli-responsive nanotools for controlled drug- or gene-delivery and with imaging capabilities.

This Special Issue of Pharmaceutics is therefore dedicated to the most recent advances in the use of mesoporous nanostructures in the field of theranostis, specifically for cancer therapy, and in advanced tissue engineering.

To have a proper overview in the specific field of mesoporous silica materials for drug delivery, targeting, and theranostics applications, the review from Prof. Maria Vallet-Regì and coworkers is very relevant [1]. Here, the authors analyze various strategies about the encapsulation and delivery of macromolecules of biological interests (i.e., enzymes, therapeutic or antibacterial proteins, growth factors, therapeutic or antibacterial peptides, glycan-based macromolecules, and nucleic acids for gene modulation and silencing, like miRNA, siRNA, and DNA). The relevant figures of merit for the correct design of mesoporous silica nanoparticles (MSNs), such as pore size and shape, nanoparticle dimensions, surface chemistry, and colloidal stability, are considered. Furthermore, the location of the biomacromolecules (either at the external surface or in the mesopores) and the bond types to the silica surface (relying on physical adsorption or chemical grafting with various chemical and sometimes triggerable bonds) are reviewed.

In the review of Dr. Sugata Barui et al. [2], a special focus is given to both the surface decoration of MSNs by ligands for active targeting of cancer cells, exploiting overexpressed receptors, and to the use of stimuli-responsive gatekeepers for the controlled release of drugs to the disease site, avoiding leakage to healthy tissues. In addition, the multimodal modifications of the MSNs for simultaneous active targeting and stimuli-responsive behavior are reviewed with the most recent applications in vitro and in vivo. Applications of MSNs in cancer diagnosis and finally in theranostics are also proposed.

Experimental results on the most recent advances in nucleic acid delivery and efficient cancer cell targeting are provided in the work of Prof. Thomas Bein and coworkers [3]. Here, multiple core-shell functionalized MSNs were used to exploit a positively charged pore surface for miRNA loading and protection of this fragile cargo in the nanoparticle interior. On the outer surface, a block copolymer was electrostatically bound with the purpose of pore gatekeeping and endosomal release triggering. Finally, a targeting peptide GE11 for the epidermal growth factor receptor (EGFR) was used to enhance the MSN uptake in bladder cancer cells in vitro and provide the miRNA delivery for gene knockdown.

Despite silica-based mesoporous materials, this Special Issue also provides recent and prominent insights in the use of mesoporous bioactive glasses (MBGs) for tissue engineering applications.

Specifically, for bone tissue engineering applications, the work of Profs. Ying Wan and Jiliang Wu [4] shows that MBGs in the form of nanoparticles can be used as carriers for insulin-like growth factor-1 and can be efficiently incorporated into an injectable hydrogel matrix. Interestingly, the sol to gel transition of the hydrogel is engineered such that it can happen at physiological temperature and pH, resulting in gel with high porosity and interconnected pores, which is thus suitable for sustaining the delivery of the cargo over weeks and maintaining its biological functions, as proven by in vitro tests with osteoblasts.

Also representative is the work of Boffito et al. [5], where MBG incorporated into a hydrogel matrix was successfully designed to simultaneously release both copper ions, with pro-angiogenic and anti-bacterial effects, and an anti-inflammatory drug. The work aims to propose a multifunctional platform for tissue healing, in particular bone healing, where, on the one hand, the thermosensitive hydrogel concentrates and maintains the MBG carriers at the diseased site and, on the other hand, the in situ and prolonged co-release of ions and drugs is achieved.

Top-down fabricated mesoporous-based nanomaterials are also presented in this Special Issue. In the work of Prof. Alessandro Grattoni and coworkers [6], a silicon nanofluidic membrane incorporating gate electrodes is presented. This nanochannel-based device is able to modulate the transport of charged molecules, here in particular methotrexate, used to treat rheumatoid arthritis, and quantum dots, which are useful for bio-imaging applications. The electrostatic gated nanochannel permeability was proven to deliver the cargos at low applied voltages in vitro, modulating the transport release performances.

In the work of Profs. Natalia Malara, Francesco Gentile, and coworkers [7], mesoporous silicon structures coated with gold nanoparticles are fabricated, allowing a high level of control over the surface at the nanoscale. These excellent characteristics enable the device to be used for theranostics purposes (i.e., first supporting the growth and proliferation of cancer cells over the nanomaterial surface, then allowing an antitumor drug uptake and subsequent delivery against cancer cells thanks to the mesopores, and finally providing imaging of the biological system by surface enhanced Raman spectroscopy (SERS) due to the presence of the gold nanoparticles).

The review of Dr. Tania Limongi, Francesca Susa, and coworkers [8] presents an innovative perspective highlighting the synthetic approaches, characteristics, and roles of 3D-printed mesoporous materials as customizable and personalized scaffolds for drug delivery studies and tissue engineering applications. Such 3D-printed mesoporous materials can provide not only a solid support for cell growth in a 3D fashion, but also and most importantly can be ad-hoc designed and customized for personalized therapy to patients, or for realistic in vitro drug delivery studies, or finally for assisting cell growth for a tissue-specific model.

As a concluding remark, with this Special Issue, we hope we have contributed to highlighting the role of mesoporous materials in cancer cell theranostics and tissue engineering, providing insights from their synthesis, surface functionalization, and characterization up to their smart and stimuli-responsive behavior with customizable properties for advanced and personalized biomedical applications.

Funding: This research received no external funding.

Conflicts of Interest: The authors declare no conflict of interest.

References

1. Castillo, R.R.; Lozano, D.; Vallet-Regí, M. Mesoporous Silica Nanoparticles as Carriers for Therapeutic Biomolecules. *Pharmaceutics* **2020**, *12*, 432. [CrossRef] [PubMed]
2. Barui, S.; Cauda, V. Multimodal Decorations of Mesoporous Silica Nanoparticles for Improved Cancer Therapy. *Pharmaceutics* **2020**, *12*, 527. [CrossRef] [PubMed]
3. Haddick, L.; Zhang, W.; Reinhard, S.; Möller, K.; Engelke, H.; Wagner, E.; Bein, T. Particle-Size-Dependent Delivery of Antitumoral miRNA Using Targeted Mesoporous Silica Nanoparticles. *Pharmaceutics* **2020**, *12*, 505. [CrossRef] [PubMed]

4. Min, Q.; Yu, X.; Liu, J.; Zhang, Y.; Wan, Y.; Wu, J. Controlled Delivery of Insulin-like Growth Factor-1 from Bioactive Glass-Incorporated Alginate-Poloxamer/Silk Fibroin Hydrogels. *Pharmaceutics* **2020**, *12*, 574. [CrossRef] [PubMed]
5. Boffito, M.; Pontremoli, C.; Fiorilli, S.; Laurano, R.; Ciardelli, G.; Vitale-Brovarone, C. Injectable Thermosensitive Formulation Based on Polyurethane Hydrogel/Mesoporous Glasses for Sustained Co-Delivery of Functional Ions and Drugs. *Pharmaceutics* **2019**, *11*, 501. [CrossRef] [PubMed]
6. Di Trani, N.; Silvestri, A.; Wang, Y.; Demarchi, D.; Liu, X.; Grattoni, A. Silicon Nanofluidic Membrane for Electrostatic Control of Drugs and Analytes Elution. *Pharmaceutics* **2020**, *12*, 679. [CrossRef]
7. Coluccio, M.L.; Onesto, V.; Marinaro, G.; Dell'Apa, M.; De Vitis, S.; Imbrogno, A.; Tirinato, L.; Perozziello, G.; Di Fabrizio, E.; Candeloro, P.; et al. Cell Theranostics on Mesoporous Silicon Substrates. *Pharmaceutics* **2020**, *12*, 481. [CrossRef]
8. Limongi, T.; Susa, F.; Allione, M.; di Fabrizio, E. Drug Delivery Applications of Three-Dimensional Printed (3DP) Mesoporous Scaffolds. *Pharmaceutics* **2020**, *12*, 851. [CrossRef]

Review

Mesoporous Silica Nanoparticles as Carriers for Therapeutic Biomolecules

Rafael R. Castillo [1,2,3,†], **Daniel Lozano** [1,2,3,†] **and María Vallet-Regí** [1,2,3,*]

1 Departamento de Química en Ciencias Farmacéuticas, Facultad de Farmacia, Universidad Complutense de Madrid, Plaza Ramón y Cajal s/n, 28040 Madrid, Spain; rafcas01@ucm.es (R.R.C.); danlozan@ucm.es (D.L.)
2 Centro de Investigación Biomédica en Red—CIBER, 28029 Madrid, Spain
3 Instituto de Investigación Sanitaria Hospital 12 de Octubre—imas12, 28041 Madrid, Spain
* Correspondence: vallet@ucm.es
† These authors contributed equally to this work.

Received: 13 April 2020; Accepted: 1 May 2020; Published: 7 May 2020

Abstract: The enormous versatility of mesoporous silica nanoparticles permits the creation of a large number of nanotherapeutic systems for the treatment of cancer and many other pathologies. In addition to the controlled release of small drugs, these materials allow a broad number of molecules of a very different nature and sizes. In this review, we focus on biogenic species with therapeutic abilities (proteins, peptides, nucleic acids, and glycans), as well as how nanotechnology, in particular silica-based materials, can help in establishing new and more efficient routes for their administration. Indeed, since the applicability of those combinations of mesoporous silica with bio(macro)molecules goes beyond cancer treatment, we address a classification based on the type of therapeutic action. Likewise, as illustrative content, we highlight the most typical issues and problems found in the preparation of those hybrid nanotherapeutic materials.

Keywords: mesoporous silica; therapeutic biomolecules; proteins; peptides; nucleic acids; glycans

1. Introduction

Since the very first example reported on nanometric mesoporous silica as a drug delivery material [1], many examples were subsequently reported including a broad number of chemical species. One of them includes biomacromolecules, which play a capital role in living beings, as they are responsible for biorecognition [2], signal transduction [3], and replication routes; hence, they are responsible for the adequate development of tissues and organs. Such is the importance of these species that even the immune system and hormone signaling are based on specific affinity interactions between biomacromolecules (BMs). From a therapeutic point of view, hijacking such ligands and receptors may be useful to regulate unbalanced systems, as well as develop new-generation therapeutic nanodevices in oncology [4,5] cancer immunotherapy [6–8], and gene therapy [9–11], among many others.

Unfortunately, most bioactive macromolecules are highly labile in vivo, as a self-regulation mechanism to avoid massive damages. Therefore, to use them, it is necessary to implement chemical modifications or to design vehicles capable of ensuring an adequate preservation and, hence, a long-lasting effect. Among known carriers, viruses have the best performance, although at the expense of having a great associated risk in handling and containment. Additionally, they are only suitable for nucleic acids (NAs), being useless for protein and peptide delivery. To overcome these limitations, nanoparticles emerged as promising vectors for nucleotides, as well as peptides and proteins. This is a consequence of two complementary aspects. On one hand, their size permits establishing intimate interactions with cell's membranes and the receptors therein. On the other hand, they exhibit the possibility to establish non-conventional interactions between particles, cargoes, and target cells.

Amongst all known materials, mesoporous silica nanoparticles (MSNs) arose as promising drug delivery platforms because of their outstanding biocompatibility, their degradability, and their great chemical and biological robustness. Moreover, the unique porous structure of MSNs permits establishing host–guest interactions of high interest for drug delivery purposes, as they allow creating protective environments for labile molecules. In addition to those, the current silica-based nanotechnology also permits creating particles with variable diameters [12,13], pore sizes [14], and structures [15], which permit fine-tuning the final application of the nanosystems, especially those intended to deliver big cargoes such as those reviewed herein. Moreover, MSNs and related hybrid particles with SiO_2 coatings also permit easily tuning the resulting outer layers of nanosystems to enhance biochemical stability and, hence, to reduce side effects and potential toxicities [16–18]. Particle diameter is one of the most critical parameters for achieving a successful therapy. Typically, the accepted window of diameters for cancer treatment comprises particles in a range between 50 and 300 nm, in which the enhanced permeation and retention effect operates. However, depending on the final purpose of the nanosystem, such values may be narrowed. For instance, in cases where a superior trafficking is desired, smaller particles would behave better, while, in nanosystems intended as biomolecule reservoirs, the diameter must unavoidably be increased. Regarding nanoparticle-based drug delivery, we recently reviewed how mesoporous silica-based nanosystems are suitable platforms to combine two or more chemical species, outranging the pharmacological profile of free species [19,20]. However, despite the possible therapeutic improvement, their efficiency and long-term stability could be compromised if the BMs are not properly protected or fully exposed to white blood cells and immune systems. The scarce protection provided by most solid nanoparticles highlights once again the importance of MSNs [21] as platforms for the development of non-viral vectors and protein carriers, whose particular porous morphology can provide a protective environment for those labile molecules, although the typical porosity (2–3 nm) of MSNs may be tuned in order to host the biggest molecules [22] (Figure 1).

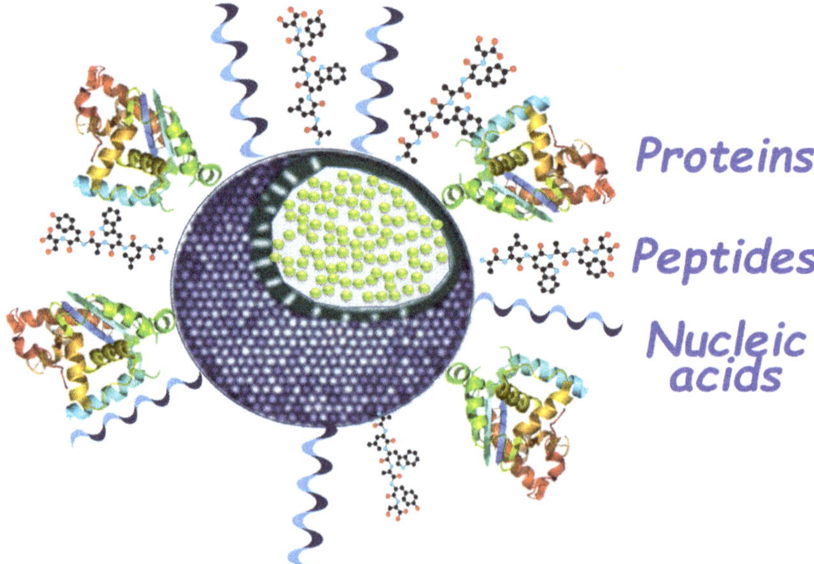

Figure 1. Main groups of therapeutic biomolecules that are possible to deliver using mesoporous nanosilica technology.

As introduced, the use of MSNs in biomedicine has a huge potential impact; in addition to acting as carriers, they also permit creating fancy structures with most functional nanomaterials. However, despite this versatility, the permeation of these materials into clinical trials is still limited [23,24]; this is not caused by poor biocompatibility, but rather by the impossibility of establishing reliable comparisons between different systems, as accounted by Florence [25]. In fact, as can be observed from the growing number of in vivo experiments using MSNs carried out by many groups worldwide, it is logical to assume that they have adequate performance in living systems; thus, MSNs will hopefully soon reach clinical practice.

2. Strategies to Deliver Biomacromolecules with Silica Nanoparticles

Loading effectiveness, cellular internalization, targeting, and cargo delivery are critical issues when developing a nanosystem with maximal therapeutic potential. As we previously mentioned, MSNs are ideal nanocarriers related to the load and delivery of biomolecules due to their unique properties, including shape, size, and surface chemistry [26]. Pore, channel, and cavity sizes can be modified in MSNs to increase the loading of therapeutic molecules [27]. The MSN surface can be functionalized with polyethyleneimine and poly-L-lysine or modified with targeting peptides or antibodies to improve cargo loading, cellular uptake, and endosomal scape rates [26]. The chemical and physical properties of MSNs are critical when designing a nanocarrier with therapeutic efficacy in biomolecule loading and delivery. In the last few years, surface nanoscale topography gained special attention due to the possibility of controlling the interactions between molecules and cells, in the process of molecular loading and cell internalization [28,29].

The chemical functionalization of silica is easily achieved through condensation processes employing functionalized alkoxysilanes. For the functionalization of surfaces, this silanization must be carried out before template removal, while the modification of mesopores could be achieved either via condensation during the template-driven synthesis of MSNs or after surface modification and template removal. In this latter case, it is important to account that pore constriction may occur if the process is not properly controlled. For the preparation of hollow MSNs (HMSNs), the most employed method is the solid template etching of an internal core, typically from solid silica, onto which a mesoporous layer is created. Pore and surface modification can be achieved in a parallel manner as done with conventional MSNs, while, for the functionalization of the internal space, this must be done prior to the formation of the mesoporous layer. Onto this slightly modified MSN, it would be possible to create additional modifications by linking chemical species through conventional chemical reactions. With this strategy, it was possible to create a huge number of functional nanosystems with pore nanogates [30], sensitive bonds [31], recognition elements [5,19], and charged surfaces able to undergo electrostatic interactions.

One common strategy to combine BMs and NPs is surface functionalization; it can be achieved either via a chemical bond or via an electrostatic interaction between charged counterparts. On MSNs, surface grafting is usually accomplished via direct condensation of the remaining Si–OH groups with functional silane reagents; these are typically modified trialkoxysilanes with aliphatic chains bearing an additional functional group. For the bonding of peptides and proteins, the most common functional groups are amino and carboxylate for amide coupling and maleimides for thiol-mediated binding [32]. Additionally, to complement the direct coupling approach, there are also many different bifunctional linkers available which are able to accomplish this task [33]. With regard to this coupling strategy, it is also important to remark the importance of controlling the bonding, which, if produced at the active region of proteins, may lead to inactivation. This is of particular importance when preparing antibody-targeted nanosystems [34,35] and sensors. On this topic, Landry and coworkers studied how the chemical linkage affected the specificity of cluster of differentiation 4 (CD4)-bonded proteins onto MSN against its target, the gp120 glycoprotein [36]. The authors proved that a conventional, unspecific, direct thiol–maleimide linkage behaved worse than a specific linkage placed far away from the active site.

The other approach for surface functionalization is electrostatic deposition. In this case, both nanocarriers and BMs need to establish strong electrostatic interactions through differently charged functional groups. This is of particular importance for NAs, whose permanent negative charge permits strong interactions with positively charged surfaces and polymers such as polyethyleneimine or chitosan among many others. Insights into this strategy are available in References [20,37]. In the most advanced models, a rational deposition of alternating cationic and anionic layers—for instance, polyethyleneimine (PEI)–small interfering RNA (siRNA)–PEI–, also permitted developing multilayered nanosystems in which targeting elements could be added in the outermost layer. This design permits placing NAs in middle layers, obtaining additional protection against nucleases. Moreover, this strategy also enables pH-driven cleavage, which permits disassembling the system in endosomal environments, as a consequence of the proton sponge effect associated with polycationic substances [38]. The main drawback of the surface loading strategy is the absence of protection for BMs, which could be deactivated if exposed to blood (opsonization, enzymatic degradation, macrophage-mediated clearance, etc.) [39] or if not properly handled during processing (non-sterile material, accidental contamination, or physicochemical decomposition).

From a protective point of view, the use of pore-expanded MSNs [40] is the most convenient strategy, but only if the cargo is adequately retained until its final destination. To this end, pore sizes should be tuned to allow cargo hosting and, eventually, pore gates may be required [30]. The typical strategies to prepare enlarged-pore MSNs consist of either using large surfactants such as Pluronic or Brij [14] or employing swelling agents able to increase the diameter of the template cetrimonium micelles during the synthesis [41]. Additionally, the use of non-surfactant species, like tannic acid, was also reported as a pore-forming agent [42]. Regarding NAs, it is also important to remark that raw pores of MSNs need to be chemically modified; otherwise, the negative charges of NAs and silanols would undergo a repulsive interaction that would hinder pore loading. This topic was previously visited by us and other groups in previous contributions [20,37]. In the case of proteins and peptides, this effect is not so relevant, as their isoelectric points are always closer to neutrality. In fact, most peptides behave as small molecules and can be loaded satisfactorily in most raw-pore MSNs.

In summary, the loading strategy must be carefully accounted for depending on the carried biomacromolecule. In this way, short peptides can be easily loaded within pores or grafted onto surfaces, while the delivery of bigger and highly charged molecules may suffer from pore rejection if the mesopores are not properly conditioned [43]. Regarding NAs, their outstanding chemical stability permits creating either fancy layered structures or pore-loaded systems [44–46]. In addition to typical porous particles, the use of hollow mesoporous silica nanoparticles (HMSNs) also receives interest because of their additional enormous internal space. However, to use them as carriers, their mesopores and surfaces must comply with all requirements outlined for in-pore loading, i.e., sufficient diameter, favorable electrostatic environment, and adequate order to permit effective diffusion processes.

In addition to cargo-related modifications, these kinds of nanodevices must also have additional modifications to increase colloidal stability and immune stealth to ensure an adequate trafficking profile (Figure 2). In the case of pore-loaded nanosystems, this can be easily achieved with a common polyethylene glycol (PEG)ylation strategy. On coated nanosystems, it may not be necessary if the particles' coatings are naturally occurring biomolecules, such as those reviewed herein. In this case, although the integrity of supported molecules is not fully ensured, surface deposition is proven to be a synthetic advantage, as it greatly reduces the number of components and synthetic steps.

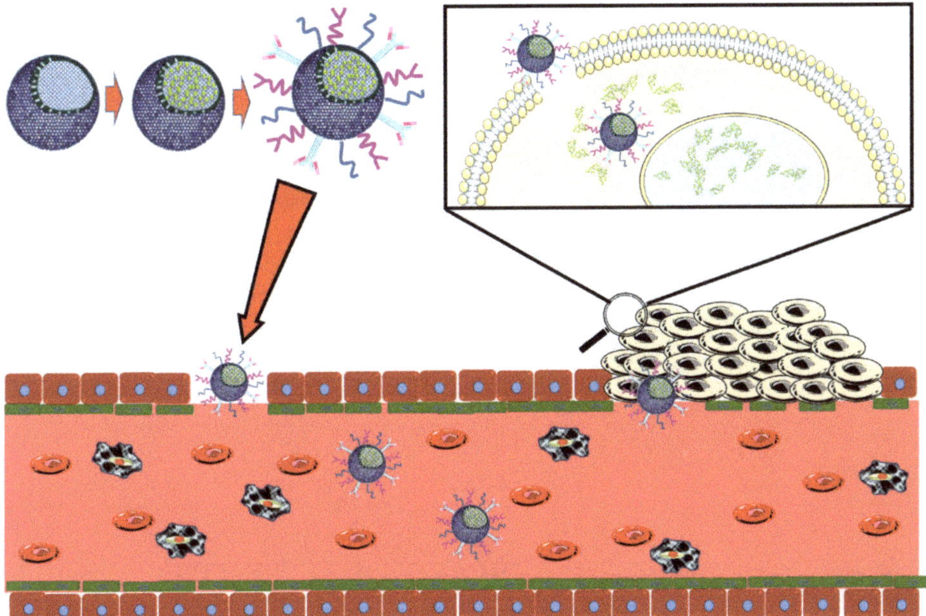

Figure 2. Blood compatibility, colloidal stability, and cell recognition are necessary on therapeutic nanosystems; otherwise, they would not reach their final destination.

3. Delivery of Proteins with Therapeutic Effect

As introduced above, the effect and potency of therapeutic proteins rely on their physiological effect and behavior. For example, cytochrome c (Cyt c) triggers a caspase-mediated apoptosis, while immunoglobulins activate the immune system and induce cell destruction. Additionally, certain enzymes and growth factors may be useful to treat certain genetic diseases based on protein malfunction. Therefore, as there is no common therapeutic effect [47], the different approaches are discussed including pro-apoptotic, immunostimulating, enzymes, growth factors, and antibacterial proteins, according to the classification shown in Table 1 and Figure 3. For the interested reader, previously published outstanding revisions dealt with insights into protein loading and delivery with MSNs [48–51].

Table 1. Examples of therapeutic proteins delivered by silica-based nanocarriers.

Protein	Carrier Type	Protein Location	Loading Strategy	Cell Line(s)	In Vivo	Reference
Anticancer proteins [52,53]						
Cytochrome C	MSNs	Mesopores	Pore filling	HeLa	None	[54,55]
	MSNs	Surface	Adsorption	None	None	[56]
	MSNs	Surface	Grafting	HeLa	None	[57]
	MSNs	Mesopores	Pore filling	SKOV3	None	[58]
Concanavalin A	MSNs	Surface	Grafting	MC3T3-E1, HOS	None	[59]

Table 1. *Cont.*

Protein	Carrier Type	Protein Location	Loading Strategy	Cell Line(s)	In Vivo	Reference
Immunostimulating proteins and vaccines						
IgG	HMSNs	Particle cavity	Cavity loading	HeLa	None	[60]
OVA	HMSNs	Mesopores	Pore filling	NIH3T3	Mice	[61,62]
OVA	DMOHS	Mesopores	Pore filling	None	Mice	[63]
CpG@OVA	MSNs	Mesopores	Pore filling	RAW264.7	Mice	[64]
CpG@OVA GM-CSF	MSNs@ MSRs	Mesopores Mesopores	Pore filling	None	Mice	[65]
Cyt c, IgG, Anti-pAkt	RSNs	Interparticles	In-pocket packing	None	None	[66]
IL-2	HMSNs	Particle cavity	Cavity loading	L929	Mice	[67]
ORF2	HMSNs	Surface	Adsorption	PK15	Mice	[68]
SWAP	MSNs	Surface	Adsorption	None	Mice	[69]
HSP700	MSNs	Surface	Adsorption	None	Mice	[70]
EspA	MSNs	Surface	Adsorption	None	Mice	[71]
rPb27	MSNs	Surface	Adsorption	HEK-293	Mice	[72]
Enzymes						
CA	MSNs	Mesopores	In-pore grafting	HeLa	None	[73]
CA or HPR	MSNs	Mesopores	In-pore grafting	None	None	[74]
β-Galactosidase	MSNs	Mesopores	Adsorption	N2a	None	[75]
SOD	MSNs	Surface	Grafting	HeLa	None	[76]
SOD or GPx	MSNs	Surface	Grafting	HeLa	None	[77]
Proteasomes	MSNs	Surface	Grafting	HEK-293, HeLa	None	[78]
Growth factors						
bFGF	MSNs	Mesopores	Microemulsion	HUVEC	None	[79]
BMP-2	MSNs	Surface	Adsorption	bMSCs	Mice	[80]
	MSN@SPION	Mesopores	Pore filling	bMSCs	None	[81]
Antibacterial proteins						
Lysozyme	MSNs	Surface	Grafting	*Escherichia coli*	Mice	[82]
	MSNs	Pores	Adsorption	*E. coli*	None	[83]
	HMSNs	Surface	Adsorption	*E. coli*	Mice	[84]
	HMSNs	Particle cavity	Cavity loading	*E. coli*	None	[85]
Concanavalin A	MSNs	Surface	Grafting	*E. coli*	None	[86]

Abbreviations: bFGF: basic fibroblast growth factor; BMP-2: bone morphogenetic protein 2; bMSCs: murine bone mesenchymal stem cells; CA: carbonic anhydrase; CpG@OVA: ovalbumin-loaded cytosine–phosphate–guanine (CpG) oligodeoxynucleotide; DMOHS: dendritic mesoporous organosilica hollow spheres; EspA: an immunogenic protein from enterohaemorrhagic *Escherichia Coli*; GM-CSF: murine granulocyte-macrophage colony-stimulating factor; GPx: glutathione peroxidase; HMSNs: hollow mesoporous silica nanoparticles; HRP: horseradish peroxidase; IgG: immunoglobulin G; IL-2: interleukin-2; ORF2: open reading frame from porcine circovirus type 2; OVA: chicken ovalbumin; RSNs: rough (non-porous and core–shell) silica nanoparticles; SOD: superoxide dismutase; SPIONs: superparamagnetic iron oxide nanoparticles.

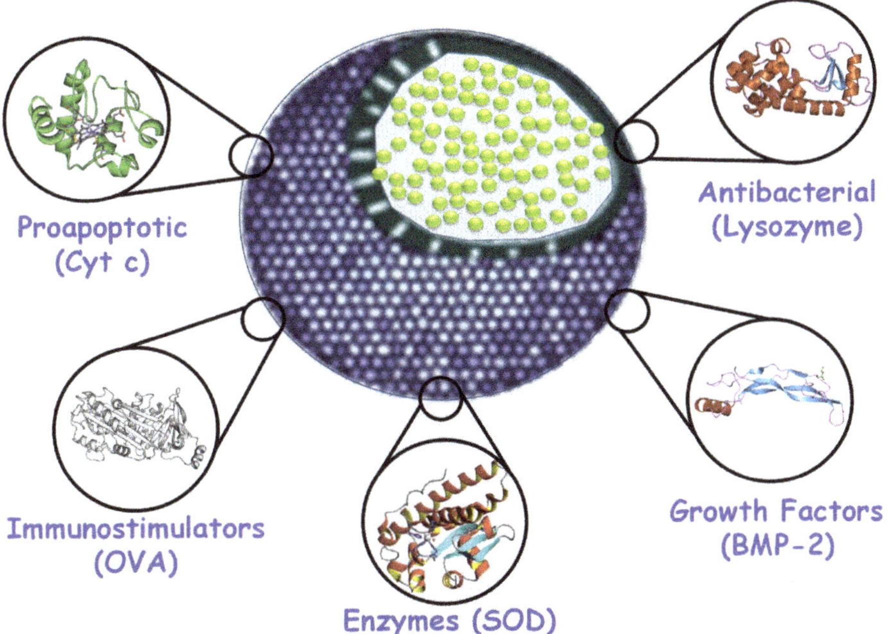

Figure 3. Roles and examples of typical therapeutic proteins delivered with silica-based nanosystems.

3.1. Anticancer Proteins

Historically, Lin's group was the first to describe the potential of MSNs to carry and deliver proteins. To achieve this, they focused on Cyt c, a small protein with a pro-apoptotic effect. In their contribution, they pioneered pore-expanded MSNs to allow protein hosting, highlighting the strategy to follow for the intracellular delivery of membrane-impermeable proteins [54]. Herein, they demonstrated that unmodified MSNs with 5.4-nm mesopores were able to host the globular 3.3-nm-width protein to produce effective intracellular delivery, although with no control. To improve this, Griebenow and coworkers evolved a system including a chemical bond able to retain Cyt c inside the mesopores. Their stimulus-responsive system was based on a redox-sensitive disulfide bond that linked Cyt c to thiol-modified mesopores [55], enabling an intracellular, glutathione-mediated release. Shang et al. also employed Cyt c as a model protein for MSN-based delivery, although, in their system, surface adsorption was preferred [56]. In this contribution, the authors did not evaluate the carrier efficiency or the stability of the protein but studied the loading capacity—protein activity—in relation to nanoparticle size. Their results showed that flatter surfaces—larger diameters—permit adsorbing more proteins and, hence, provide higher activities. Another example employing Cyt c was reported by Davis and coworkers, who studied the best-performing linker to connect the protein onto the particles' surface [57]. In their research, they systematically tested several custom-made linkers against the most critical parameters on optimal delivery for surface-grafted proteins: suitable charging capacity (>40 mV on ξ-potential), cationic character at acidic pH, ability to undergo endosomal escape, and capacity to permanently retain Cyt c immobilized on the surface. They found that MSNs modified with 1 mol.% primary amine were the only material able to satisfactorily accomplish all these tasks, providing a fantastic know-how for subsequent investigations into surface grafting. More recently, Choi et al. explored another possibility to deliver Cyt c, employing eroded MSNs with rougher surfaces and enlarged pores [58]. Herein, these matured MSNs permitted loading the Cyt c instead of obtaining the surface deposition that occurred onto particles bearing conventional (2–3 nm) pores. As a result of the increased in-pore loading, the particles showed an enhanced release compared with both the free Cyt c and the nanosystem with higher surface deposition.

Apart from Cyt c, there were reported numerous anticancer proteins [52,53], albeit few examples in combination with MSNs. A relevant example on the topic was reported with concanavalin A, a lectin with anticancer and antibacterial properties [59,86]. Regarding the anticancer example, the lectin also proved to be an efficient targeting element against human osteosarcoma (HOS) and murine prosteoblast (MC3T3-E1) cells. In our system, MSNs were firstly loaded with the typical chemotherapeutic doxorubicin (DOX) and then coated with a polymeric layer linked to the silica surface through pH-sensitive linkers. To conclude, the concanavalin A (ConA) was grafted onto the surface through amide bonds. The resulting system was able to efficiently deliver the drug intracellularly, but only when the pH dropped enough to cleave the bis-acetal linker. The evaluation of the system demonstrated an enhanced killing effect when both species (ConA and DOX) were co-delivered. This suggested a potent adjuvant effect of the protein, which was able to drop the cell viability from 50% with DOX alone to almost 100%.

3.2. Immunostimulating Proteins

Immunostimulation is one of the most popular strategies for improving cancer treatment, as it promises to recruit the patient's immune system to fight tumors. However, any indiscriminate activation may lead to disastrous results. Therefore, there is interest in developing immunostimulant nanosystems able to induce local immune responses thanks to the enhanced permeability and retention (EPR) effect. MSN-based immunotherapy also has an additional advantage, as these particles were proven to be efficient coadjutants [87–89]. Despite these promising features, the delivery of immune-regulating proteins with silica-based carriers is still in its infancy.

Regarding already reported examples, proteins and MSNs can be combined via either surface deposition—grafted or not—or loading into enlarged pores, although those are not the only options. One such exception was the contribution by Lim et al. who delivered immunoglobulin (IgG) to HeLa cells employing a very special silica particle [60]. These were called unconventional perforated HMSNs. Those particles were able to efficiently deliver large membrane-impermeable cargoes, although without any study on immune response. However, their contribution established the basis for local immunostimulation, opening the way to new therapeutic strategies. In more recent examples, several research groups studied immune inductions in mice using nanoparticles. Along this line, Wang, Ito, Tsuji, and coworkers reported a series of articles which firstly studied the immunostimulating behavior of HMSNs [87–89] in murine models, and then studied how the surface modification of such HMSNs with the T cell-dependent antigen, chicken ovalbumin (OVA), stimulated the overall response [61]. The authors paid special attention to which markers were upregulated when treating the animals with these devices [62]. They found a four-action pathway: an anticancer effect through the use of HMSNs themselves, an effector memory on the $CD4^+$ and $CD8^+$ T-cell population, an overexpression on T helper 1 (Th1) and Th2 cytokines, and an enhanced secretion of immunoglobulin antibodies [62]. Employing another silica nanostructure, a multi-shelled dendritic mesoporous organosilica hollow sphere (DMOHS), Yang et al. also obtained similar stimulation patterns when they employed OVA-loaded particles [63]. Herein, a parallel upregulation on $CD4^+$, $CD8^+$, and Th1 immunoproteins was also reported, characterized by the secretion of interleukin-12 (IL-12), interferon-γ (IFN-γ), and tumor necrosis factor-α (TNF-α). More recently, Cha et al. evolved this system by adding an additional immunostimulating entity. In their model, large-pore MSNs (20–30 nm) were sequentially loaded with OVA plus an additional "danger signal", an agonist for Toll-like receptor 9 (TLR9) [64]. As a result, an increased immunostimulation was obtained, which again suggests the importance of combined therapies for the development of more effective cancer treatments and vaccines. More recently, these authors implemented their immunostimulant nanosystem by designing a combination of these MSNs embedded in large chemokine-loaded mesoporous nanorods suitable for injection [65]. With this system, the authors overcame the limitations of the intravenous dosage of their previous nanosystem, as, with this approach, they could effectively recruit and mature the dendritic cells in order to achieve a more efficient cancer vaccination. Their final formulation demonstrated a significant tumor progression reduction together with great survival rates.

In addition to the typical strategies to link proteins onto nanoparticles, Niu et al. reported another possible approach for protein delivery. They designed different porous systems with variable cavities by controlling a core–shell assembly between solid silica nanoparticles. In their designs, core particles had a preset size, while encircling particles had smaller but variable diameters. The resulting disposition created different interparticle voids able to host different sized proteins [66]. The resulting nanostructures showed variable roughness (14, 21, and 38 nm, respectively) that permitted satisfactorily loading a wide variety of proteins, such as Cyt c, monoclonal rabbit antibodies (IgG), and even antibody fragments (horseradish peroxidase (HRP)-linked anti-rabbit IgG antibody), preserving in all cases their activities, as demonstrated by surface plasmon resonance. From a therapeutic point of view, the most interesting model was the 38-nm hydrophobically modified rough (non-porous and core-shell) silica nanoparticles (RSNs), which were able to deliver the anti-pAkt antibody into MCF-7 breast cancer cells. In this case, the therapeutic effect occurred because the antibody created a significant reduction of cell proliferation, together with a downregulated expression of the anti-apoptotic B-cell lymphoma 2 (Bcl-2) protein.

In addition to immunogenic proteins, another interesting therapeutic possibility enabled by nanotechnology is non-viral-based vaccination [90,91]. Therefore, this could be achieved if antigens are delivered to immune cells, thereby avoiding a general immune response. One of the first examples of vaccination employing silica nanoparticles was reported by Guo et al., who employed HMSNs to deliver the open reading frame of the porcine circovirus type 2 (ORF2) protein using a murine model [68]. In this case, the nanovaccine was prepared via direct adsorption of the protein onto raw nanoparticles. In mice, the MSN-based delivery of ORF2 induced overexpression of the typical markers: IFN-γ, Th1, CD4, and CD8, suggesting an immune activation. This strategy was also successfully employed to create other specific MSN-based non-viral vaccines for mice: (1) against *Schistosoma mansoni* employing homogenates from the parasite [69]; (2) against porcine enzootic pneumonia by using a recombinant heat-shock protein 70 (HSP70) antigen fragment ($HSP70_{212–600}$) from *Mycoplasma hyopneumoniae* [70]; (3) against enterohaemorrhagic *Escherichia coli* by using a recombinant fragment of filamentous immunogenic protein from enterohaemorrhagic *E. coli* (EspA) protein [71]; (4) against the pathogenic fungus *Paracoccidioides brasiliensis* employing the antigenic protein rPb27 [72]. In light of these results, it is noteworthy to highlight that these nanoparticle-based vaccinations permit immunizing against different types of pathogens, such as parasites, fungi, or bacteria, with a better profile than the free antigen or other known pharmacological formulations in which the delivery toward macrophages is not so efficient.

3.3. Enzymes

The deregulation of normal enzymatic levels has a fundamental role in many diseases. For example, toxic or metabolic syndromes are a direct consequence of this protein malfunction. In cancer, the deregulation of homeostasis is also a consequence of an abnormal expression of proteins and enzymes. However, the use of enzymes in therapy clearly goes beyond cancer treatment based on the delivery of cell-damaging proteins. Unfortunately, like all proteins, enzymes usually do not nicely tolerate systemic administration unless properly modified [92]. Such is the case of the p53 anticancer protein capsule designed by Zhao et al. [93] or the collagenase capsule reported by Villegas et al. [94], which performed best only when they had a degradable protective shell. With regard to the enzymes delivered by MSNs, it is important to remark that most examples are located at the surface; otherwise, the diffusion of both substrates and products may be difficult. One of the most studied examples is carbonic anhydrase (CA), which, despite not having a significant therapeutic effect, was widely employed as a model for in-pore loading approaches [73,74]. The advantage of CA as a model is a direct consequence of the facile determination of its remaining activity, as both substrate (CO_2)

and final product (HCO_3^-) are small enough to diffuse through the pores with freedom and can be easily determined.

Regarding therapeutic models, the mitigation of toxic syndromes caused by genetic disorders is one of the most promising research fields. One interesting example was reported by Xu et al. [75], who employed MSNs to deliver β-galactosidase (β-Gal) to treat Morquio B syndrome. This disease appears when the enzyme is not able to properly cleave the glycoside bond on oligosaccharides, thus producing the accumulation of substrates. In this case, contrary to other models in which the enzyme was located on the surface, the authors decided to load it into the mesopores to obtain additional protection and long-term stability. In this case, the significant size (119 kDa) of β-Gal made it mandatory to prepare ultra-large-pore MSNs. For this, the authors created core–cone structured MSNs with dahlia-like mesopores that were able to host, preserve, and deliver catalytically active β-Gal to N2a cells. In this case, the large size of mesopores, together with the small sizes of both enzymatic substrates and products, permitted supporting the enzyme within the pores; however, in other cases, in which the substrates have diffusion barriers because of the size or nature, this approach may not be so convenient.

With regard to supported proteins, Mou and coworkers reported the use of two antioxidant enzymes, superoxide dismutase (SOD) and glutathione peroxidase (GPx), to intracellularly deplete reactive oxygen species (ROS). In this case, the different nature of possible substrates demanded an adequate exposure of proteins on the nanosystem; otherwise, they would not be able to prevent oxidative stress. In their first example, the authors developed a mesoporous silica-based multifunctional nanocarrier for SOD [76]. This device employed an Ni^{+2}-chelated nitrilotriacetic acid (Ni-NTA)-modified silane, which was able to connect through a histidine moiety to a TAT-containing peptide sequence bound to the enzyme. At this point, the authors denatured the enzyme with urea in order to have only intracellular refolding and to avoid extracellular ROS depletion. The resulting system was able to internalize into HeLa cells and, therein, refold the enzyme into its active form and reduce the oxidative stress induced by paraquat, a well-known superoxide anion generator. More recently, the authors implemented the antioxidant performance of this model by employing two differently loaded nanodevices: one with SOD and the other with GPx [77]. With this combination, the authors found a complementary synergic effect in which the co-delivery of both antioxidant proteins improved the effect of single treatments.

Another interesting example, reported by Han et al., focused on the delivery of human proteasomes to delay tau aggregation associated with Alzheimer's disease [78]. In this model, the authors employed the same building strategy reported by Mou's group: Ni-NTA able to bind histidine moieties from active human 26S proteasomes isolated from a HEK293-derived cell line. Again, the system proved to deliver proteasomes without a significant proteotoxic effect associated with its enzymatic activity. At this point, the authors claimed that their system was not intended to permeate the blood-brain barrier, although it could be implemented onto permeable systems, opening the way to novel strategies to prevent and treat Alzheimer's.

3.4. Growth Factors

Growth factors are biomacromolecules responsible for inducing cellular growth, proliferation, and differentiation; they also play important roles in wound healing, bone formation, and vascularization. The most known growth factor is somatotropin, the human growth factor, which is widely employed to promote impaired growth in healthy individuals. The main drawback of this therapy is the need to accomplish a continuous—intramuscular or subcutaneous—dosage in order to achieve a sustained effect. Because of this, it is a treatment with very low adherence and, therefore, it needs improvement. Among the new strategies developed, the most promising are those that permit reaching long-lasting formulations with noninvasive administration routes [95,96]. With regard to the remaining growth factors, mostly cytokines and steroid hormones, it is typical to have compounds with very little aqueous solubility. As a result, they also require form delivery agents to perform their action properly.

This need could be a niche for non-biogenic carriers, especially those of high loading capacity such as silica-based platforms.

Regarding the examples available in the literature, ten years ago, Zhang et al. already reported the potential of MSNs as carriers for growth factors [79]. In this contribution, a water-in-oil microemulsion strategy was employed to create ad hoc MSNs for the 18-kDa basic fibroblast growth factor (bFGF) protein. The resulting system was able to release the protein for over 20 days, and it could be satisfactorily taken up by HUVEC cells. Moreover, the cytotoxicity assay showed low toxicity even when administered at high concentrations (50 μg/mL). Despite this promising result, the remaining examples with therapeutic growth factors are limited to two contributions by Gan et al., who employed two complementary approaches for the delivery of the bone morphogenetic protein 2 (BMP-2). In their first contribution, they prepared chitosan-coated MSNs to deposit proteins onto the particles' surfaces [80], while, in their second work, SBA-15 microparticles with iron oxide nanocaps were employed to load the protein within the 6.2-nm mesopores [81]. In the first example, the authors included dexamethasone, a corticoid, into the mesopores to increment bone growth due to co-delivery of two osteogenic species. Unfortunately, although a significant osteogenesis was achieved, it was completely uncontrolled, and significant ectopic bone formation occurred. To prevent this, the authors changed the model to deliver only the growth factor. To do so, they employed large-pore silica microparticles to carry the BMP-2 and Fe_3O_4 nanocaps to prevent a premature release. The resulting pH-driven carrier was able to deliver the protein while exhibiting excellent biocompatibility, but only in vitro, where the huge size of the chosen silica particles did not affect the guest survival.

3.5. Antibacterial Proteins

In addition to the previous examples, it is also possible to transport proteins bearing antimicrobial effects. On the delivery of antibacterial proteins with MSNs, most contributions were made with lysozyme, a 14.4-kDa enzyme known to damage bacterial cells by hydrolyzing the major component of Gram-positive bacterial walls, i.e., the β-1,4 bonds between *N*-acetylmuramic acid and *N*-acetyl-D-glucosamine. The first example, reported by Lin et al., employed negatively charged MCM-41 MSNs to bind lysozyme units through electrostatic interactions to provide a stable corona [82]. This model permitted having an antibacterial effect even on Gram-negative bacteria such as *E. coli*. Such an effect was a consequence of the high concentration of lysozyme released in bacterial surroundings. Although forthcoming models improved the antibacterial performance of this device, in this work, the authors set two important advances: the possibility of treating bacterial infections with nanotechnology, and the finding that even Gram-negative bacteria could be highly damaged by these glycoside hydrolase enzymes.

In light of these previous results, Yu's group attempted to improve the antibacterial capacity by using higher-loading carriers. In one contribution, they differently evaluated dendritic pore MSNs, finding that a bigger pore led to a faster release and, thus, higher antibacterial effects [83]. In the second contribution, they developed silica nanopollens, which are engineered non-porous hollow nanoparticles coated with nanosized silica spikes to provide a rough surface [84]. Similarly, to the former model, they evaluated different roughnesses, finding that a rougher surface led to a better loading and the associated antibacterial effect. A comparison between models seemed to indicate that, in the case of large-pore MSNs, the antibacterial effect of lysozyme was a consequence of a faster release, while, in the case of nanopollens, it was a combination of a more sustained release together with an unknown nanoparticle-induced effect. Unfortunately, these models were only tested on planktonic bacteria. To address a more complex situation in which the bacteria form a biofilm, Ye and coworkers studied hollow MSNs as high-capacity nanocarriers (up to 350 mg per gram in the case of their enlarged pore HMSNs) [85]. In this work, the authors reported a threshold for free lysozyme activity (400 μg/mL), over which there was no therapeutic improvement. However, the use of MSN-based delivery raised that maximal effect, as they showed a more sustained release pattern able to induce a long-lasting effect.

Apart from lysozyme, antibacterial effects with other proteins were also reported such as ConA (in this case, in combination with levofloxacin) [86]. In this example, carboxylate-modified MCM-41 MSNs were functionalized with ConA upon drug loading. Herein, the combined action of both species permitted achieving complete biofilm destruction even at minimal concentrations (10 μg MSNs per mL), improving the precedent models due to combination therapy. In addition to these reported models, there are many more proteins that lead to successful combinations [97]. Such is the case of lactoferrin, known for having antibacterial properties [98], which was extensively employed in the preparation of glioblastoma-targeted nanodevices [99,100].

4. Delivery of Peptides with Therapeutic Effect

Many of known bioactive peptides are simplified amino-acid sequences that replicate the active sites of proteins and enzymes. The success of these peptides is a consequence of two complementary aspects: (1) the ability to retain features from their parent proteins, and (2) an outstanding chemical profile that arises from their facile synthesis, significantly low cost, and chemical robustness. Those aspects permitted discovering a large number of peptides of different nature and specificity. For example, regarding targeting, those sequences are selected for their affinity toward eukaryotic [101,102] or prokaryotic [103,104] cell receptors. However, applications of peptides go beyond targeting [105], as they are able to accomplish different tasks in regulating metabolic cycles and signaling. In this section, we focus on the potential of MSN-carried peptides for the development of new therapeutic devices. As shown in Table 2 and Figure 4, their applications range from cancer treatment [106–108] to antibacterial effects [107,109], cell regulation processes [110], and immunostimulation.

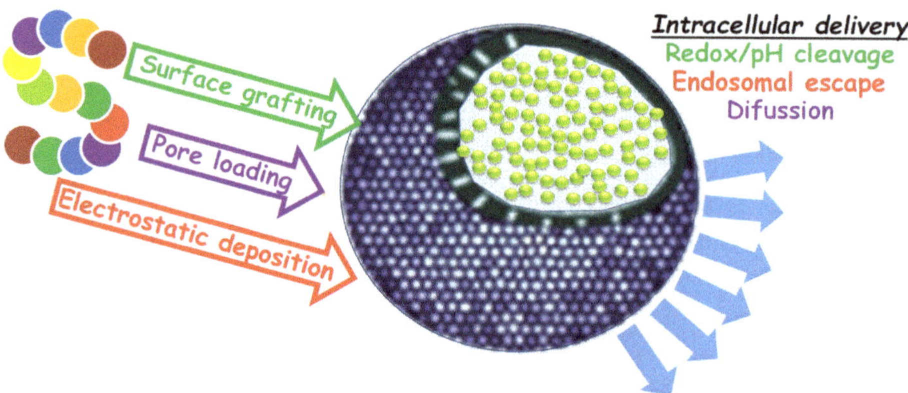

Figure 4. Strategies for delivery of therapeutic peptides employing silica-based carriers.

Table 2. Examples of different types of therapeutic peptides delivered by mesoporous silica-based nanocarriers.

Peptide	Carrier	Pore Cargo	Release Mechanism	Cell Line(s)	In Vivo	Reference
Anticancer Peptides						
K_8-Citraconate $K_8(RGD)_2$	MSNs	Doxorubicin	Electrostatic	COS7, U87 MG	None	[111]
TPP-K-$(KLAKLAK)_2$-	MSNs	Topotecan	Electrostatic/Redox cleavage	KB	None	[112]
C-$GRK_2R_2QR_3P_2$Q-RGDS C-GKGG-$D(KLAKLAK)_2$	MSNs	Doxorubicin	Redox cleavage	HeLa, COS7	None	[113]
$(RGDWWW)_2KC$	MSNs	Doxorubicin	Redox cleavage	COS7, U87 MG	Mice	[114]
$(KLAKLAK)_2$	MSNs	Doxorubicin	Enzymatic degradation Redox cleavage	HeLa	None	[115]
ε-poly-L-lysine C9h	MSNs	C9h	Enzymatic degradation Pore release	HeLa	None	[116]
RDG-Hylin a1	MSNs	RDG-Hylin a1	Pore release	HeLa Hep2	Mice	[117]
Pepstatin A	HMSNs	Pepstatin A	Release from cavity	MCF-7	None	[118]
NuBCP9	MSNs	NuBCP9	Pore release	HeLa, HEK293	Zebrafish	[119]
NuBCP9	PAMAM@MSNs	NuBCP9	Pore release upon PAMAM detachment	HepG2,H292, MCF-7,HeLa	Mice	[120]
Immunostimulating Peptides						
HGP100 TRP2	HMSNs	HGP100 TRP2	Lipid layer disassembly Release from cavity	BMDCs	None	[121]
Antibacterial Peptides						
LL37	SiO_2/MSNs	None/LL37	Surface adsorption vs. pore release	*E. coli*	None	[122]
NZX	MSNs	NZX	Pore release	*Mycobacterium tuberculosis*	Mice	[123]
Melittin	MSNs@ MagMSNs	Melittin Ofloxacin	Pore release upon hyperthermia triggering	*Pseudomonas aeruginosa*	Mice	[124]

Table 2. *Cont.*

Peptide	Carrier	Pore Cargo	Release Mechanism	Cell Line(s)	In Vivo	Reference
Hormones and Growth Factors						
Insulin	MSNs	cAMP	Glucose-mediated displacement	RIN-5F	None	[125]
	MSNs	Insulin	Glucose-sensitive polymer shell	None	None	[126]
	MSNs	Insulin	Pore release	Caco-2	None	[127]
Osteostatin	MSNs	Osteostatin	Pore release	MC3T3-E1	Rabbit	[128–130]
OGP	MSNs MS-HANs	OGP	Pore release	None	None	[131]
BMP-2	MSNs	Surface	None	BMSCs	Rat	[132]

For an illustrative revision of the coupling protocols employed for linking peptides onto nanocarriers, please check Reference [133]. Abbreviations: BMDCs: murine bone-marrow-derived dendritic cells; BMP-2: bone morphogenetic protein 2; BMSCs: rat bone mesenchymal stem cells; cAMP: cyclic adenosine monophosphate; MagMSNs: SPION@MSNs core–shell nanoparticles; MS-HANs: mesoporous silica–hydroxyapatite nanoparticles; OGP: osteogenic growth peptide; PAMAM: polyamidoamine dendrimer.

4.1. Anticancer Peptides

Typically, anticancer effects of peptides occur through two possible mechanisms. One of these arises when the peptide is able to disrupt the normal function of the membrane, while the other occurs when it triggers pro-apoptotic pathways. The membrane-disrupting mechanism operates when the peptides are enriched with cationic amino acids: lysine (Lys, K; amine), arginine (Arg, R; guanidine), and histidine (Hys, H; imidazole). Although this mechanism is not fully clear, these peptides demonstrated a certain membrane-lytic effect similar to other cationic species [134]. The first evidence on the topic was observed with TAT-targeted nanosystems [135], as well with other cell-membrane-penetrating peptides [136].

Regarding the use of cationic peptides as therapeutic agents, it is highly desirable to hide their cationic character to prevent premature damages such as hemolysis and embolisms during trafficking. Apart from pore loading, which preserves cargoes from undesired interactions, another interesting possibility arises from the use of additional negative components to balance the resulting surface charge. One elegant strategy, reported by Zhang and coworkers, employed citraconic anhydride to transform an octa-lysine into a negatively charged peptide able to balance the cationic $K_8(RGD)_2$ functionalized MSNs [111]. The resulting outermost layer was able to accomplish three tasks: target integrin receptors through the RDG moiety, show a highly biocompatible neutral charge, and carry a detachable but toxic K_8 sequence. Unfortunately, the reported studies mainly focused on assessing the effect of carried DOX than on studying the effect of the peptide release, overlooking any potential anticancer effect. In a later contribution by these authors, their system was evolved by including a triphenyl phosphonium group into the KLA peptide sequence (KLAKLAKKLAKLAK) in order to target the mitochondrion. Mitochondria appeared as a key target in cancer treatment because of their roles in regulating cell apoptosis and metabolism [137]. The delivery of antitumoral therapeutic molecules to mitochondria may improve their therapeutic efficacy while avoiding resistance pathways. Furthermore, stimulation of apoptosis mediated by mitochondria could also improve the efficacy of cancer therapy. The resulting organelle-targeted compound was bound to MSNs through disulfide bonds, known to be broken intracellularly by glutathione. Finally, to prevent undesired damages, the authors created a PEGylated anionic polymer shell aimed at increasing biocompatibility and reducing the leakage of topotecan [112]. In this system, the non-loaded model was able to exert a significant pro-apoptotic even at low concentrations, proving the potential of peptides in anticancer therapies; however, as expected, the highest antiproliferative effect occurred when topotecan was loaded within the mesopores.

Encouraged by these results, this group also addressed the synthesis of a multiple pro-apoptotic nanosystem: delivery of DOX, in combination with the delivery of two therapeutic peptides with glutathione-mediated cleavage [113]. In this model, the employed peptides contained different membrane-disrupting sequences, together with two different targeting elements: one of them specific to mitochondrion (C-GKGG-DKLAKLAKKLAKLAK) and the other specific to membranes (C-GRKKRRQRRRPPQ-RGDS). Additionally, to increase the potential cellular damage, the mesopores were loaded with DOX while the peptides were bonded to particles through disulfide bonds, in order to induce oxidative stress upon glutathione depletion. As result, the complete system showed high efficiency against HeLa and COS7 cells, although this effect decreased significantly in the absence of the drug, which highlights the limited effect of membrane-lytic peptides. More recently, this group also employed the glutathione-mediated disulfide cleavage to prepare a drug delivery system for DOX and a membrane-targeted therapeutic peptide rich in tryptophan ($(RGDWWW)_2KC$) [114]. In this case, this peptide was designed to have a DNA-intercalant effect due to the high concentration of indoles. This postulate seems to be justified based on the toxicity obtained for such a nanosystem; however, as expected, the best effect was obtained when DOX was co-delivered. More recently, Feng's group also employed this coating approach to co-deliver DOX and the anticancer peptide KLA. In this model, the final capping was done with a bovine serum albumin (BSA) corona [115], which permitted achieving a double effect: creating a diffusion barrier for both therapeutic agents, and enabling a protease/glutathione-mediated intracellular release. An interesting aspect of the system is that BSA was employed in its wild-type form, which might trigger an additional cellular response when in combination with the remaining multi-apoptotic effects.

Regarding the delivery of peptides that trigger pro-apoptotic routes, the reported examples focused on threading such peptides into the mesopores. For example, Martínez-Máñez's group reported the use of ε-polylysine as a coating layer to prevent the leakage of the pro-apoptotic C9h (YVETLDDIFEQWAHSEDLK) peptide [116]. In this model, the polylysine coating had multiple roles, as it favored cellular uptake due to its cationic character, while maintaining the C9h peptide within the mesopores until proteases cleaved this protective layer. The in vitro testing of this model proved that the encapsulated peptide showed a better therapeutic profile that its free form. However, this effect reached a maximum plateau over which higher dosages did not augment apoptosis. This points out once again the limited anticancer effect of peptides, which demand additional chemotherapeutic agents to obtain satisfactory results. In another similar example employing pore modifications, Cao et al. reported the delivery of a different proapoptotic peptide. In this model, they employed large-pore MSNs to deliver a bifunctional RGD-containing Hylin a1 peptide (IFGAILPLALGALKNLIK) able to target and kill cancerous cells [117]. To accommodate the peptide, the authors functionalized the internal facets of mesopores to favor threading and enable a pH-dependent release. In vitro studies of this system showed that encapsulation drastically reduced the hemolytic rate shown by the free peptide without affecting the potent cytotoxic effect against HeLa and Hep2 cells in vitro. Published in vivo studies with this system showed a clear tumor growth arrest in murine models, although, unfortunately, complete tumor remission could not be achieved.

As previously outlined, the anticancer effect of peptides and proteins is relatively low; thus, their use is generally limited to therapeutic adjuvants of more active species. However, the overall therapeutic effect may be increased if higher doses can be delivered; in this case, high-loading carriers gain interest. Among silica-based carriers, HMSNs are the most suitable candidates, since they theoretically allow loading in larger quantities than their porous analogues. The diverging aspects of MSNs vs. HMSNs in peptide delivery were studied by Rahmani et al., who focused on pepstatin A, a cathepsin D inhibitor peptide [118]. Surprisingly, the authors found two unexpected behaviors: HMSNs loaded less peptide than typical large-pore MSNs, but their effect was higher. Therein, the authors justified such behavior based on the release patterns observed from HMSNs, which provided a more sustained release (longer therapeutic effect), in comparison to regular pore-expended MSNs, which showed a burst-like release.

Another promising anticancer peptide is NuBCP9 (FSRSLHSLL), which is able to bind the Bcl-2 protein, highly overexpressed in cancer cells, turning a cell protector into an apoptosis inductor [138]. Along this line, Wu et al. reported folate-targeted, large-pore MSNs able to deliver this NuBCP9 peptide into HeLa tumors in zebrafish [119]. The resulting ca. 35-nm-width MSNs with pores in the range of 20 nm showed effective internalization into folate-positive HeLa cells, reaching up to 70% reduction of viability when loaded with the peptide. On the other hand, these nanoparticles showed fantastic biocompatibility, as they permitted obtaining fish survival above 80% in concentrations up to 200 μg/mL. In addition to this targeted example, this peptide was also employed in combination with a typical chemotherapeutic. In this contribution, the authors employed ca. 30-nm MSNs with large pores to load the peptide and an outer coating of a fifth-generation polyamidoamine (PAMAM) dendrimer able to load DOX within the structure [120]. As a result, the nanosystem was able to co-deliver both pro-apoptotic species to several cancer cell lines, achieving almost complete cell destruction in concentrations up to 1 μg/mL. To validate the potential of such a combination, the authors tested the efficiency of their nanosystem against resistant cell lines, obtaining outstanding results except for the case of the DOX-resistant MCF-7 line. Nevertheless, these results must be carefully accounted for, as the outermost PAMAM cationic coating may produce a decrease in overall biocompatibility.

4.2. Immunostimulating Peptides

Although most examples of immunostimulation were reported with fully functional proteins, the use of peptides is also possible. Along this line, Xie et al. reported the use of hollow mesoporous silica-based nanocarriers to deliver two melanoma-derived antigen peptides: the hydrophobic H2-K^b peptide $TRP2_{180–188}$ (SVYDFFVWL) and the hydrophilic H2-D^b peptide $HGP100_{25–33}$ (KVPRNQWL) [121]. In order to achieve the desired double loading, the authors modified their HMSNs with amino groups at the internal space and the mesopores with carboxylates in order to create two different preferential locations for peptides within the HMSNs. Then, to provide adequate retention and colloidal stability, the system was further coated with a lipid bilayer. Furthermore, on this lipid layer, another therapeutic species was included to increase the overall effect: monophosphoryl lipid A (MPLA), a Toll-like receptor 4 (TLR4) agonist. The resulting hollow protocells showed time-dependent uptake by murine bone marrow-derived dendritic cells and, according to data shown, induced cell maturation as seen by the overexpression of CD86, TNF-α, IFN-γ, IL-12, and IL-4 proteins. When the system was tested against melanoma lung metastasis in murine models, the vaccinated animals showed less tumor growth and creation of metastatic lymph nodes, demonstrating that the system was able to induce an effective anticancer response in mice.

4.3. Antibacterial Peptides

Although antimicrobial peptides are one of the most promising research lines to fight multi-resistant bacteria, their incorporation into nanocarriers was poorly studied [139]. Among the few systems reported, the contributions made by Braun et al. can be highlighted, who studied the best loading strategy and silica composition for the cationic antibacterial peptide LL37 (LLGDFFRKSKEKIGKEFKRIVQRIKDFLRNLVPRTES). In their first contribution, these authors focused on the membrane interactions between *E. coli* and several silica-based (non-porous, calcined mesoporous, and amino-capped mesoporous) nanoparticles loaded with the LL37 peptide [122]. As expected, the best loading profiles were obtained for the porous calcined—most negatively charged—nanoparticles. In addition, as expected, MSNs provided an adequate protective environment for the peptide and, hence, reduced the associated hemo- and proteolysis. At this point, it is also interesting to remark that these authors also studied how the porous structure affected the delivery, in this case, by comparing regular sized-MSNs with large-pore HMSNs [140]. They found that ca. 2.5-nm-pore MSNs produced a burst release, while the HMSNs showed a more sustained release, in concordance with data obtained by Rahmani et al. [118], suggesting that HMSNs behave better for peptide delivery.

Apart from the activation of immune cells and vaccination, the treatment of infected cells is another big issue that could be solved by applying nanotechnology. One inspiring example was recently reported by Tenland et al., who employed MSNs to deliver an anti-tuberculosis peptide to infected macrophages [123]. In their work, the NZX (GFGCNGPWSEDDIQCHNHCKSIKGYKGG YCARGGFVCKCY) peptide with a proven anti-tuberculosis effect was employed [141]. In this case, the system was assembled by threading the peptide into the mesopores, employing previously optimized nanoparticles [140]. The system demonstrated effective internalization into macrophages and produced peptide release once internalized. As a result, the infecting mycobacterium could the killed without significantly affecting host macrophages; moreover, the MSN-carried peptide showed a longer therapeutic effect than its free form. This effect, seen in other reviewed examples, was justified by the accumulation of nanocarriers within intracellular vacuoles, which created therapeutic reservoirs by maintaining peptide integrity.

Apart from the delivery of antibacterial peptides, combination therapies were also recently reported. Along this line, Zink and coworkers reported the simultaneous delivery of the melittin (MEL, GIGAVLKVLTTGLPALISWIKRKRQQ) peptide and the antibiotic ofloxacin (OFL) with a mesoporous silica-based assembled nanosystem [124]. In this system, the authors employed a very elegant strategy to co-deliver the medium-size peptide together with the small antibiotic. To do so, they prepared large-pore MSNs which were loaded with MEL and capped with a β-cyclodextrin-modified polyethyleneimine and ofloxacin-loaded, adamantane-modified SPION@MSNs capped by cucurbit[6]uril units. The resulting nanoparticles were able to self-assemble and create nanogates able to maintain both drugs within the closed pores. The system was proven to disassemble under magnetothermal induction, thereby releasing both species, which demonstrated efficient destruction of planktonic bacteria and biofilms. Indeed, this system demonstrated its efficiency even when embedded into implants, which were able to prevent bacterial infection in mice.

4.4. Growth Factors

As in many other cases, peptides could also have anabolic properties. These features make them interesting candidates for their use as cargoes in new-generation therapeutic devices aimed at tissue growth induction or hormone therapy. The first example in this field, reported by Lin's group, employed MSNs to intracellularly deliver cyclic adenosine monophosphate (cAMP). To obtain chemical responsiveness, the authors modified the MSNs with a boronic acid –$B(OH)_2$ moiety able to coordinate glucuronic acid moieties. The mesopores were then loaded with cAMP and closed through glucuronate-modified insulin nanocaps [125]. The resulting system was sensitive to glucose, as it could be disassembled by chemical displacement. As a result, this system permitted regulating hyperglycemic levels through a dual release mechanism: Insulin to trigger glucose assimilation by liver, and cAMP delivery to induce glycolysis. Unfortunately, this proof of concept was not further evaluated in vivo.

More recently, Sun et al. revisited insulin delivery with MSNs. To do so, they developed a glucose-sensitive nanodevice employing rod-like MSNs and a polymer shell able to change its conformation upon the presence of glucose [126]. In their investigations, the authors optimized the composition of this polymeric shell by using three different monomers: *N*-isopropylaminomethacrylate (NIPAM) to provide the needed morphological shift, 3-acrylamidophenylboronic acid as a glucose-sensitive fragment, and dextran maleate as an optional hydrophilic component. The resulting system could be loaded with insulin at low temperatures (4 °C), in which the polymer adopted an expanded conformation and closed upon heating (37 °C). Both reported compositions showed effective insulin release when glucose was present; however, when pH was below 7, the polymer was not able to shift morphology, avoiding insulin release. Although only in vitro tests were reported, the possibilities of this model are considerable, as the incorporation of dextran may favor colloidal stability and enhance biocompatibility, as suggested by the high cell viability values obtained. Insulin delivery was also investigated by Zakeri-Siavashani et al., who employed SBA-15 mesoporous microparticles [127]; however, therein, the authors profited from the bigger pore size provided by these particles to achieve higher loading and more sustained release. Although their system showed a nice insulin release profile, the typical bigger sizes of SBA-15 particles (generally above 300 nm) limit their application in intravenous formulations, although they may be used to create biocompatible reservoirs.

The facile aggregation of SBA-15 particles caused by their size, which makes their trafficking through alveoli and capillaries extremely difficult, makes these materials ideal candidates for non-systemic therapies, for example, bone replacement and tissue regeneration. In our first contribution to the topic, we found that the osteostatin peptide, derived from parathyroid hormone-related protein ($PTHrP_{107-111}$), could be efficiently delivered using this kind of material [128]. A clear regenerative effect was observed, which permitted recovering bone mass in peri-implant bone regeneration in cavitary defects [129], as well as in osteoporotic subjects [130]. More recently, our investigations focused on the combination of this peptide with osteogenic elements embedded in bioactive glasses [142,143], finding an enhanced bone formation consequence of the exerted combination therapy.

In addition to these examples, other research groups also focused on the combination of osteoinductive peptides with bone-forming agents. Such is the case of the model reported by Mendes et al., which combined an osteogenic growth peptide (OGP, ALKRQGRTLYGFGG) with hydroxyapatite in an SBA-15-based formulation [131]. Therein, the authors found a substantial behavior difference when hydroxyapatite was present, while deeper studies on the possibilities of such a combination were not addressed. Another interesting combination for remineralizing bone was reported by He and coworkers, who employed MCM-41 to prepare a nanosystem able to deliver a BMP-2-derived peptide (KIPKASSVPTELSAISTLYL) together with dexamethasone, a potent synthetic glucocorticoid [132]. Although, in this case, the peptide was not intended to be released, experimental results showed that its presence exerted a clear osteogenic induction. Moreover, the system was able to promote osteogenic differentiation on mesenchymal stem cells in vitro, as indicated by the upregulation of alkaline phosphatase (ALP) and other bone-forming proteins. Additionally, an increased calcium deposition was observed, demonstrating that mesoporous silica nanoparticles are adequate materials for bone engineering.

5. Delivery of Nucleic Acids: Gene Modulation and Silencing

Gene modulation and silencing are promising strategies to efficiently treat cancer, as well as cardiovascular or inflammatory-related diseases [20]. NAs are rapidly degraded by endonucleases and barely internalized by cells. Therefore, the biggest challenge today is to develop an effective gene delivery vector to transport and introduce NAs into the gene therapy target cells. These carriers must be easy to manipulate and show non-toxic properties in vitro and in vivo.

There are two main carriers for gene delivery therapy: viral and non-viral vector platforms [144]. For non-pathogenic viruses (retro and adenovirus), the most used platforms are viral vectors, but they present some limitations such as high cost and manipulation, immune response, or inadequate accessibility of genetic cargo internalized by cells [144]. Currently, research groups are focusing their efforts on the improvement of cheaper and easier preparation of non-viral platforms as cation polymers, liposomes, or inorganic porous nanoparticles such as MSNs [145]. In this respect, PEI is a widely used polycation in gene delivery systems due to its amino protonation improving gene delivery [20,144,145]. Liposomes are also commonly applied to enhance the cell internalization of NAs. Both polycations and liposomes present some limitations for therapeutic applications related to their molecular weight or disadvantageous gene escape and physicochemical instability, respectively [20,144,145]. Nevertheless, MSNs showed optimal properties for intracellular delivery of NAs due to their easy manipulation, high loading capacity into the mesopores, surface functionalization, and non-toxicity characteristics [20]. In addition, MSNs can be combined with polycations and other therapeutic drugs for personalized treatment related to cancer and other diseases.

In this regard, plasmid DNAs (pDNAs) are the most widely used gene therapy material to treat different diseases [20,144,145], although there is increasing interest in small interfering RNAs (siRNAs) and microRNAs (miRNAs) for therapeutic application. In this section, we focus on NA-loaded (DNA, siRNA, and miRNA) MSN-based drug delivery systems, as shown in Table 3 and Figure 5.

Table 3. Examples of different therapeutic nucleic acids delivered by silica-based nanocarriers.

Nucleic Acid	Carrier Type	NA Location	Loading Strategy	Cell Line(s)	In Vivo	Application	Reference
DNA							
pDNA	MSNs	Surface	Adsorption	Neuro-2A	None	Cancer	[146]
pDNA	MSNs	Surface	Adsorption	HEPA-1	None	Cancer	[147]
dsDNA	MSNs	Surface	Adsorption	HeLa	None	Cancer	[148]
dsDNA	MSNs	Surface	Adsorption	HL-60	None	Cancer	[149]
pDNA	IDMSMs	Surface	Adsorption	QGY-7703	Mice	Cancer	[150]
ssDNA	MSNs	Surface	Disulfide bond Maleimide	MCF-7	None	Cancer	[151]
dsDNA	HMSNs	Surface	Adsorption	MCF-7, A549, HepG2	None	Cancer	[152]
ssDNA	MSNs	Surface	Adsorption	MDA-MB-231	Mice	Cancer	[153]
ssDNA	MSNs	Surface	Adsorption	MCF-7, K-562, U2OS	Mice	Cancer	[154]
pDNA	MSNs	Surface	Adsorption	293T, RAW264.7	Mice	Hepatitis B	[155]
siRNA							
PKM2	MSNs	Surface	Adsorption	MDA-MB-231	Mice	Cancer	[156]
Cell-killing	MSNs	Mesopores	Adsorption	MDA-MB-231	None	Cancer	[157]
HER2	MSNs	Surface	Adsorption	BT474	None	Cancer	[158]
TGFβR-1	MSNs	Surface	Adsorption	RT3	Mice	Cancer	[159]
MDR1	MSNs	Surface	Adsorption	CAL27	None	Cancer	[160]
MDR1	MSNs	Surface	Adsorption	KBV	Mice	Cancer	[161]
Bcl-2	MSNs	Surface	Adsorption	SKOV-3, MCF-7	None	Cancer	[162]
Bcl-2	MSNs	Surface	Adsorption	SKOV-3	Mice	Cancer	[163]

Table 3. *Cont.*

Nucleic Acid	Carrier Type	NA Location	Loading Strategy	Cell Line(s)	In Vivo	Application	Reference
siRNA							
TWIST	MSNs	Surface	Adsorption	F2, Ovcar8	Mice	Cancer	[164]
VEGF	MSNs	Surface	Adsorption	HeLa, HepG2	None	Cancer	[165]
VEGF	MSNs	Surface	Adsorption	HepG2, Huh7	None	Cancer	[166]
HSP47	MSNs	Surface	Adsorption	PMDF	Mice	Fibrosis	[167]
TnC	MSNs	Surface	Adsorption	rHSCs, mHSCs	None	Fibrosis	[168]
COLL1A1	MSNs	Surface	Adsorption	HDFs	Rat	Tissue regeneration	[169]
SOST	MSNs	Surface	Adsorption	MEF, HeLa	Mice	Osteoporosis	[170]
miRNA							
miR-155	MSNs	Mesopores	Adsorption	SW480	Mice	Cancer	[171]
miR-328	MSNs	Mesopores	Adsorption	SW480, SW620, HT-29, Lovo, Caco-2	Mice	Cancer	[172]
miR-34a miR-10b	MSNs	Mesopores	Adsorption	MDA-MB-231, MDA-MB-468	Mice	Cancer	[173]
miR-26a	MSNs	Mesopores	Adsorption	rBMSCs	None	Tissue regeneration	[174]
miR-200c siPlk1	MSNs	Mesopores	Adsorption	MDA-MB-231	Mice	Cancer	[175]

Abbreviations: IDMSN: imidazole dendritic mesoporous silica nanoparticles; PKM2: glycolytic enzyme pyruvate kinase; HER2: human epidermal growth factor receptor 2; TGFβR-1: transforming growth factor, beta receptor I; MDR1: multidrug resistance protein 1; Bcl-2: B-cell lymphoma 2; VEGF: vascular endothelial growth factor; HSP47: heat-shock protein 47; TnC: tenascin-C; COLL1A1: collagen type I; PMDFs: primary mouse dermal fibroblasts; HSCs: human stem cells; HDFs: human dermal fibroblasts; MEF: mouse embryonic fibroblastic cells; rBMSCs: rat bone marrow stromal cells.

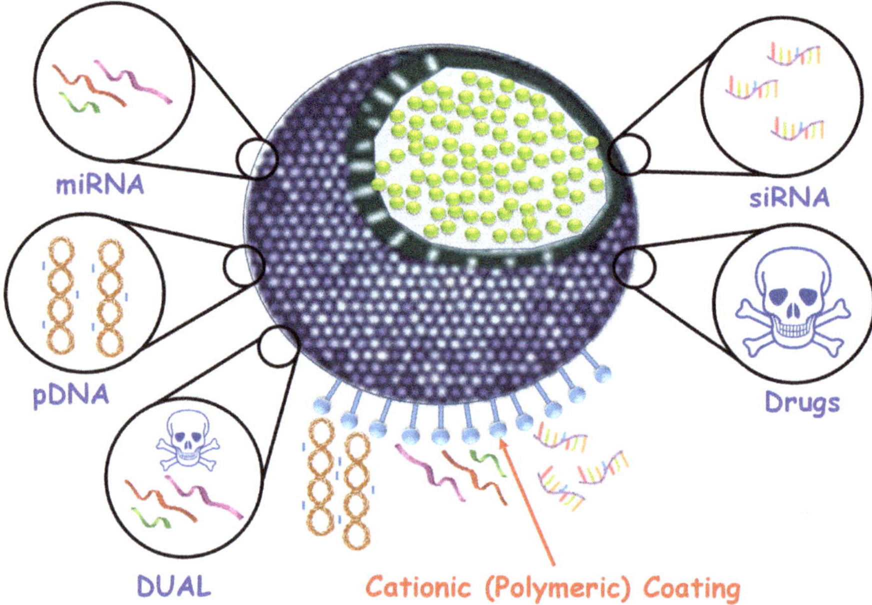

Figure 5. Types of nucleic acids and the possible loading strategies into MSNs for tuning gene expression on cells.

5.1. DNA

As we previously mentioned, MSNs can be functionalized with PEIs for better DNA molecule adsorption and, thus, for a more effective intracellular delivery, protecting NAs from endonuclease degradation due to their positive charge. These positive charges interact with negatively charged cell membranes, inducing higher cell internalization rates and cell death. For example, Zarei et al. [146] developed a functionalized nanosystem based on phosphonate MSNs-PEI loaded with a lysosomotropic factor, chloroquine (CQ), complexed with plasmid DNA. The resulting cell internalization and viability of this pDNA-MSN-PEI was analyzed with the fluorescent protein plasmid (pGFP (green fluorescent protein)), showing a significant increase in transfection of pGFP into the mouse neuroblastoma cell line Neuro-2A. In this respect, Xia et al. [147] proposed the use of a cationic MSN-PEI nanocarrier with a potential therapeutic application. The resulting cationic surface of MSNs facilitated the DNA attachment in order to balance the cationic 10-kDa PEI and, hence, achieve nontoxic effects and higher cellular uptake rates. This nanosystem was loaded with a plasmid DNA (pEGFP) and siRNA construct that was capable of knocking down GFP expression in hepatocellular carcinoma mouse (HEPA-1) cells, with a fluorescent GFP expression of 70% in these cells. In addition, the authors loaded the cationic MSN-PEI nanocarrier with paclitaxel, a hydrophobic anti-tumoral drug used in pancreatic tumor treatment, increasing the cell internalization and delivery of this drug in HEPA-1 cells. On the other hand, Song et al. [29] demonstrated that controlling the nanotopography of MSNs as pDNA vectors had an influence on the transfection efficacy. For example, ambutan-like MSNs-PEI with spiky surfaces showed pDNA-binding capability and a transfection efficacy of 88% in HEK-293T cells, which was a higher rate compared with other MSNs systems. In addition, the authors demonstrated that these types of surface spikes of MSNs induced a continuous open space with two objectives: binding DNA chains through multivalent interactions and protecting the gene cargo covered in the spiky layer against enzymatic degradation, without negative effects on transfection rates. These results indicated that this nanosystem is an interesting approach as a non-viral vector for effective gene delivery.

In a study by Wang et al. [148], an MSN based nanoplatform functionalized with a DNA gate was evaluated in vitro. In this report, the authors developed a fluorometric method for adenosine triphosphate (ATP) recognition using rolling circle amplification (RCA) consisting of proximity ligation-mediated amplification. In addition, the nanosystem was functionalized with graphene oxide modified with folic acid (FA) as a DNA vehicle and loaded with DOX, to evaluate the control release efficacy in HeLa tumoral cells. Following the RCA process, long DNA chains that contained a complementary strand to the DNA at the gate permitted dehybridizing such a nanogate and producing DOX release into tumoral cells. This nanoplatform was effectively internalized by the FA receptor, upregulated in those tumoral cells, while the DOX release induced increased cell toxicity compared with control systems, improving the targeted cancer therapeutics on these HeLa cells. In another study by Wang et al. [149], a DNA-capped MSN nanocarrier loaded with DOX for tumor marker-triggered on-demand drug release was developed and evaluated in vitro. As a first step, a DNA biomolecular gate adsorbed on the MSN–NH_2 was developed at neutral pH via electrostatic adsorption. In the absence of a stimulus, the pores were locked, and it was not possible to release the drug. However, when the nanosystem recognized the internal stimulus (survivin messenger RNA (mRNA) or miR-21, overexpressed biomarkers in cancer), the DNA caps were removed from the MSNs, allowing DOX release. This stimulus-responsive behavior was confirmed by different microscopy techniques in cell-acute myeloblastic leukemia (HL-60) cells. This nanocarrier was effectively internalized by these cells, releasing DOX cargo into the cytoplasm and inducing significant cell death. Therefore, this nanoplatform is a very useful novel system for both imaging diagnosis and therapeutic controlled drug delivery applications. In this regard, Li et al. [150] designed and evaluated another anti-tumoral and gene drug co-delivery nanosystem with on-demand release properties in vitro and in vivo. For this purpose, the authors developed dendritic MSNs modified with imidazole groups employing a Schiff-base imine linkage, which permitted loading DOX within the pores and electrostatically depositing the survivin short hairpin RNA (shRNA)-expressing plasmid (iSur-pDNA). The imidazole functionalization triggered efficient endosomal escape, improving the accumulation of iSur-pDNA and gene knockdown efficacy. Such nanosystems were successfully internalized by QGY-7703 hepatoma cells and decreased tumoral cell viability in vitro, due to the pH-sensitive co-delivery of DOX and iSur-pDNA. In addition, this nanoplatform induced tumor growth reduction in H-22 tumor-bearing mice, indicating that it could be a promising nanocarrier for co-delivery cancer treatment.

On the other hand, aptamers are short single-stranded oligonucleotides with high affinity and specificity to several molecule targets, and they exhibit potential properties as therapeutic and diagnostic factors in different diseases [176]. In this regard, Sun et al. [151] developed an MSN-based nanocarrier with different DNA molecules on the surface for self-assembly. For the controlled delivery of anti-tumoral drug cargo (DOX), the authors used an aptamer oligonucleotide as a gatekeeper and other oligonucleotides on the MSN surface to allow DNA-guided immobilization bearing single-stranded capture oligonucleotides. This nanovalve induced an increase in cell adhesion rates in MCF-7 adenocarcinoma cells, with an efficient triggered release of DOX drug in these cells. The DNA-directed self-assembly aptamer-based nanosystem is efficient in surface-bound monolayers and can be used as a delivery system for different applications and treatments, allowing site-specific delivery of anti-tumoral drugs. In the same way, Li et al. [152] developed a dual multi-locked DNA valve HMSN-based nanosystem to intracellular cancer-related mRNAs for controlled DOX drug release. The upregulated endogenous targeted mRNAs (K1 and GalNAc-T mRNAs) were able to dehybridize the multi-locked DNA valves and open the pores to release DOX. As a result, a synergic effect was obtained, caused by the intratumoral delivery of the drug together with the depletion of TK1 mRNA, implicated in cell division as a tumor growth factor, and GalNAc-T, upregulated in several tumoral cells. The nanocarrier was evaluated in vitro in different cancer cells, inducing higher rates of cytotoxicity compared to non-tumoral cells. In an in vivo study, Pascual et al. [153] designed another novel non-viral drug delivery and diagnostic nanocarrier based on a capped MSN-NH_2 hybrid platform, which was loaded with DOX and gated with Mucin 1 (MUC1) aptamer (S-apMUC1). MUC1 is a protein present in the cell surface and upregulated in breast cancer cells. The S-apMUC1-MSNs nanosystem was efficiently and specifically internalized only by tumoral MDA-MB-231 cells related to MUC1 receptor overexpression of these cells. After S-apMUC1-MSNs internalization by tumoral cells, the DOX cargo was released into the cytoplasm of tumoral cells, inducing a decrease in cell viability. This nanoplatform exhibited reduced cargo release when DNAse I was not present. In vivo, S-apMUC1-MSNs were radionuclide-labeled with ^{99}Tc for a radio-imaging study in MDA-MB-231 tumor-bearing Balb/c mice. S1-apMUC1-Tc displayed significant tumor signaling and accumulation in these mice. These results suggest that MSN nanosystems capped with aptamers showing radiopharmaceutical properties are promising hybrid nanomaterials in the clinical context.

Of particular interest is a study by Srivastava et al. [154], where the authors proposed a novel telomerase-responsive delivery of DOX loaded in an MSN nanosystem against telomerase-positive tumoral cells. DOX loading was performed using a telomeric repeat complementary sequence and a telomerase substrate primer sequence, allowing controlled release in MCF-7 cancer cells. This functionalized nanocarrier was efficiently internalized, and it significantly reduced the tumoral cell viability, with an inhibition of function when a specific telomerase inhibitor was present. This oligo-wrapped nanoprobe specific for telomerase was used in a telomerase-positive Dalton's lymphoma mice model. The nanosystem induced a significant inhibition of tumors, enhanced survival, and re-established histopathological parameters, including neo-angiogenesis.

Furthermore, DNA vaccination was proven to provide an immune response to viruses or infectious diseases [177]. In this regard, Wang et al. [155] developed a layered double hydroxide (LDH) MSN nanocarrier as a vaccine delivery platform and immune stimulant, using a GFP expression plasmid as model DNA. The pDNA-MSN-LDH nanosystem showed high internalization rates in monocyte-derived dendritic cells (MDDCs) and stimulated macrophage activation via nuclear factor kappa B (NF-κB) signaling pathway, increasing IFN-γ, IL-6, CD86, and major histocompatibility complex class I (MHC-I). In vivo, the nanocarrier immunized BALB/c mice, indicating that the DNA vaccine-MSN-LDH system increased the serum antibody response and promoted T-cell proliferation, skewing T helper to Th1 polarization. Taken together, these results indicated that MSN-LDH nanosystems could operate as a potential non-viral gene delivery platform.

5.2. *siRNAs*

Small interfering RNA (siRNA) are highly potent, short, double-stranded RNA molecules (20–25 bp) used in gene-based therapy of different diseases that can be delivered in a sustained manner [178]. Moreover, siRNAs are anti-cancer targets because they can modify the expression of pro-apoptotic oncogenes and cell-cycle key regulators. However, siRNAs have a very short half-life and poor cell internalization capability [178]. In this regard, MSNs attracted pronounced consideration for the intracellular delivery of NAs due to their high loading capacity into the mesopores and surfaces. Polycations such as PEI combined onto the surface of MSNs permit a more stable encapsulation, loading, controlled release, and intracellular internalization of potential siRNA cargoes [178]. In addition, positively charged amine-rich nanosystems are required for a better result in siRNA/MSN cellular uptake rates [178]. Future in vivo efforts are necessary to validate the tissue-specific siRNA therapeutic efficacy using imaging techniques to analyze MSN biodistribution at the whole-body level. On the other hand, it is mandatory to find nanosystems able to efficiently deliver those therapeutic siRNAs in vivo with enough efficiency, as well as with protective features against nucleases, able to enhance their gene-silencing capabilities at the target tissues.

In this respect, Shen et al. [156] developed a universal siRNA multi-component system carrier based on MSNs functionalized with cyclodextrin-grafted polyethyleneimine (CP). CP offers a positive charge to improve the siRNA loading on MSNs through electrostatic interaction, allowing endosomal escape. To analyze the effectiveness of the proposed delivery system, an siRNA of the M2 isoform of the glycolytic enzyme pyruvate kinase (PKM2) was used, due to it being overexpressed in several cancer types. Firstly, siRNA-loaded CP-MSNs were positively evaluated in vitro in MDA-MB-231 human breast cancer cells with higher rates of cellular internalization and gene silencing capability. The nanosystem showed specific tumor accumulation and gene silencing in an orthotopic mouse model of MDA-MB-231 cells, inhibiting tumor cell growth, invasion, and migration. Prabhakar et al. [157] designed a new nanosystem based on MSNs with expanded mesopores and pore–surface-hyperbranched PEI tethered with redox-cleavable organic linkers that could carry high concentrations of siRNAs. This stimulus-responsive nanosystem was highly internalized by MDA-MB-231 cells, releasing the cell-killing siRNA as cargo into the cytoplasm, followed by endosomal escape triggered by the intracellular redox situation, with high efficiency in long-term gene knockdown for several days. In addition, long-term stability of therapeutic nanosystems is mandatory to provide personalized medicine. In this respect, lyophilization or freeze-drying are useful and common methods to ensure this aim. Ngamcherdtrakul et al. [158] successfully reported a lyophilized nanosystem consisting of MSNs modified with cationic polymers (PEI), PEG, and antibody (trastuzumab), which was loaded with a human epidermal growth factor receptor 2 (HER2) siRNA upon reconstitution. This nanosystem was successfully internalized by $HER2^+$ human breast cancer cells, BT474, decreasing their viability; it maintained a cake-like structure and retained hydrodynamic size, charge, siRNA loading ability, and silencing efficacy. With the tested method, the lyophilized nanocarrier can be stored stably for eight weeks at 4 °C and at least six months at −20 °C.

Oral cancer is the seventh most frequent cancer and the ninth most frequent cause of death in the world [179]. About 90% of oral cancer is of squamous cell carcinoma type (SCC) [179]. Currently, surgery and radiation, combined or not with chemotherapy, are the main treatments, but this approach presents several limitations due to multidrug resistance; thus, new therapeutic drugs must be evaluated. Lio et al. [159] developed a topical formulation for the transdermal delivery of siRNA built on MSNs for the treatment of skin disorders. In this regard, transforming growth factor beta receptor I (TGFβR-1) is overexpressed in patients with skin SCC, and the anti-tumor effects of TGFβR-1 inhibitors were proven in several types of cancer. The nanosystem was built by loading TGFβR-1 siRNA into the pores of MSNs and further coating the resulting particle with a layer of poly-L-lysine (PLL) to improve transdermal delivery. The tested platform demonstrated positive cellular uptake rates and detection of specific gene biomarkers in human SCC cells in vitro. This system induced a topical delivery of TGFβR-1 siRNA targeting the SCC in a mouse xenograft model, decreasing the TGFβR-1 siRNA levels and reducing tumor volume. These results indicated that this platform is a promising non-invasive transdermal drug delivery system. Moreover, multidrug resistance plays an important role in the failure of oral cancer chemotherapy. In this sense, in a study by Wang et al. [160], a novel MSN loaded with both multidrug resistance protein 1 (MDR1) siRNA and DOX was evaluated in vitro and in vivo to directly kill tumor cells without the effect of multidrug resistance. MSNP-PEI@siRNA loaded with DOX was efficiently transfected into human oral squamous carcinoma DOX-resistant (KBV) cells, decreasing the gene expression of MDR1 and increasing the apoptosis of these cells. This nanocarrier significantly reduced the tumor size and decelerated tumor growth rate in vivo, showing a huge potential therapeutic application for multidrug-resistant cancer. On the other hand, Shi et al. [161] developed hyaluronic acid (HA)-assembled MSNs nanosystems loaded with MDR1 siRNA and protein MutT homolog 1 (MTH1) inhibitor TH287. Downregulation of MTH1 protein expression increased oxidative stress levels and significantly decreased tumor survival and proliferation. TH287 is a potent small molecule that inhibits the MTH1 protein, and it could act as a potential chemotherapeutic drug. The HA-MSNs nanosystem was successfully internalized by CAL27 cancer cells, inducing a high rate of cell apoptosis when TH287 and MDR1 siRNA were present. In vivo, this nanocarrier reduced the tumor growth compared with untreated control and free TH287 in male Balb/c mice. These results suggest that HA-MSN-based nanosystems are a promising vehicle to control the delivery of MDR1 siRNA and TH287, with a possible combination with other therapeutic agents for oral cancer treatment.

Regarding multidrug resistance as a persistent limitation for cancer treatment, Pan et al. [162] proposed a novel pH-responsive delivery nanosystem based on carboxylated MSNs (MSN-COOH) with a zeolitic imidazole framework-8 (ZIF-8) film synthesized for pore blocking and efficient loading of anti-apoptotic B-cell lymphoma 2 (Bcl-2) siRNA. The nanosystem developed was efficiently internalized by MCF-7/ADR and SKOV-3/ADR tumoral cells. The ZIF-8 film can be degraded in the acidic endo-lysosome and trigger the intracellular release of both siRNAs and potential antitumor drugs, decreasing cancer cell proliferation. In this regard, Choi et al. [163] developed a new biodegradable and biocompatible amine-free, neutral MSN-based nanocarrier for siRNA delivery. In this study, Bcl-2 siRNA was loaded via a calcium ion (Ca^{2+})-mediated interconnection between phosphates of siRNA and surface silicates of MSNs, reducing the risk of amine-related cytotoxicity and immunogenicity. This nanosystem showed positive results in internalization rates in tumoral SKOV-3 cells, inducing a significant decrease in cell viability when the siRNA was released, compared with MSNs or siRNA alone. This reduction in tumoral cell proliferation was dramatically enhanced when the nanocarrier was loaded with DOX. The therapeutic potential of the co-delivery system was also analyzed in SKOV3 xenografts in nude mice. The authors found that co-delivery of Bcl-2-siRNA and DOX MSNs-Ca^{2+} reduced tumor volume with an inhibition rate of 72%, higher than MSNs-Ca^{2+} loaded with DOX and MSNs-Ca^{2+} with siRNA. The results indicated that Ca^{2+}-bound MSNs can act as an siRNA nanocarrier with a less toxic, more effective, and safer siRNA potential clinical application. On the other hand, patients with epithelial ovarian cancer (EOC) metastases relapse after initial effective chemotherapy treatment due to multiple drug resistance. In order to resolve this problem, Shahin et al. [164] proposed a novel nanoparticle delivery platform to reverse chemoresistance based on HA-MSNs loaded with TWIST siRNA. Hyaluronic acid (HA) is a ligand for CD44, which is overexpressed and correlates with worse prognosis in EOC. TWIST is a key protein implicated in epithelial–mesenchymal transition, cancer metastasis, angiogenesis, and drug resistance. The developed nanocarrier was successfully internalized in vitro by F2 and Ovcar8 cell lines, inducing a sustained TWIST gene knockdown. MSN-HA@siRNA showed safety and specific tumor targeting in an EOC mouse model. Mice treated with the complete nanosystem induced significant tumor reduction, and, when the authors combined this nanocarrier with cisplatin, the reduction of tumor growth was greater compared with the corresponding control groups.

For the metastatic spread of cancer, growth of the vascular network is a critical event [29]. Vascular endothelial growth factor (VEGF) is a glycoprotein and key mediator of angiogenesis in tumoral cells [180]. Keeping this in mind, Zheng et al. [165] proposed a new nanocarrier based on MSNs functionalized with FA-targeted tumoral cells and co-loading of VEGF siRNA and ursolic acid (UA), a pentacyclic triterpenoid with anticancer efficacy. HeLa (FA receptor-overexpressed) and HepG2 tumoral cell lines were used to analyze the MSN-FA-UA@siRNA nanosystems in vitro. The MSN-FA nanocarrier enhanced transfection efficiency and siVEGF stability, improving the targeted antitumoral efficacy of siVEGF and UA, with a significant decrease in cancer-related VEGF protein levels in HeLa cells. In this sense, the same authors [166] developed an alternative MSN nanosystem consisting of an asialoglycoprotein receptor (ASGPR)-targeting drug delivery system for co-delivery of siVEGF and sorafenib (SO), a multi-kinase inhibitor that can target VEGF for the inhibition of tumor growth and neo-angiogenesis in hepatocellular carcinoma (HCC). Lactobionic acid (LA) is derived from lactose oxidation, and it was able to effectively target human HCC. In addition, ASGPR is overexpressed in several types of cancer cells, facilitating hepatic infection, and it is specifically recognized by glycoproteins which bind with it. The MSN-LA-SO@siVEGF nanocarrier induced synthesis (S) phase cell-cycle arrest, increasing the anti-cancer efficacy (cell apoptosis) of SO and siVEGF through the active targeting of LA in ASGPR-overexpressing Huh7 tumoral cells. Furthermore, this nanosystem enhanced the siVEGF transfection efficiency and inhibited the expression of angiogenesis-related VEGF proteins in these tumoral cells. Taking both studies together, these types of functionalized MSN-based nanosystems could be potential nanocarriers for targeted delivery of anti-tumoral drugs and VEGF siRNAs to improve anti-tumor capability.

On the other hand, fibrotic diseases are related to augmented oxidative stress and upregulation of pro-fibrotic genes [181]. Oxygen species (ROS), nicotinamide adenine dinucleotide phosphate

(NADPH) oxidase 4 (NOX4) and heat-shock protein 47 (HSP47) are implicated in this disease [181], producing disproportionate collagen synthesis. Morry et al. [167] proposed an MSN platform with PEI and PEG coating that effectively delivered HSP47 siRNA in vitro and in vivo. This nanocarrier induced a knockdown of the HSP47 gene and a decrease in ROS and NOX4 levels in an in vitro model of fibrosis based on TGF-β-stimulated fibroblasts. Intradermal administration of MSNs@siHSP47 successfully decreased HSP47 protein levels in a bleomycin-induced scleroderma mouse model in vivo, reducing ROS production and different pro-fibrotic markers. These nanocarriers could be a very interesting platform to treat both fibrotic and inflammation diseases. On the other hand, it is necessary to develop anti-fibrotic drugs for chronic liver fibrotic diseases. Hepatic stellate cell (HSC) activation is present in the progression of liver fibrosis, secreting extracellular matrix proteins as tenascin-C (TnC). Vivero-Escoto et al. [168] proposed an MSN@siTnC based nanosystem as a favorable alternative to evaluate the siRNA therapy of chronic liver disease in preclinical trials. This novel nanocarrier was successfully internalized by human stem cells in vitro and produced a downregulation in TnC mRNA and protein levels, inducing a decrease in inflammatory cytokine expression by macrophages and hepatoma migration. These promising results must be tested in a liver fibrosis in vivo model.

As demonstrated in this section, siRNA-MSN based nanosystems were widely studied and showed a high efficiency in silencing genes and reducing tumoral growth related to cancer processes in vivo. Nevertheless, in tissue engineering applications, scaffold-mediated delivery is desired to achieve local and sustained release of drugs. In this respect, Pinese et al. [169] developed an MSN-PEI@siRNA nanocarrier delivered from electrospun scaffolds via surface adsorption and nanofiber encapsulation. These nanosystems were either combined with the surfaces of nanofiber substrates or directly encapsulated. MSN-PEI@siRNA coated scaffolds offered sustained availability of siRNA for at least 30 days and five months in the case of MSN-PEI@siRNA-encapsulated scaffolds within the electrospun fibers in human dermal fibroblasts cells in vitro. The scaffolds showed excellent bifunctionality in vitro and in vivo by targeting collagen type I (COLL1A1) as a target gene to reduce fibrous capsule formation. These platforms were subcutaneously implanted in a rat model in vivo and COLL1A1 siRNA/MSN-PEI induced a reduction of 45.8% in fibrous capsule formation after four weeks compared to negative siRNA treatment. This method improved siRNA internalization rates and sustained targeted protein silencing in vitro and in vivo. The nanoplatform proposed by these authors can be used to release and internalize siRNA/miRNA combined with other specific low-molecular-weight drugs in long-term tissue engineering applications and cancer. In addition, the current pharmacological therapy of bone diseases such as osteoporosis presents several limitations related to bioavailability and toxicity. In this sense, gene knockdown through siRNA delivery received great attention as a promising treatment in osteoporosis [182]. In a recent study by Mora-Raimundo et al. [170], a nanosystem based on MSN-PEI, which can effectively deliver SOST siRNA and osteostatin inside cells, was evaluated in vitro and in vivo. The SOST gene inhibits the Wnt signaling pathway, decreasing osteoblast differentiation, whereas osteostatin is a pentapeptide with osteogenic and anti-osteoclastic properties [183]. SOST siRNA/MSN-PEI loaded with osteostatin induced a dramatic suppression of SOST gene expression in mouse embryonic fibroblastic cells with the consequent upregulation of several osteogenic markers with a synergistic effect. In vivo, the complete nanosystem was injected in the femoral bone marrow of ovariectomized mice, and the results were in concordance with in vitro studies, knocking down SOST and upregulating the osteogenic marker levels in a synergistic manner. The nanoplatform proposed by these authors was able to transport, co-deliver, and internalize SOST siRNA and osteostatin in cells, preserving its activity and achieving an effective silencing effect.

5.3. miRNAs

MiRNAs are small endogenous noncoding RNA molecules (17–25 nucleotides) that act as regulators of gene expression and downregulate target proteins via miRNA degradation or translational inhibition [184]. The aberrant expression of miRNAs in several cancers is associated with several mechanisms such as cellular differentiation, proliferation, survival, and apoptosis [184,185]. MiRNAs can accomplish a gene knockdown effect by regulating multiple miRNAs compared to siRNA, which is very useful when treating complex multifactorial diseases such as cancer [184,185]. Several drugs based on miRNAs were used in clinical trials in cancer [186]; however, the effective and safe delivery of anti-miRNAs or miRNA mimics is a key challenge for clinical applications.

Regarding these objectives, Li et al. [171] proposed a novel nanosystem based on MSN functionalized with polymerized dopamine (PDA) and AS1411 aptamer loaded with microRNA-155 (anti-miRNA-loaded MSN@PDA-Apt) for the specific treatment of colorectal cancer (CRC). Upregulation of miR-155 (oncogenic microRNA) was found in several cancer-related pathways, including in CRC. Firstly, the authors demonstrated that nuclear factor kappa B (NF-κB), an important transcription factor with a crucial role in the process of tumor growth, has a positive feedback loop with miR-155 in vitro and in CRC tissues. MSNs-anti-miR-155@PDA-Apt decreased miR-155 expression in SW480 cells and successfully targeted the CRC tumor, leading to gene knockdown and tumor growth reduction, thanks to both active targeting of AS1411 aptamer and passive targeting of the EPR effect. In addition, the same authors [172] explored the miR-328 pathway and developed a new pH-responsive nanoplatform consisting of MSNs-miRNA-328 decorated with PDA, a cell adhesion molecule aptamer, and bevacizumab (MSNs-miR-328@PDA-PEG-Apt-Bev) for the dual-targeting treatment of CRC. MiR-328 is a tumor suppressor downregulated in several human cancers, including CRC, and it is correlated with drug resistance. The authors demonstrated for the first time that CPTP, a ubiquitously expressed lipid transfer protein associated with inflammation and CRC, is a direct target of miR-328. MSNs-miR-328@PDA-PEG-Apt-Bev increased miR-328 levels and inhibited CPTP expression in SW480 cells, increasing binding ability and showing much higher cytotoxicity in vitro and in vivo. In addition, this nanoplatform showed efficient gene silencing and tumor growth inhibition of the target tumor in a SW480 xenograft mouse tumor model in vivo. Taken together, these results indicate that the MSN-based functionalized nanoplatforms reported herein show promise for miRNA therapy in cancer, specifically CRC. In a recent study by Ahir et al. [173], a nanosystem based on MSN-HA@miRNA with an appended PEG–poly(lactic-*co*-glycolic acid) (PLGA) polymer to target triple-negative breast cancer (TNBC) was evaluated in vitro and in vivo. In this context, miR-34a is downregulated and miR-10b is upregulated in TNBC disease, inducing tumorigenesis and metastatic dissemination. These authors proposed the MSN nanocarrier for co-delivery of miR-34a-mimic and antisense-miR-10b, targeting the CD44 receptor. In vitro, this nanosystem showed positive results in terms of cellular internalization rates, release profile, and a subsequent pro-apoptosis effect in human mammary carcinoma cell lines (MDAMB-231 and MDAMB-468). In vivo studies exhibited a high specificity in TNBC tumor targeting, leading to effective tumor growth reduction and metastasis delay in mice.

As previously mentioned, MSN nanosystems emerged as a promising bone regeneration methodology in bone tissue engineering. In this regard, Yan et al. [174] developed a novel miR-26a delivery nanosystem based on MSNs. MiR-26a was confirmed to regulate several pathways of osteogenic differentiation and promote bone regeneration. The authors demonstrated the protection effectiveness of the vectors to the miRNA and the positive internalization rates in rat bone marrow stromal cells (BMSCs) in vitro, releasing miR-26a into the cytoplasm without cytotoxicity effects. This nanosystem promoted stem-cell osteogenic differentiation, inducing an increase in alkaline phosphatase activity and mineralization, as well as in several genes and proteins implicated in osteogenesis in BMSCs. The nanocarrier proposed by these authors provides new methods and strategies for the delivery of mRNAs in bone tissue engineering, but it is necessary to further validate them in vivo.

On the other hand, a combined siRNA/miRNA therapy would be a very interesting approach, especially for cancer treatment, targeting multiple disease-related pathways and silencing specific genes. Currently, the use of this combination in clinical studies shows several limitations such as the availability of safe and effective systemic delivery nanocarriers with efficient tumor penetration. In this respect, Wang et al. [175] developed multifunctional tumor-penetrating MSNs for co-delivery of siRNA (siPlk1) and miRNA (miR-200c), upregulated in several types of tumor and implicated in cancer development and progression. In addition, the authors functionalized the nanosystem to facilitate endosomal scape using a photosensitizer indocyanine green (ICG) and surface conjugation of the iRGD peptide to allow deep tumor penetration. The complete nanosystem induced an increase in internalization rates in MDA-MB-231 cells and three-dimensional (3D) tumor spheroids in vitro, whereas ROS produced by ICG upon light irradiation induced the endosomal scape of the siRNA/miRNA into the cytoplasm with a deleterious effect on cell viability. In vivo, a significant reduction in tumor growth and metastasis upon short-light irradiation was found when the nanosystem was intravenously administrated in mice with metastatic breast cancer.

6. Delivery of Glycan-Based Biomolecules

Beyond the use of proteins, peptides, and nucleotides, glycan-based structures were also successfully employed in MSN-based delivery. Among them, the most recurrent examples are reported with heparin, chondroitin sulfate (CS), chitosan, and HA. Those compounds, apart from being important components of structural matrices, are known to affect many biochemical processes. For instance, HA is a valuable targeting element for the development of new therapeutic tools, as it is able to interact with CD44 and CD168, which are upregulated in many cancerous cells and are closely related to cell adhesion and migration in metastases [187,188].

Similarly, chitosan [188–190] and chondroitin sulfate [191–193] were employed for the development of high colloidally stable and biocompatible nanosystems with certain targeting capabilities. Moreover, as both substances are important structural components of extracellular matrices on vertebrates and crustaceans, they can act as bioactive components for tissue regeneration. Such was the case in the contributions by Kavya et al. [194] and Porgham Daryasari et al. [195], who employed those elements to induce osteogenesis through two different approaches. In the first approach, Kavya et al. designed a crosslinked chitosan/CS scaffold reinforced with nanometric SiO_2. In this system, the presence of silica permitted improving the mechanical properties and obtaining slower degradation, while the glycans permitted obtaining an adequate moisture swelling for creating a favorable environment for cell proliferation [194]. In the article by Porgham Daryasari et al., the authors employed chitosan-coated, dexamethasone-loaded MSNs as a bioactive component in a poly-L-lactic acid scaffold. The system was proven to induce osteogenic differentiation of human adipose stem cells according to the ALP activity [195]. Although both models offered promising results, the absence of osteoregeneration experiments in vivo makes these materials a mere design exercise, in which the presence of glycans is not completely justified. Apart from its use as a stabilizer, CS was also successfully employed as a functional coating able to interact with blood lipoproteins through a double effect: electrostatic interaction plus saccharide recognition [196]. Although these CS-modified dendritic MSNs permitted selectively binding and isolating low-density lipoproteins (LDL) for analytical purposes, the potential of such a nanosystem as a therapeutic agent is of enormous interest, as it may enable an interesting therapeutic effect against hypercholesterolemia and blood vessel atheromatosis.

Heparin is a highly sulfated, negatively charged glycosaminoglycan with outstanding properties as an anticoagulant. Beyond injectable applications, heparin is employed as an anticoagulant surface in medical devices like test tubes and dialysis machines. Regarding additional uses in combination with mesoporous silica, its capacity as a targeting moiety was also described against human hepatocyte carcinoma HepG2 [197,198] and HUVEC cell lines [198].

Additionally, as a consequence of its anticoagulant effect, some investigations focused on improving its release profile from silica-based carriers as nanodepots. Along this line, the research articles authored by Zhu and coworkers performed two systematic studies on how different types of mesoporous silica, with varying pore sizes and functionalization, affected the loading-release process [199,200]. Based on their results, the authors established that a slight pore enlargement of MCM-41 yielded better adsorption and release. Indeed, they also demonstrated that the presence of amino groups in the mesopores increased the amount of heparin adsorbed, albeit at the expense of slowing down the release [199]. With regard to SBA-15, they found that, apart from pore size and charge, the presence of rough pore surfaces caused by carbonaceous deposits upon calcination boosted both the retention and the release of heparin, improving the performance of such a material and enabling access to better-performing nanodepots.

Apart from its use as a drug delivery agent, there are examples of heparin-loaded silica-based materials used as anticoagulant coatings for medical devices. For example, Wei et al. reported the use of SBA-15 to prepare a heparin-releasing anticoagulant coating [201]. Their system employed amino-modified mesopores to favor heparin loading and delay its release. To incorporate these particles into chosen substrates (silicon wafer, glass, or polyvinylchloride), the authors coated the particles with polydopamine and the substrate with a catechol-modified chitosan polymer. This configuration permitted depositing the negatively charged, heparin loaded MSNs onto chitosan, which was further bound to the catechol–dopamine coating of substrates. Finally, to prevent undesired detachment, this dopamine–catechol bond was oxidized with $NaIO_4$ in order to create a non-degradable polymeric layer. When the authors tested this coating against blood, they found a very low fibrinogen adsorption, as well as platelet adhesion and hemolysis, which is in concordance with a sustained anticoagulant release and good biocompatibility. More recently, Wu et al. reported a similar approach but with an added feature: antibacterial adhesion. To achieving this, the authors loaded the mesopores of MSNs with two glycan-based structures: agarose for an antibacterial effect and heparin as an anticoagulant [202]. The agarose-heparin loaded MCM-41 nanoparticles were electrostatically immobilized onto the prepared device, i.e., an amino-modified silicone film. Therefore, the final coating presented a negative charge and exposed mesopores, which were able to produce effective agarose and heparin release. The system demonstrated a low hemolytic effect comparable with that of the previous model but with an additional antibacterial effect against *E. coli*, gaining additional applicability. However, regarding the design, the possibility of detaching SBA-15 particles within the bloodstream has to be considered as a potential risk and must be accounted for in the further development of such devices.

7. Challenges and Future Perspectives

As reviewed, the number of developing therapeutic hybrid nanosystems employing biomolecules together with mesoporous silica nanoparticles is enormous, as reported over the last few years. In general, and according to the literature here visited, cancer therapy is by far the most active research field and the one that exploits the most up-to-date technological advances such as PEGylation, controlled drug delivery, or the implementation of hybrid nanomaterials, among many others. However, such technology permitted developing alternative therapies to cancer, applied to other not-so-common diseases with a promising prognosis. This fact is largely due to the particular structure of MSNs, with a high capacity to load and release therapeutic molecules within their porous matrix. However, the advancement of knowledge and the need to adapt these nanomaterials to advanced preclinical in vivo models, with new aspects in loading and release systems, must be considered when designing these medical nanosystems. In this context, short therapeutic peptides and small molecules seem to be the ideal target for transfer to preclinical studies because they can be easily designed, developed, functionalized, and coupled. Among these new emerging therapeutic strategies with biomolecules that came to light recently, the most promising for implementation together with mesoporous silica materials seem to be immune therapy, gene modulation, and anti-infective nano-agents, as can be determined from the number and quality of related papers. In any case, there are different issues to be addressed regarding the nanocarriers themselves such as the absence of reliable in vivo pre-clinical studies, methods of clearance of the nanoplatforms, and how different particles properties (shape, size, pore symmetry and connectivity, and chemical composition) affect their application in nanomedicine.

Author Contributions: D.L. and R.R.C. contributed equally to this work, M.V.-R. coordinated the work. All authors have read and agreed to the published version of the manuscript.

Funding: The authors want to acknowledge financial support from the European Research Council ERC-2015-AdG-694160. R.R.C especially acknowledges Centro de Investigación Biomédica en Red for a postdoctoral contract.

Conflicts of Interest: The authors declare no conflicts of interest.

Abbreviations

ATP	Adenosine triphosphate
Bcl-2	B-cell lymphoma 2
bFGF	Basic fibroblast growth factor
BM	Biomacromolecule
BMP-2	Bone morphogenetic protein 2
bMSCs	Murine bone mesenchymal stem cells
BSA	Bovine serum albumin
CA	Carbonic anhydrase
cAMP	Cyclic adenosine monophosphate
COLL1A1	Collagen type 1
ConA	Concanavalin A
CS	Chondroitin sulfate
DMOHS	Dendritic mesoporous organosilica hollow spheres
DNA	Deoxyribonucleic acid
Cyt c	Cytochrome c
DOX	Doxorubicin
FA	Folic acid/folate
gG	Immunoglobulin G
GFP	Green fluorescent protein
GM-CSF	Murine granulocyte-macrophage colony-stimulating factor
GPx	Glutathione peroxidase
HMSNs	Hollow mesoporous silica nanoparticles
HA	Hyaluronic acid
HDFs	Human dermal fibroblasts
HOS	Human osteosarcoma
HRP	Horseradish peroxidase
HSCs	Human stem cells
HSP47	Heat-shock protein 47
IDMSN	Imidazole dendritic mesoporous silica nanoparticles
IL-2	Interleukin-2
kDa	Kilodaltons
LA	Lactobionic acid
LDL	Low-density lipoprotein
MCM-41	Mobil composition of matter formulation 41
MDR1	Multidrug resistance protein 1
MEF	Mouse embryonic fibroblastic cells
MPLA	Monophosphoryl lipid A
miRNA	Micro ribonucleic acid
mRNA	messenger ribonucleic acid
MSNs	Mesoporous silica nanoparticles
NA	Nucleic acid
NOX4	NADPH oxidase 4
OGP	Osteogenic growth peptide
OVA	Chicken ovalbumin
PAMAM	Polyamidoamine dendrimer
PDA	Polydopamine
PEG	Polyethylene glycol
PEI	Polyethyleneimine
pGFP	Fluorescent protein plasmid

PKM2	Glycolytic enzyme pyruvate kinase
PMDFs	Primary mouse dermal fibroblast
ROS	Oxygen species
RSNs	Rough (non-porous and core–shell) silica nanoparticles
SBA-15	Santa Barbara amorphous formulation 15
rBMSCs	Rat bone marrow stromal cells
RCA	Rolling circle amplification
shRNA	Short hairpin ribonucleic acid
siRNA	Small interfering ribonucleic acid
SO	Sorafenib
SOD	Superoxide dismutase
SPIONs	Superparamagnetic iron oxide nanoparticles.
TAT	Transactivator of transcription
TnC	Tenascin-C
TGFβR-1	Transforming growth factor beta receptor I
VEGF	Vascular endothelial growth factor

References

1. Vallet-Regi, M.; Rámila, A.; del Real, R.P.; Pérez-Pariente, J. A New Property of MCM-41: Drug Delivery System. *Chem. Mater.* **2001**, *13*, 308–311. [CrossRef]
2. Yang, J.; Li, L.; Kopeček, J. Biorecognition: A key to drug-free macromolecular therapeutics. *Biomaterials* **2019**, *190*, 11–23. [CrossRef]
3. Sever, R.; Brugge, J.S. Signal Transduction in Cancer. *Cold Spring Harb. Perspect. Med.* **2015**, *5*, a006098. [CrossRef]
4. Núñez-Lozano, R.; Cano, M.; Pimentel, B.; de la Cueva-Méndez, G. 'Smartening' anticancer therapeutic nanosystems using biomolecules. *Curr. Opin. Biotechnol.* **2015**, *35*, 135–140. [CrossRef]
5. Castillo, R.R.; Lozano, D.; González, B.; Manzano, M.; Izquierdo-Barba, I.; Vallet-Regí, M. Advances in mesoporous silica nanoparticles for targeted stimuli-responsive drug delivery: An update. *Expert Opin. Drug Deliv.* **2019**, *16*, 415–439. [CrossRef]
6. Dempke, W.C.M.; Fenchel, K.; Uciechowski, P.; Dale, S.P. Second-and third-generation drugs for immuno-oncology treatment—The more the better? *Eur. J. Cancer* **2017**, *74*, 55–72. [CrossRef]
7. Topalian, S.L.; Taube, J.M.; Anders, R.A.; Pardoll, D.M. Mechanism-driven biomarkers to guide immune checkpoint blockade in cancer therapy. *Nat. Rev. Cancer* **2016**, *16*, 275–287. [CrossRef]
8. Ribas, A.; Wolchok, J.D. Cancer immunotherapy using checkpoint blockade. *Science* **2018**, *359*, 1350–1355. [CrossRef]
9. Ahuja, N.; Sharma, A.R.; Baylin, S.B. Epigenetic Therapeutics: A New Weapon in the War Against Cancer. *Annu. Rev. Med.* **2016**, *67*, 73–89. [CrossRef]
10. Das, S.K.; Menezes, M.E.; Bhatia, S.; Wang, X.-Y.; Emdad, L.; Sarkar, D.; Fisher, P.B. Gene Therapies for Cancer: Strategies, Challenges and Successes. *J. Cell. Physiol.* **2015**, *230*, 259–271. [CrossRef]
11. Fukuhara, H.; Ino, Y.; Todo, T. Oncolytic virus therapy: A new era of cancer treatment at dawn. *Cancer Sci.* **2016**, *107*, 1373–1379. [CrossRef]
12. Yamada, H.; Urata, C.; Aoyama, Y.; Osada, S.; Yamauchi, Y.; Kuroda, K. Preparation of colloidal mesoporous silica nanoparticles with different diameters and their unique degradation behavior in static aqueous systems. *Chem. Mater.* **2012**, *24*, 1462–1471. [CrossRef]
13. Yamada, H.; Urata, C.; Ujiie, H.; Yamauchi, Y.; Kuroda, K. Preparation of aqueous colloidal mesostructured and mesoporous silica nanoparticles with controlled particle size in a very wide range from 20 nm to 700 nm. *Nanoscale* **2013**, *5*, 6145–6153. [CrossRef] [PubMed]
14. Knežević, N.Ž.; Durand, J.-O. Large pore mesoporous silica nanomaterials for application in delivery of biomolecules. *Nanoscale* **2015**, *7*, 2199–2209. [CrossRef] [PubMed]
15. Wu, S.H.; Lin, H.P. Synthesis of mesoporous silica nanoparticles. *Chem. Soc. Rev.* **2013**, *42*, 3862–3875. [CrossRef]

16. Lindén, M. Biodistribution and Excretion of Intravenously Injected Mesoporous Silica Nanoparticles: Implications for Drug Delivery Efficiency and Safety. In *The Enzymes*; Elsevier: Amsterdam, The Netherlands, 2018; Volume 43, pp. 155–180.
17. Croissant, J.G.; Fatieiev, Y.; Khashab, N.M. Degradability and Clearance of Silicon, Organosilica, Silsesquioxane, Silica Mixed Oxide, and Mesoporous Silica Nanoparticles. *Adv. Mater.* **2017**, *29*, 1604634. [CrossRef]
18. Castillo, R.R.; Vallet-Regí, M. Functional Mesoporous Silica Nanocomposites: Biomedical applications and Biosafety. *Int. J. Mol. Sci.* **2019**, *20*, 929. [CrossRef]
19. Castillo, R.R.; Colilla, M.; Vallet-Regí, M. Advances in mesoporous silica-based nanocarriers for co-delivery and combination therapy against cancer. *Expert Opin. Drug Deliv.* **2017**, *14*, 229–243. [CrossRef]
20. Castillo, R.R.; Baeza, A.; Vallet-Regí, M. Recent applications of the combination of mesoporous silica nanoparticles with nucleic acids: Development of bioresponsive devices, carriers and sensors. *Biomater. Sci.* **2017**, *5*, 353–377. [CrossRef]
21. Narayan, R.; Nayak, U.; Raichur, A.; Garg, S. Mesoporous Silica Nanoparticles: A Comprehensive Review on Synthesis and Recent Advances. *Pharmaceutics* **2018**, *10*, 118. [CrossRef]
22. Vallet-Regí, M.; Balas, F.; Colilla, M.; Manzano, M. Bone-regenerative bioceramic implants with drug and protein controlled delivery capability. *Prog. Solid State Chem.* **2008**, *36*, 163–191. [CrossRef]
23. Anselmo, A.C.; Mitragotri, S. A Review of Clinical Translation of Inorganic Nanoparticles. *AAPS J.* **2015**, *17*, 1041–1054. [CrossRef] [PubMed]
24. Anselmo, A.C.; Mitragotri, S. Nanoparticles in the clinic: An update. *Bioeng. Transl. Med.* **2019**, *4*, e10143. [CrossRef]
25. Florence, A.T. Nanotechnologies for site specific drug delivery: Changing the narrative. *Int. J. Pharm.* **2018**, *551*, 1–7. [CrossRef]
26. Baeza, A.; Manzano, M.; Colilla, M.; Vallet-Regí, M. Recent advances in mesoporous silica nanoparticles for antitumor therapy: Our contribution. *Biomater. Sci.* **2016**, *4*, 803–813. [CrossRef]
27. Singh, R.K.; Knowles, J.C.; Kim, H.-W. Advances in nanoparticle development for improved therapeutics delivery: Nanoscale topographical aspect. *J. Tissue Eng.* **2019**, *10*, 1–9. [CrossRef]
28. Niu, Y.; Yu, M.; Meka, A.; Liu, Y.; Zhang, J.; Yang, Y.; Yu, C. Understanding the contribution of surface roughness and hydrophobic modification of silica nanoparticles to enhanced therapeutic protein delivery. *J. Mater. Chem. B.* **2016**, *4*, 212–219. [CrossRef]
29. Song, H.; Yu, M.; Lu, Y.; Gu, Z.; Yang, Y.; Zhang, M.; Fu, J.; Yu, C. Plasmid DNA Delivery: Nanotopography Matters. *J. Am. Chem. Soc.* **2017**, *139*, 18247–18254. [CrossRef]
30. Aznar, E.; Oroval, M.; Pascual, L.; Murguía, J.R.; Martínez-Máñez, R.; Sancenón, F. Gated Materials for On-Command Release of Guest Molecules. *Chem. Rev.* **2016**, *116*, 561–718. [CrossRef]
31. Thi, T.T.H.; Nguyen, T.N.Q.; Hoang, D.T.; Nguyen, D.H. Functionalized mesoporous silica nanoparticles and biomedical applications. *Mater. Sci. Eng. C* **2019**, *99*, 631–656. [CrossRef]
32. Hu, J.J.; Xiao, D.; Zhang, X.Z. Advances in Peptide Functionalization on Mesoporous Silica Nanoparticles for Controlled Drug Release. *Small* **2016**, *12*, 3344–3359. [CrossRef] [PubMed]
33. Bruno, B.J.; Miller, G.D.; Lim, C.S. Basics and recent advances in peptide and protein drug delivery. *Ther. Deliv.* **2013**, *4*, 1443–1467. [CrossRef] [PubMed]
34. Baker, M. Reproducibility crisis: Blame it on the antibodies. *Nature* **2015**, *521*, 274–276. [CrossRef] [PubMed]
35. Montenegro, J.-M.; Grazu, V.; Sukhanova, A.; Agarwal, S.; de la Fuente, J.M.; Nabiev, I.; Greiner, A.; Parak, W.J. Controlled antibody/(bio-) conjugation of inorganic nanoparticles for targeted delivery. *Adv. Drug Deliv. Rev.* **2013**, *65*, 677–688. [CrossRef] [PubMed]
36. Cheng, K.; El-Boubbou, K.; Landry, C.C. Binding of HIV-1 gp120 Glycoprotein to Silica Nanoparticles Modified with CD4 Glycoprotein and CD4 Peptide Fragments. *ACS Appl. Mater. Interfaces* **2012**, *4*, 235–243. [CrossRef]
37. Kamegawa, R.; Naito, M.; Miyata, K. Functionalization of silica nanoparticles for nucleic acid delivery. *Nano Res.* **2018**, *11*, 5219–5239. [CrossRef]
38. Benjaminsen, R.V.; Mattebjerg, M.A.; Henriksen, J.R.; Moghimi, S.M.; Andresen, T.L. The Possible "Proton Sponge" Effect of Polyethylenimine (PEI) Does Not Include Change in Lysosomal pH. *Mol. Ther.* **2013**, *21*, 149–157. [CrossRef]

39. Boraschi, D.; Duschl, A. (Eds.) *Nanoparticles and the Immune System*; Academic Press: Cambridge, MA, USA, 2014; ISBN 9780124080850.
40. Tu, J.; Boyle, A.L.; Friedrich, H.; Bomans, P.H.H.; Bussmann, J.; Sommerdijk, N.A.J.M.; Jiskoot, W.; Kros, A. Mesoporous Silica Nanoparticles with Large Pores for the Encapsulation and Release of Proteins. *ACS Appl. Mater. Interfaces* **2016**, *8*, 32211–32219. [CrossRef]
41. Kruk, M. Access to Ultralarge-Pore Ordered Mesoporous Materials through Selection of Surfactant/ Swelling-Agent Micellar Templates. *Acc. Chem. Res.* **2012**, *45*, 1678–1687. [CrossRef]
42. Gao, Z.; Zharov, I. Large Pore Mesoporous Silica Nanoparticles by Templating with a Nonsurfactant Molecule, Tannic Acid. *Chem. Mater.* **2014**, *26*, 2030–2037. [CrossRef]
43. Balas, F.; Manzano, M.; Horcajada, P.; Vallet-Regi, M. Confinement and controlled release of bisphosphonates on ordered mesoporous silica-based materials. *J. Am. Chem. Soc.* **2006**, *128*, 8116–8117. [CrossRef] [PubMed]
44. Wu, M.; Meng, Q.; Chen, Y.; Du, Y.; Zhang, L.; Li, Y.; Zhang, L.; Shi, J. Large-Pore Ultrasmall Mesoporous Organosilica Nanoparticles: Micelle/Precursor Co-templating Assembly and Nuclear-Targeted Gene Delivery. *Adv. Mater.* **2015**, *27*, 215–222. [CrossRef] [PubMed]
45. Na, H.-K.; Kim, M.-H.; Park, K.; Ryoo, S.-R.; Lee, K.E.; Jeon, H.; Ryoo, R.; Hyeon, C.; Min, D.-H. Efficient Functional Delivery of siRNA using Mesoporous Silica Nanoparticles with Ultralarge Pores. *Small* **2012**, *8*, 1752–1761. [CrossRef] [PubMed]
46. Ashley, C.E.; Carnes, E.C.; Epler, K.E.; Padilla, D.P.; Phillips, G.K.; Castillo, R.E.; Wilkinson, D.C.; Wilkinson, B.S.; Burgard, C.A.; Kalinich, R.M.; et al. Delivery of Small Interfering RNA by Peptide-Targeted Mesoporous Silica Nanoparticle-Supported Lipid Bilayers. *ACS Nano* **2012**, *6*, 2174–2188. [CrossRef]
47. Akash, M.S.H.; Rehman, K.; Tariq, M.; Chen, S. Development of therapeutic proteins: Advances and challenges. *Turk. J. Biol.* **2015**, *39*, 343–358. [CrossRef]
48. Liu, H.-J.; Xu, P. Smart Mesoporous Silica Nanoparticles for Protein Delivery. *Nanomaterials* **2019**, *9*, 511. [CrossRef]
49. Xu, C.; Lei, C.; Yu, C. Mesoporous Silica Nanoparticles for Protein Protection and Delivery. *Front. Chem.* **2019**, *7*, 290. [CrossRef]
50. Yu, M.; Gu, Z.; Ottewell, T.; Yu, C. Silica-based nanoparticles for therapeutic protein delivery. *J. Mater. Chem. B* **2017**, *5*, 3241–3252. [CrossRef]
51. Deodhar, G.V.; Adams, M.L.; Trewyn, B.G. Controlled release and intracellular protein delivery from mesoporous silica nanoparticles. *Biotechnol. J.* **2017**, *12*, 1600408. [CrossRef]
52. Karpiński, T.; Adamczak, A. Anticancer Activity of Bacterial Proteins and Peptides. *Pharmaceutics* **2018**, *10*, 54. [CrossRef]
53. Liu, X.; Wu, F.; Ji, Y.; Yin, L. Recent Advances in Anti-cancer Protein/Peptide Delivery. *Bioconjug. Chem.* **2019**, *30*, 305–324. [CrossRef] [PubMed]
54. Slowing, I.I.; Trewyn, B.G.; Lin, V.S.-Y. Mesoporous Silica Nanoparticles for Intracellular Delivery of Membrane-Impermeable Proteins. *J. Am. Chem. Soc.* **2007**, *129*, 8845–8849. [CrossRef] [PubMed]
55. Méndez, J.; Morales Cruz, M.; Delgado, Y.; Figueroa, C.M.; Orellano, E.A.; Morales, M.; Monteagudo, A.; Griebenow, K. Delivery of chemically glycosylated cytochrome c immobilized in mesoporous silica nanoparticles induces apoptosis in HeLa cancer cells. *Mol. Pharm.* **2014**, *11*, 102–111. [CrossRef] [PubMed]
56. Shang, W.; Nuffer, J.H.; Muñiz-Papandrea, V.A.; Colón, W.; Siegel, R.W.; Dordick, J.S. Cytochrome c on silica nanoparticles: Influence of nanoparticle size on protein structure, stability, and activity. *Small* **2009**, *5*, 470–476. [CrossRef]
57. Huang, W.-Y.; Davies, G.-L.; Davis, J.J. Engineering Cytochrome-Modified Silica Nanoparticles To Induce Programmed Cell Death. *Chem. A Eur. J.* **2013**, *19*, 17891–17898. [CrossRef]
58. Choi, E.; Lim, D.-K.; Kim, S. Hydrolytic surface erosion of mesoporous silica nanoparticles for efficient intracellular delivery of cytochrome c. *J. Colloid Interface Sci.* **2020**, *560*, 416–425. [CrossRef]
59. Martínez-Carmona, M.; Lozano, D.; Colilla, M.; Vallet-Regí, M. Lectin-conjugated pH-responsive mesoporous silica nanoparticles for targeted bone cancer treatment. *Acta Biomater.* **2018**, *65*, 393–404. [CrossRef]
60. Lim, J.-S.; Lee, K.; Choi, J.-N.; Hwang, Y.-K.; Yun, M.-Y.; Kim, H.-J.; Won, Y.S.; Kim, S.-J.; Kwon, H.; Huh, S. Intracellular protein delivery by hollow mesoporous silica capsules with a large surface hole. *Nanotechnology* **2012**, *23*, 085101. [CrossRef]

61. Wang, X.; Li, X.; Ito, A.; Yoshiyuki, K.; Sogo, Y.; Watanabe, Y.; Yamazaki, A.; Ohno, T.; Tsuji, N.M. Silica Nanospheres: Hollow Structure Improved Anti-Cancer Immunity of Mesoporous Silica Nanospheres In Vivo. *Small* **2016**, *12*, 3510–3515. [CrossRef]
62. Wang, X.; Li, X.; Yoshiyuki, K.; Watanabe, Y.; Sogo, Y.; Ohno, T.; Tsuji, N.M.; Ito, A. Comprehensive Mechanism Analysis of Mesoporous-Silica-Nanoparticle-Induced Cancer Immunotherapy. *Adv. Healthc. Mater.* **2016**, *5*, 1169–1176. [CrossRef]
63. Yang, Y.; Lu, Y.; Abbaraju, P.L.; Zhang, J.; Zhang, M.; Xiang, G.; Yu, C. Multi-shelled Dendritic Mesoporous Organosilica Hollow Spheres: Roles of Composition and Architecture in Cancer Immunotherapy. *Angew. Chem. Int. Ed.* **2017**, *56*, 8446–8450. [CrossRef] [PubMed]
64. Cha, B.G.; Jeong, J.H.; Kim, J. Extra-Large Pore Mesoporous Silica Nanoparticles Enabling Co-Delivery of High Amounts of Protein Antigen and Toll-like Receptor 9 Agonist for Enhanced Cancer Vaccine Efficacy. *ACS Cent. Sci.* **2018**, *4*, 484–492. [CrossRef] [PubMed]
65. Nguyen, T.L.; Cha, B.G.; Choi, Y.; Im, J.; Kim, J. Injectable dual-scale mesoporous silica cancer vaccine enabling efficient delivery of antigen/adjuvant-loaded nanoparticles to dendritic cells recruited in local macroporous scaffold. *Biomaterials* **2020**, *239*, 119859. [CrossRef] [PubMed]
66. Niu, Y.; Yu, M.; Zhang, J.; Yang, Y.; Xu, C.; Yeh, M.; Taran, E.; Hou, J.J.C.; Gray, P.P.; Yu, C. Synthesis of silica nanoparticles with controllable surface roughness for therapeutic protein delivery. *J. Mater. Chem. B* **2015**, *3*, 8477–8485. [CrossRef] [PubMed]
67. Kong, M.; Tang, J.; Qiao, Q.; Wu, T.; Qi, Y.; Tan, S.; Gao, X.; Zhang, Z. Biodegradable Hollow Mesoporous Silica Nanoparticles for Regulating Tumor Microenvironment and Enhancing Antitumor Efficiency. *Theranostics* **2017**, *7*, 3276–3292. [CrossRef]
68. Guo, H.C.; Feng, X.M.; Sun, S.Q.; Wei, Y.Q.; Sun, D.H.; Liu, X.T.; Liu, Z.X.; Luo, J.X.; Yin, H. Immunization of mice by hollow mesoporous silica nanoparticles as carriers of porcine circovirus type 2 ORF2 protein. *Virol. J.* **2012**, *9*, 1–10. [CrossRef]
69. Oliveira, D.C.d.P.; de Barros, A.L.B.; Belardi, R.M.; de Goes, A.M.; de Oliveira Souza, B.K.; Soares, D.C.F. Mesoporous silica nanoparticles as a potential vaccine adjuvant against Schistosoma mansoni. *J. Drug Deliv. Sci. Technol.* **2016**, *35*, 234–240. [CrossRef]
70. Virginio, V.G.; Bandeira, N.C.; dos Anjos Leal, F.M.; Lancellotti, M.; Zaha, A.; Ferreira, H.B. Assessment of the adjuvant activity of mesoporous silica nanoparticles in recombinant Mycoplasma hyopneumoniae antigen vaccines. *Heliyon* **2017**, *3*, e00225. [CrossRef]
71. Hajizade, A.; Salmanian, A.H.; Amani, J.; Ebrahimi, F.; Arpanaei, A. EspA-loaded mesoporous silica nanoparticles can efficiently protect animal model against enterohaemorrhagic E. coli O157: H7. *Artif. Cells Nanomed. Biotechnol.* **2018**, *46*, S1067–S1075. [CrossRef]
72. Ferreira Soares, D.C.; Soares, L.M.; Miranda de Goes, A.; Melo, E.M.; Branco de Barros, A.L.; Alves Santos Bicalho, T.C.; Leao, N.M.; Tebaldi, M.L. Mesoporous SBA-16 silica nanoparticles as a potential vaccine adjuvant against Paracoccidioides brasiliensis. *Microporous Mesoporous Mater.* **2020**, *291*, 109676. [CrossRef]
73. Méndez, J.; Monteagudo, A.; Griebenow, K. Stimulus-responsive controlled release system by covalent immobilization of an enzyme into mesoporous silica nanoparticles. *Bioconjug. Chem.* **2012**, *23*, 698–704. [CrossRef] [PubMed]
74. Gößl, D.; Singer, H.; Chiu, H.Y.; Schmidt, A.; Lichtnecker, M.; Engelke, H.; Bein, T. Highly active enzymes immobilized in large pore colloidal mesoporous silica nanoparticles. *N. J. Chem.* **2019**, *43*, 1671–1680. [CrossRef]
75. Xu, C.; Yu, M.; Noonan, O.; Zhang, J.; Song, H.; Zhang, H.; Lei, C.; Niu, Y.; Huang, X.; Yang, Y.; et al. Core-Cone Structured Monodispersed Mesoporous Silica Nanoparticles with Ultra-large Cavity for Protein Delivery. *Small* **2015**, *11*, 5949–5955. [CrossRef] [PubMed]
76. Chen, Y.P.; Chen, C.T.; Hung, Y.; Chou, C.M.; Liu, T.P.; Liang, M.R.; Chen, C.T.; Mou, C.Y. A new strategy for intracellular delivery of enzyme using mesoporous silica nanoparticles: Superoxide dismutase. *J. Am. Chem. Soc.* **2013**, *135*, 1516–1523. [CrossRef] [PubMed]
77. Lin, Y.H.; Chen, Y.P.; Liu, T.P.; Chien, F.C.; Chou, C.M.; Chen, C.T.; Mou, C.Y. Approach to Deliver Two Antioxidant Enzymes with Mesoporous Silica Nanoparticles into Cells. *ACS Appl. Mater. Interfaces* **2016**, *8*, 17944–17954. [CrossRef] [PubMed]

78. Han, D.H.; Na, H.-K.; Choi, W.H.; Lee, J.H.; Kim, Y.K.; Won, C.; Lee, S.-H.; Kim, K.P.; Kuret, J.; Min, D.-H.; et al. Direct cellular delivery of human proteasomes to delay tau aggregation. *Nat. Commun.* **2014**, *5*, 5633. [CrossRef]
79. Zhang, J.; Postovit, L.M.; Wang, D.; Gardiner, R.B.; Harris, R.; Abdul, M.M.; Thomas, A.A. In situ loading of basic fibroblast growth factor within porous silica nanoparticles for a prolonged release. *Nanoscale Res. Lett.* **2009**, *4*, 1297–1302. [CrossRef]
80. Gan, Q.; Zhu, J.; Yuan, Y.; Liu, H.; Qian, J.; Li, Y.; Liu, C. A dual-delivery system of pH-responsive chitosan-functionalized mesoporous silica nanoparticles bearing BMP-2 and dexamethasone for enhanced bone regeneration. *J. Mater. Chem. B* **2015**, *3*, 2056–2066. [CrossRef]
81. Gan, Q.; Zhu, J.; Yuan, Y.; Liu, C. pH-Responsive Fe3O4 Nanopartilces-Capped Mesoporous Silica Supports for Protein Delivery. *J. Nanosci. Nanotechnol.* **2016**, *16*, 5470–5479. [CrossRef]
82. Li, L.-L.; Wang, H. Enzyme-coated mesoporous silica nanoparticles as efficient antibacterial agents in vivo. *Adv. Healthc. Mater.* **2013**, *2*, 1351–1360. [CrossRef]
83. Wang, Y.; Nor, Y.A.; Song, H.; Yang, Y.; Xu, C.; Yu, M.; Yu, C. Small-sized and large-pore dendritic mesoporous silica nanoparticles enhance antimicrobial enzyme delivery. *J. Mater. Chem. B* **2016**, *4*, 2646–2653. [CrossRef] [PubMed]
84. Song, H.; Ahmad Nor, Y.; Yu, M.; Yang, Y.; Zhang, J.; Zhang, H.; Xu, C.; Mitter, N.; Yu, C. Silica Nanopollens Enhance Adhesion for Long-Term Bacterial Inhibition. *J. Am. Chem. Soc.* **2016**, *138*, 6455–6462. [CrossRef] [PubMed]
85. Xu, C.; He, Y.; Li, Z.; Ahmad Nor, Y.; Ye, Q. Nanoengineered hollow mesoporous silica nanoparticles for the delivery of antimicrobial proteins into biofilms. *J. Mater. Chem. B* **2018**, *6*, 1899–1902. [CrossRef]
86. Martínez-Carmona, M.; Izquierdo-Barba, I.; Colilla, M.; Vallet-Regí, M. Concanavalin A-targeted mesoporous silica nanoparticles for infection treatment. *Acta Biomater.* **2019**, *96*, 547–556. [CrossRef]
87. Wang, X.; Li, X.; Ito, A.; Sogo, Y.; Ohno, T. Particle-size-dependent toxicity and immunogenic activity of mesoporous silica-based adjuvants for tumor immunotherapy. *Acta Biomater.* **2013**, *9*, 7480–7489. [CrossRef]
88. Li, X.; Wang, X.; Sogo, Y.; Ohno, T.; Onuma, K.; Ito, A. Mesoporous Silica-Calcium Phosphate-Tuberculin Purified Protein Derivative Composites as an Effective Adjuvant for Cancer Immunotherapy. *Adv. Healthc. Mater.* **2013**, *2*, 863–871. [CrossRef]
89. Wang, X.; Li, X.; Ito, A.; Watanabe, Y.; Sogo, Y.; Tsuji, N.M.; Ohno, T. Stimulation of In Vivo Antitumor Immunity with Hollow Mesoporous Silica Nanospheres. *Angew. Chem. Int. Ed.* **2016**, *55*, 1899–1903. [CrossRef]
90. Kim, M.-G.; Park, J.Y.; Shon, Y.; Kim, G.; Shim, G.; Oh, Y.-K. Nanotechnology and vaccine development. *Asian J. Pharm. Sci.* **2014**, *9*, 227–235. [CrossRef]
91. Pati, R.; Shevtsov, M.; Sonawane, A. Nanoparticle Vaccines Against Infectious Diseases. *Front. Immunol.* **2018**, *9*, 2224. [CrossRef]
92. Villegas, M.R.; Baeza, A.; Vallet-Regí, M. Nanotechnological Strategies for Protein Delivery. *Molecules* **2018**, *23*, 1008. [CrossRef]
93. Zhao, M.; Liu, Y.; Hsieh, R.S.; Wang, N.; Tai, W.; Joo, K.-I.; Wang, P.; Gu, Z.; Tang, Y. Clickable Protein Nanocapsules for Targeted Delivery of Recombinant p53 Protein. *J. Am. Chem. Soc.* **2014**, *136*, 15319–15325. [CrossRef] [PubMed]
94. Villegas, M.R.; Baeza, A.; Usategui, A.; Ortiz-Romero, P.L.; Pablos, J.L.; Vallet-Regí, M. Collagenase nanocapsules: An approach to fibrosis treatment. *Acta Biomater.* **2018**, *74*, 430–438. [CrossRef] [PubMed]
95. Mohammed, G.K.; Obaidat, R.M.; Assaf, S.; Khanfar, M.; Al-Taani, B. Formulations and Technologies in Growth Hormone Delivery. *Int. J. Pharm. Pharm. Sci.* **2017**, *9*, 1–12. [CrossRef]
96. Rohrer, T.R.; Horikawa, R.; Kappelgaard, A.-M. Growth hormone delivery devices: Current features and potential for enhanced treatment adherence. *Expert Opin. Drug Deliv.* **2017**, *14*, 1253–1264. [CrossRef] [PubMed]
97. Kuijpers, A.J.; Engbers, G.H.; van Wachem, P.B.; Krijgsveld, J.; Zaat, S.A.; Dankert, J.; Feijen, J. Controlled delivery of antibacterial proteins from biodegradable matrices. *J. Control. Release* **1998**, *53*, 235–247. [CrossRef]
98. Jenssen, H.; Hancock, R.E.W. Antimicrobial properties of lactoferrin. *Biochimie* **2009**, *91*, 19–29. [CrossRef]
99. Zhu, Y.; Feijen, J.; Zhong, Z. Dual-targeted nanomedicines for enhanced tumor treatment. *Nano Today* **2018**, *18*, 65–85. [CrossRef]

100. Song, Y.; Du, D.; Li, L.; Xu, J.; Dutta, P.; Lin, Y. In Vitro Study of Receptor-Mediated Silica Nanoparticles Delivery across Blood–Brain Barrier. *ACS Appl. Mater. Interfaces* **2017**, *9*, 20410–20416. [CrossRef]
101. Kalmouni, M.; Al-Hosani, S.; Magzoub, M. Cancer targeting peptides. *Cell. Mol. Life Sci.* **2019**, *76*, 2171–2183. [CrossRef]
102. Zhao, N.; Qin, Y.; Liu, H.; Cheng, Z. Tumor-Targeting Peptides: Ligands for Molecular Imaging and Therapy. *Anticancer Agents Med. Chem.* **2018**, *18*, 74–86. [CrossRef]
103. Robinson, J.A. Folded Synthetic Peptides and Other Molecules Targeting Outer Membrane Protein Complexes in Gram-Negative Bacteria. *Front. Chem.* **2019**, *7*, 45. [CrossRef] [PubMed]
104. Malanovic, N.; Lohner, K. Antimicrobial Peptides Targeting Gram-Positive Bacteria. *Pharmaceuticals* **2016**, *9*, 59. [CrossRef] [PubMed]
105. Jiang, Z.; Guan, J.; Qian, J.; Zhan, C. Peptide ligand-mediated targeted drug delivery of nanomedicines. *Biomater. Sci.* **2019**, *7*, 461–471. [CrossRef]
106. Marqus, S.; Pirogova, E.; Piva, T.J. Evaluation of the use of therapeutic peptides for cancer treatment. *J. Biomed. Sci.* **2017**, *24*, 21. [CrossRef]
107. Gaspar, D.; Salomé Veiga, A.; Castanho, M.A.R.B. From antimicrobial to anticancer peptides. A review. *Front. Microbiol.* **2013**, *4*, 294. [CrossRef]
108. Kurrikoff, K.; Aphkhazava, D.; Langel, Ü. The future of peptides in cancer treatment. *Curr. Opin. Pharmacol.* **2019**, *47*, 27–32. [CrossRef]
109. Mahlapuu, M.; Håkansson, J.; Ringstad, L.; Björn, C. Antimicrobial Peptides: An Emerging Category of Therapeutic Agents. *Front. Cell Infect. Microbiol.* **2016**, *6*, 194. [CrossRef]
110. Lau, J.L.; Dunn, M.K. Therapeutic peptides: Historical perspectives, current development trends, and future directions. *Bioorg. Med. Chem.* **2018**, *26*, 2700–2707. [CrossRef]
111. Luo, G.-F.; Chen, W.-H.; Liu, Y.; Zhang, J.; Cheng, S.-X.; Zhuo, R.-X.; Zhang, X.-Z. Charge-reversal plug gate nanovalves on peptide-functionalized mesoporous silica nanoparticles for targeted drug delivery. *J. Mater. Chem. B* **2013**, *1*, 5723–5732. [CrossRef]
112. Luo, G.F.; Chen, W.H.; Liu, Y.; Lei, Q.; Zhuo, R.X.; Zhang, X.Z. Multifunctional enveloped mesoporous silica nanoparticles for subcellular co-delivery of drug and therapeutic peptide. *Sci. Rep.* **2014**, *4*, 6064. [CrossRef]
113. Cheng, Y.J.; Zeng, X.; Cheng, D.B.; Xu, X.D.; Zhang, X.Z.; Zhuo, R.X.; He, F. Functional mesoporous silica nanoparticles (MSNs) for highly controllable drug release and synergistic therapy. *Colloids Surfaces B Biointerfaces* **2016**, *145*, 217–225. [CrossRef] [PubMed]
114. Xiao, D.; Hu, J.-J.; Zhu, J.-Y.; Wang, S.-B.; Zhuo, R.-X.; Zhang, X.-Z. A redox-responsive mesoporous silica nanoparticle with a therapeutic peptide shell for tumor targeting synergistic therapy. *Nanoscale* **2016**, *8*, 16702–16709. [CrossRef] [PubMed]
115. Zhang, J.; Shen, B.; Chen, L.; Chen, L.; Meng, Y.; Feng, J. A dual-sensitive mesoporous silica nanoparticle based drug carrier for cancer synergetic therapy. *Colloids Surfaces B Biointerfaces* **2019**, *175*, 65–72. [CrossRef] [PubMed]
116. De la Torre, C.; Domínguez-Berrocal, L.; Murguía, J.R.; Marcos, M.D.; Martínez-Máñez, R.; Bravo, J.; Sancenón, F. ϵ-Polylysine-Capped Mesoporous Silica Nanoparticles as Carrier of the C9h Peptide to Induce Apoptosis in Cancer Cells. *Chem. A Eur. J.* **2018**, *24*, 1890–1897. [CrossRef] [PubMed]
117. Cao, J.; Zhang, Y.; Shan, Y.; Wang, J.; Liu, F.; Liu, H.; Xing, G.; Lei, J.; Zhou, J. A pH-dependent Antibacterial Peptide Release Nano-system Blocks Tumor Growth in vivo without Toxicity. *Sci. Rep.* **2017**, *7*, 1–13. [CrossRef] [PubMed]
118. Rahmani, S.; Budimir, J.; Sejalon, M.; Daurat, M.; Aggad, D.; Vivès, E.; Raehm, L.; Garcia, M.; Lichon, L.; Gary-Bobo, M.; et al. Large pore mesoporous silica and organosilica nanoparticles for pepstatin A delivery in breast cancer cells. *Molecules* **2019**, *24*, 332. [CrossRef]
119. Wu, Y.; Ge, P.; Xu, W.; Li, M.; Kang, Q.; Zhang, X.; Xie, J. Cancer-targeted and intracellular delivery of Bcl-2-converting peptide with functional macroporous silica nanoparticles for biosafe treatment. *Mater. Sci. Eng. C* **2020**, *108*, 110386. [CrossRef]
120. Xie, J.; Xu, W.; Wu, Y.; Niu, B.; Zhang, X. Macroporous organosilicon nanocomposites co-deliver Bcl2-converting peptide and chemotherapeutic agent for synergistic treatment against multidrug resistant cancer. *Cancer Lett.* **2020**, *469*, 340–354. [CrossRef]

121. Xie, J.; Yang, C.; Liu, Q.; Li, J.; Liang, R.; Shen, C.; Zhang, Y.; Wang, K.; Liu, L.; Shezad, K.; et al. Encapsulation of Hydrophilic and Hydrophobic Peptides into Hollow Mesoporous Silica Nanoparticles for Enhancement of Antitumor Immune Response. *Small* **2017**, *13*, 1701741. [CrossRef]
122. Braun, K.; Pochert, A.; Lindén, M.; Davoudi, M.; Schmidtchen, A.; Nordström, R.; Malmsten, M. Membrane interactions of mesoporous silica nanoparticles as carriers of antimicrobial peptides. *J. Colloid Interface Sci.* **2016**, *475*, 161–170. [CrossRef]
123. Tenland, E.; Pochert, A.; Krishnan, N.; Rao, K.U.; Kalsum, S.; Braun, K.; Glegola-Madejska, I.; Lerm, M.; Robertson, B.D.; Lindén, M.; et al. Effective delivery of the anti-mycobacterial peptide NZX in mesoporous silica nanoparticles. *PLoS ONE* **2019**, *14*, e0212858. [CrossRef]
124. Yu, Q.; Deng, T.; Lin, F.-C.; Zhang, B.; Zink, J.I. Supramolecular Assemblies of Heterogeneous Mesoporous Silica Nanoparticles to Co-deliver Antimicrobial Peptides and Antibiotics for Synergistic Eradication of Pathogenic Biofilms. *ACS Nano* **2020**. [CrossRef]
125. Zhao, Y.; Trewyn, B.G.; Slowing, I.I.; Lin, V.S.-Y. Mesoporous Silica Nanoparticle-Based Double Drug Delivery System for Glucose-Responsive Controlled Release of Insulin and Cyclic AMP. *J. Am. Chem. Soc.* **2009**, *131*, 8398–8400. [CrossRef] [PubMed]
126. Sun, L.; Zhang, X.; Zheng, C.; Wu, Z.; Li, C. A pH gated, glucose-sensitive nanoparticle based on worm-like mesoporous silica for controlled insulin release. *J. Phys. Chem. B* **2013**, *117*, 3852–3860. [CrossRef] [PubMed]
127. Zakeri Siavashani, A.; Haghbin Nazarpak, M.; Fayyazbakhsh, F.; Toliyat, T.; McInnes, S.J.P.; Solati-Hashjin, M. Effect of amino-functionalization on insulin delivery and cell viability for two types of silica mesoporous structures. *J. Mater. Sci.* **2016**, *51*, 10897–10909. [CrossRef]
128. Lozano, D.; Manzano, M.; Doadrio, J.C.; Salinas, A.J.; Vallet-Regí, M.; Gómez-Barrena, E.; Esbrit, P. Osteostatin-loaded bioceramics stimulate osteoblastic growth and differentiation. *Acta Biomater.* **2010**, *6*, 797–803. [CrossRef]
129. Trejo, C.G.; Lozano, D.; Manzano, M.; Doadrio, J.C.; Salinas, A.J.; Dapía, S.; Gómez-Barrena, E.; Vallet-Regí, M.; García-Honduvilla, N.; Buján, J.; et al. The osteoinductive properties of mesoporous silicate coated with osteostatin in a rabbit femur cavity defect model. *Biomaterials* **2010**, *31*, 8564–8573. [CrossRef]
130. Lozano, D.; Trejo, C.G.; Gómez-Barrena, E.; Manzano, M.; Doadrio, J.C.; Salinas, A.J.; Vallet-Regí, M.; García-Honduvilla, N.; Esbrit, P.; Buján, J. Osteostatin-loaded onto mesoporous ceramics improves the early phase of bone regeneration in a rabbit osteopenia model. *Acta Biomater.* **2012**, *8*, 2317–2323. [CrossRef]
131. Mendes, L.S.; Saska, S.; Martines, M.A.U.; Marchetto, R. Nanostructured materials based on mesoporous silica and mesoporous silica/apatite as osteogenic growth peptide carriers. *Mater. Sci. Eng. C* **2013**, *33*, 4427–4434. [CrossRef]
132. Zhou, X.; Feng, W.; Qiu, K.; Chen, L.; Wang, W.; Nie, W.; Mo, X.; He, C. BMP-2 Derived Peptide and Dexamethasone Incorporated Mesoporous Silica Nanoparticles for Enhanced Osteogenic Differentiation of Bone Mesenchymal Stem Cells. *ACS Appl. Mater. Interfaces* **2015**, *7*, 15777–15789. [CrossRef]
133. Du, A.W.; Stenzel, M.H. Drug carriers for the delivery of therapeutic peptides. *Biomacromolecules* **2014**, *15*, 1097–1114. [CrossRef] [PubMed]
134. Mader, J.S.; Hoskin, D.W. Cationic antimicrobial peptides as novel cytotoxic agents for cancer treatment. *Expert Opin. Investig. Drugs* **2006**, *15*, 933–946. [CrossRef] [PubMed]
135. Torchilin, V.P. Tat peptide-mediated intracellular delivery of pharmaceutical nanocarriers. *Adv. Drug Deliv. Rev.* **2008**, *60*, 548–558. [CrossRef]
136. Boohaker, R.J.; Lee, M.W.; Vishnubhotla, P.; Perez, J.L.M.; Khaled, A.R. The Use of Therapeutic Peptides to Target and to Kill Cancer Cells. *Curr. Med. Chem.* **2012**, *19*, 3794–3804. [CrossRef]
137. Naz, S.; Wang, M.; Han, Y.; Hu, B.; Teng, L.; Zhou, J.; Zhang, H.; Chen, J. Enzyme-responsive mesoporous silica nanoparticles for tumor cells and mitochondria multistage-targeted drug delivery. *Int. J. Nanomed.* **2019**, *14*, 2533–2542. [CrossRef]
138. Kolluri, S.K.; Zhu, X.; Zhou, X.; Lin, B.; Chen, Y.; Sun, K.; Tian, X.; Town, J.; Cao, X.; Lin, F.; et al. A Short Nur77-Derived Peptide Converts Bcl-2 from a Protector to a Killer. *Cancer Cell* **2008**, *14*, 285–298. [CrossRef]
139. Nordström, R.; Malmsten, M. Delivery systems for antimicrobial peptides. *Adv. Colloid Interface Sci.* **2017**, *242*, 17–34. [CrossRef]
140. Braun, K.; Pochert, A.; Gerber, M.; Raber, H.F.; Lindén, M. Influence of mesopore size and peptide aggregation on the adsorption and release of a model antimicrobial peptide onto/from mesoporous silica nanoparticles in vitro. *Mol. Syst. Des. Eng.* **2017**, *2*, 393–400. [CrossRef]

141. Tenland, E.; Krishnan, N.; Rönnholm, A.; Kalsum, S.; Puthia, M.; Mörgelin, M.; Davoudi, M.; Otrocka, M.; Alaridah, N.; Glegola-Madejska, I.; et al. A novel derivative of the fungal antimicrobial peptide plectasin is active against Mycobacterium tuberculosis. *Tuberculosis* **2018**, *113*, 231–238. [CrossRef]
142. Heras, C.; Sanchez-Salcedo, S.; Lozano, D.; Peña, J.; Esbrit, P.; Vallet-Regi, M.; Salinas, A.J. Osteostatin potentiates the bioactivity of mesoporous glass scaffolds containing Zn 2+ ions in human mesenchymal stem cells. *Acta Biomater.* **2019**, *89*, 359–371. [CrossRef]
143. Pérez, R.; Sanchez-Salcedo, S.; Lozano, D.; Heras, C.; Esbrit, P.; Vallet-Regí, M.; Salinas, A.J. Osteogenic effect of ZnO-mesoporous glasses loaded with osteostatin. *Nanomaterials* **2018**, *8*, 592. [CrossRef] [PubMed]
144. Nayerossadat, N.; Ali, P.; Maedeh, T. Viral and nonviral delivery systems for gene delivery. *Adv. Biomed. Res.* **2012**, *1*, 27. [CrossRef] [PubMed]
145. Cha, W.; Fan, R.; Miao, Y.; Zhou, Y.; Qin, C.; Shan, X.; Wan, X.; Li, J. Mesoporous Silica Nanoparticles as Carriers for Intracellular Delivery of Nucleic Acids and Subsequent Therapeutic Applications. *Molecules* **2017**, *22*, 782. [CrossRef] [PubMed]
146. Zarei, H.; Kazemi Oskuee, R.; Hanafi-Bojd, M.Y.; Gholami, L.; Ansari, L.; Malaekeh-Nikouei, B. Enhanced gene delivery by polyethyleneimine coated mesoporous silica nanoparticles. *Pharm. Dev. Technol.* **2019**, *24*, 127–132. [CrossRef]
147. Xia, T.; Kovochich, M.; Liong, M.; Meng, H.; Kabehie, S.; George, S.; Zink, J.I.; Nel, A.E. Polyethyleneimine Coating Enhances the Cellular Uptake of Mesoporous Silica Nanoparticles and Allows Safe Delivery of siRNA and DNA Constructs. *ACS Nano* **2009**, *3*, 3273–3286. [CrossRef]
148. Wang, Y.; Shang, X.; Liu, J.; Guo, Y. ATP mediated rolling circle amplification and opening DNA-gate for drug delivery to cell. *Talanta* **2018**, *176*, 652–658. [CrossRef]
149. Wang, S.; Liu, F.; Li, X.-L. Monitoring of "on-demand" drug release using dual tumor marker mediated DNA-capped versatile mesoporous silica nanoparticles. *Chem. Commun.* **2017**, *53*, 8755–8758. [CrossRef]
150. Li, Z.; Zhang, L.; Tang, C.; Yin, C. Co-Delivery of Doxorubicin and Survivin shRNA-Expressing Plasmid Via Microenvironment-Responsive Dendritic Mesoporous Silica Nanoparticles for Synergistic Cancer Therapy. *Pharm. Res.* **2017**, *34*, 2829–2841. [CrossRef]
151. Sun, P.; Leidner, A.; Weigel, S.; Weidler, P.G.; Heissler, S.; Scharnweber, T.; Niemeyer, C.M. Biopebble Containers: DNA-Directed Surface Assembly of Mesoporous Silica Nanoparticles for Cell Studies. *Small* **2019**, *15*, 1900083. [CrossRef]
152. Li, Y.; Chen, Y.; Pan, W.; Yu, Z.; Yang, L.; Wang, H.; Li, N.; Tang, B. Nanocarriers with multi-locked DNA valves targeting intracellular tumor-related mRNAs for controlled drug release. *Nanoscale* **2017**, *9*, 17318–17324. [CrossRef]
153. Pascual, L.; Cerqueira-Coutinho, C.; García-Fernández, A.; de Luis, B.; Bernardes, E.S.; Albernaz, M.S.; Missailidis, S.; Martínez-Máñez, R.; Santos-Oliveira, R.; Orzaez, M.; et al. MUC1 aptamer-capped mesoporous silica nanoparticles for controlled drug delivery and radio-imaging applications. *Nanomed. Nanotechnol. Biol. Med.* **2017**, *13*, 2495–2505. [CrossRef]
154. Srivastava, P.; Hira, S.K.; Sharma, A.; Kashif, M.; Srivastava, P.; Srivastava, D.N.; Singh, R.A.; Manna, P.P. Telomerase Responsive Delivery of Doxorubicin from Mesoporous Silica Nanoparticles in Multiple Malignancies: Therapeutic Efficacies against Experimental Aggressive Murine Lymphoma. *Bioconjug. Chem.* **2018**, *29*, 2107–2119. [CrossRef] [PubMed]
155. Wang, J.; Zhu, R.; Gao, B.; Wu, B.; Li, K.; Sun, X.; Liu, H.; Wang, S. The enhanced immune response of hepatitis B virus DNA vaccine using SiO2@LDH nanoparticles as an adjuvant. *Biomaterials* **2014**, *35*, 466–478. [CrossRef] [PubMed]
156. Shen, J.; Kim, H.-C.; Su, H.; Wang, F.; Wolfram, J.; Kirui, D.; Mai, J.; Mu, C.; Ji, L.-N.; Mao, Z.-W.; et al. Cyclodextrin and Polyethylenimine Functionalized Mesoporous Silica Nanoparticles for Delivery of siRNA Cancer Therapeutics. *Theranostics* **2014**, *4*, 487–497. [CrossRef] [PubMed]
157. Prabhakar, N.; Zhang, J.; Desai, D.; Casals, E.; Gulin-Sarfraz, T.; Näreoja, T.; Westermarck, J.; Rosenholm, J. Stimuli-responsive hybrid nanocarriers developed by controllable integration of hyperbranched PEI with mesoporous silica nanoparticles for sustained intracellular siRNA delivery. *Int. J. Nanomed.* **2016**, *11*, 6591–6608. [CrossRef] [PubMed]
158. Ngamcherdtrakul, W.; Sangvanich, T.; Reda, M.; Gu, S.; Bejan, D.; Yantasee, W. Lyophilization and stability of antibody-conjugated mesoporous silica nanoparticle with cationic polymer and PEG for siRNA delivery. *Int. J. Nanomed.* **2018**, *13*, 4015–4027. [CrossRef]

159. Lio, D.C.S.; Liu, C.; Oo, M.M.S.; Wiraja, C.; Teo, M.H.Y.; Zheng, M.; Chew, S.W.T.; Wang, X.; Xu, C. Transdermal delivery of small interfering RNAs with topically applied mesoporous silica nanoparticles for facile skin cancer treatment. *Nanoscale* **2019**, *11*, 17041–17051. [CrossRef]
160. Wang, D.; Xu, X.; Zhang, K.; Sun, B.; Wang, L.; Meng, L.; Liu, Q.; Zheng, C.; Yang, B.; Sun, H. Codelivery of doxorubicin and MDR1-siRNA by mesoporous silica nanoparticles-polymerpolyethylenimine to improve oral squamous carcinoma treatment. *Int. J. Nanomed.* **2017**, *13*, 187–198. [CrossRef]
161. Shi, X.-L.; Li, Y.; Zhao, L.-M.; Su, L.-W.; Ding, G. Delivery of MTH1 inhibitor (TH287) and MDR1 siRNA via hyaluronic acid-based mesoporous silica nanoparticles for oral cancers treatment. *Colloids Surfaces B Biointerfaces* **2019**, *173*, 599–606. [CrossRef]
162. Pan, Q.-S.; Chen, T.-T.; Nie, C.-P.; Yi, J.-T.; Liu, C.; Hu, Y.-L.; Chu, X. In Situ Synthesis of Ultrathin ZIF-8 Film-Coated MSNs for Codelivering Bcl 2 siRNA and Doxorubicin to Enhance Chemotherapeutic Efficacy in Drug-Resistant Cancer Cells. *ACS Appl. Mater. Interfaces* **2018**, *10*, 33070–33077. [CrossRef]
163. Choi, E.; Lee, J.; Kwon, I.C.; Lim, D.-K.; Kim, S. Cumulative directional calcium gluing between phosphate and silicate: A facile, robust and biocompatible strategy for siRNA delivery by amine-free non-positive vector. *Biomaterials* **2019**, *209*, 126–137. [CrossRef] [PubMed]
164. Shahin, S.A.; Wang, R.; Simargi, S.I.; Contreras, A.; Parra Echavarria, L.; Qu, L.; Wen, W.; Dellinger, T.; Unternaehrer, J.; Tamanoi, F.; et al. Hyaluronic acid conjugated nanoparticle delivery of siRNA against TWIST reduces tumor burden and enhances sensitivity to cisplatin in ovarian cancer. *Nanomed. Nanotechnol. Biol. Med.* **2018**, *14*, 1381–1394. [CrossRef]
165. Zheng, G.; Shen, Y.; Zhao, R.; Chen, F.; Zhang, Y.; Xu, A.; Shao, J. Dual-Targeting Multifuntional Mesoporous Silica Nanocarrier for Codelivery of siRNA and Ursolic Acid to Folate Receptor Overexpressing Cancer Cells. *J. Agric. Food Chem.* **2017**, *65*, 6904–6911. [CrossRef] [PubMed]
166. Zheng, G.; Zhao, R.; Xu, A.; Shen, Z.; Chen, X.; Shao, J. Co-delivery of sorafenib and siVEGF based on mesoporous silica nanoparticles for ASGPR mediated targeted HCC therapy. *Eur. J. Pharm. Sci.* **2018**, *111*, 492–502. [CrossRef] [PubMed]
167. Morry, J.; Ngamcherdtrakul, W.; Gu, S.; Goodyear, S.M.; Castro, D.J.; Reda, M.M.; Sangvanich, T.; Yantasee, W. Dermal delivery of HSP47 siRNA with NOX4-modulating mesoporous silica-based nanoparticles for treating fibrosis. *Biomaterials* **2015**, *66*, 41–52. [CrossRef] [PubMed]
168. Vivero-Escoto, J.L.; Vadarevu, H.; Juneja, R.; Schrum, L.W.; Benbow, J.H. Nanoparticle mediated silencing of tenascin C in hepatic stellate cells: Effect on inflammatory gene expression and cell migration. *J. Mater. Chem. B* **2019**, *7*, 7396–7405. [CrossRef]
169. Pinese, C.; Lin, J.; Milbreta, U.; Li, M.; Wang, Y.; Leong, K.W.; Chew, S.Y. Sustained delivery of siRNA/mesoporous silica nanoparticle complexes from nanofiber scaffolds for long-term gene silencing. *Acta Biomater.* **2018**, *76*, 164–177. [CrossRef]
170. Mora-Raimundo, P.; Lozano, D.; Manzano, M.; Vallet-Regí, M. Nanoparticles to Knockdown Osteoporosis-Related Gene and Promote Osteogenic Marker Expression for Osteoporosis Treatment. *ACS Nano* **2019**, *13*, 5451–5464. [CrossRef]
171. Li, Y.; Duo, Y.; Bi, J.; Zeng, X.; Mei, L.; Bao, S.; He, L.; Shan, A.; Zhang, Y.; Yu, X. Targeted delivery of anti-miR-155 by functionalized mesoporous silica nanoparticles for colorectal cancer therapy. *Int. J. Nanomed.* **2018**, *13*, 1241–1256. [CrossRef]
172. Li, Y.; Duo, Y.; Zhai, P.; He, L.; Zhong, K.; Zhang, Y.; Huang, K.; Luo, J.; Zhang, H.; Yu, X. Dual targeting delivery of miR-328 by functionalized mesoporous silica nanoparticles for colorectal cancer therapy. *Nanomedicine* **2018**, *13*. [CrossRef]
173. Ahir, M.; Upadhyay, P.; Ghosh, A.; Sarker, S.; Bhattacharya, S.; Gupta, P.; Ghosh, S.; Chattopadhyay, S.; Adhikary, A. Delivery of dual miRNA through CD44-targeted mesoporous silica nanoparticles for enhanced and effective triple-negative breast cancer therapy. *Biomater. Sci.* **2020**. [CrossRef] [PubMed]
174. Yan, J.; Lu, X.; Zhu, X.; Hu, X.; Wang, L.; Qian, J.; Zhang, F.; Liu, M. Effects of miR-26a on Osteogenic Differentiation of Bone Marrow Mesenchymal Stem Cells by a Mesoporous Silica Nanoparticle-PEI-Peptide System. *Int. J. Nanomed.* **2020**, *15*, 497–511. [CrossRef] [PubMed]
175. Wang, Y.; Xie, Y.; Kilchrist, K.V.; Li, J.; Duvall, C.L.; Oupický, D. Endosomolytic and Tumor-Penetrating Mesoporous Silica Nanoparticles for siRNA/miRNA Combination Cancer Therapy. *ACS Appl. Mater. Interfaces* **2020**, *12*, 4308–4322. [CrossRef] [PubMed]

176. Maimaitiyiming, Y.; Hong, D.F.; Yang, C.; Naranmandura, H. Novel insights into the role of aptamers in the fight against cancer. *J. Cancer Res. Clin. Oncol.* **2019**, *145*, 797–810. [CrossRef] [PubMed]
177. Ingolotti, M.; Kawalekar, O.; Shedlock, D.J.; Muthumani, K.; Weiner, D.B. DNA vaccines for targeting bacterial infections. *Expert Rev. Vaccines* **2010**, *9*, 747–763. [CrossRef] [PubMed]
178. Darvishi, B.; Farahmand, L.; Majidzadeh-A, K. Stimuli-Responsive Mesoporous Silica NPs as Non-viral Dual siRNA/Chemotherapy Carriers for Triple Negative Breast Cancer. *Mol. Ther. Nucleic Acids* **2017**, *7*, 164–180. [CrossRef]
179. Lopes, C.F.B.; de Angelis, B.B.; Prudente, H.M.; de Souza, B.V.G.; Cardoso, S.V.; de Azambuja Ribeiro, R.I.M. Concomitant consumption of marijuana, alcohol and tobacco in oral squamous cell carcinoma development and progression: Recent advances and challenges. *Arch. Oral Biol.* **2012**, *57*, 1026–1033. [CrossRef]
180. Nishida, N.; Yano, H.; Nishida, T.; Kamura, T.; Kojiro, M. Angiogenesis in cancer. *Vasc. Health Risk Manag.* **2006**, *2*, 213–219. [CrossRef]
181. Richter, K.; Konzack, A.; Pihlajaniemi, T.; Heljasvaara, R.; Kietzmann, T. Redox-fibrosis: Impact of TGFβ1 on ROS generators, mediators and functional consequences. *Redox Biol.* **2015**, *6*, 344–352. [CrossRef]
182. Mora-Raimundo, P.; Manzano, M.; Vallet-Regí, M. Nanoparticles for the treatment of osteoporosis. *AIMS Bioeng.* **2017**, *4*, 259–274. [CrossRef]
183. Lozano, D.; Sánchez-Salcedo, S.; Portal-Núñez, S.; Vila, M.; López-Herradón, A.; Ardura, J.A.; Mulero, F.; Gómez-Barrena, E.; Vallet-Regí, M.; Esbrit, P. Parathyroid hormone-related protein (107–111) improves the bone regeneration potential of gelatin–glutaraldehyde biopolymer-coated hydroxyapatite. *Acta Biomater.* **2014**, *10*, 3307–3316. [CrossRef] [PubMed]
184. Ansari, A.S.; Santerre, P.J.; Uludağ, H. Biomaterials for polynucleotide delivery to anchorage-independent cells. *J. Mater. Chem. B* **2017**, *5*, 7238–7261. [CrossRef] [PubMed]
185. Gambari, R.; Brognara, E.; Spandidos, D.A.; Fabbri, E. Targeting oncomiRNAs and mimicking tumor suppressor miRNAs: New trends in the development of miRNA therapeutic strategies in oncology (Review). *Int. J. Oncol.* **2016**, *49*, 5–32. [CrossRef] [PubMed]
186. Bonneau, E.; Neveu, B.; Kostantin, E.; Tsongalis, G.J.; De Guire, V. How close are miRNAs from clinical practice? A perspective on the diagnostic and therapeutic market. *Electron. J. Int. Fed. Clin. Chem. Lab. Med.* **2019**, *30*, 114–127.
187. Song, G.; Wang, Q.; Wang, Y.; Lv, G.; Li, C.; Zou, R.; Chen, Z.; Qin, Z.; Huo, K.; Hu, R.; et al. A low-toxic multifunctional nanoplatform based on Cu9S5@mSiO2core-shell nanocomposites: Combining photothermal- and chemotherapies with infrared thermal imaging for cancer treatment. *Adv. Funct. Mater.* **2013**, *23*, 4281–4292. [CrossRef]
188. Salis, A.; Fanti, M.; Medda, L.; Nairi, V.; Cugia, F.; Piludu, M.; Sogos, V.; Monduzzi, M. Mesoporous Silica Nanoparticles Functionalized with Hyaluronic Acid and Chitosan Biopolymers. Effect of Functionalization on Cell Internalization. *ACS Biomater. Sci. Eng.* **2016**, *2*, 741–751. [CrossRef]
189. Popat, A.; Liu, J.; Lu, G.Q.; Qiao, S.Z. A pH-responsive drug delivery system based on chitosan coated mesoporous silica nanoparticles. *J. Mater. Chem.* **2012**, *22*, 11173–11178. [CrossRef]
190. Nairi, V.; Medda, S.; Piludu, M.; Casula, M.F.; Vallet-Regì, M.; Monduzzi, M.; Salis, A. Interactions between bovine serum albumin and mesoporous silica nanoparticles functionalized with biopolymers. *Chem. Eng. J.* **2018**, *340*, 42–50. [CrossRef]
191. Xi, J.; Qin, J.; Fan, L. Chondroitin sulfate functionalized mesostructured silica nanoparticles as biocompatible carriers for drug delivery. *Int. J. Nanomed.* **2012**, *7*, 5235–5247. [CrossRef]
192. Xu, X.; Lü, S.; Gao, C.; Bai, X.; Feng, C.; Gao, N.; Liu, M. Multifunctional drug carriers comprised of mesoporous silica nanoparticles and polyamidoamine dendrimers based on layer-by-layer assembly. *Mater. Des.* **2015**, *88*, 1127–1133. [CrossRef]
193. Radhakrishnan, K.; Tripathy, J.; Datey, A.; Chakravortty, D.; Raichur, A.M. Mesoporous silica—Chondroitin sulphate hybrid nanoparticles for targeted and bio-responsive drug delivery. *N. J. Chem.* **2015**, *39*, 1754–1760. [CrossRef]
194. Kavya, K.C.; Dixit, R.; Jayakumar, R.; Nair, S.V.; Chennazhi, K.P. Synthesis and characterization of chitosan/chondroitin sulfate/nano- SiO_2 composite scaffold for bone tissue engineering. *J. Biomed. Nanotechnol.* **2012**, *8*, 149–160. [CrossRef] [PubMed]

195. Porgham Daryasari, M.; Dusti Telgerd, M.; Hossein Karami, M.; Zandi-Karimi, A.; Akbarijavar, H.; Khoobi, M.; Seyedjafari, E.; Birhanu, G.; Khosravian, P.; SadatMahdavi, F. Poly-l-lactic acid scaffold incorporated chitosan-coated mesoporous silica nanoparticles as pH-sensitive composite for enhanced osteogenic differentiation of human adipose tissue stem cells by dexamethasone delivery. *Artif. Cells Nanomed. Biotechnol.* **2019**, *47*, 4020–4029. [CrossRef] [PubMed]
196. Cao, J.-F.; Xu, W.; Zhang, Y.-Y.; Shu, Y.; Wang, J.-H. Chondroitin sulfate-functionalized 3D hierarchical flower-type mesoporous silica with a superior capacity for selective isolation of low density lipoprotein. *Anal. Chim. Acta* **2020**, *1104*, 78–86. [CrossRef]
197. Argyo, C.; Cauda, V.; Engelke, H.; Rädler, J.; Bein, G.; Bein, T. Heparin-coated colloidal mesoporous silica nanoparticles efficiently bind to antithrombin as an anticoagulant drug-delivery system. *Chem. A Eur. J.* **2012**, *18*, 428–432. [CrossRef]
198. Dai, L.; Li, J.; Zhang, B.; Liu, J.; Luo, Z.; Cai, K. Redox-responsive nanocarrier based on heparin end-capped mesoporous silica nanoparticles for targeted tumor therapy in vitro and in vivo. *Langmuir* **2014**, *30*, 7867–7877. [CrossRef]
199. Wan, M.M.; Yang, J.Y.; Qiu, Y.; Zhou, Y.; Guan, C.X.; Hou, Q.; Lin, W.G.; Zhu, J.H. Sustained release of heparin on enlarged-pore and functionalized MCM-41. *ACS Appl. Mater. Interfaces* **2012**, *4*, 4113–4122. [CrossRef]
200. Qian, W.J.; Wan, M.M.; Lin, W.G.; Zhu, J.H. Fabricating a sustained releaser of heparin using SBA-15 mesoporous silica. *J. Mater. Chem. B* **2014**, *2*, 92–101. [CrossRef]
201. Wei, H.; Han, L.; Ren, J.; Jia, L. Anticoagulant surface coating using composite polysaccharides with embedded heparin-releasing mesoporous silica. *ACS Appl. Mater. Interfaces* **2013**, *5*, 12571–12578. [CrossRef]
202. Wu, F.; Xu, T.; Zhao, G.; Meng, S.; Wan, M.; Chi, B.; Mao, C.; Shen, J. Mesoporous Silica Nanoparticles-Encapsulated Agarose and Heparin as Anticoagulant and Resisting Bacterial Adhesion Coating for Biomedical Silicone. *Langmuir* **2017**, *33*, 5245–5252. [CrossRef]

Review

Multimodal Decorations of Mesoporous Silica Nanoparticles for Improved Cancer Therapy

Sugata Barui and Valentina Cauda *

Department of Applied Science and Technology, Politecnico di Torino, Corso Duca degli Abruzzi 24, 10129 Turin, Italy; sugata.barui@polito.it

* Correspondence: valentina.cauda@polito.it; Tel.: +39-011-090-7389

Received: 7 May 2020; Accepted: 4 June 2020; Published: 8 June 2020

Abstract: The presence of leaky vasculature and the lack of lymphatic drainage of small structures by the solid tumors formulate nanoparticles as promising delivery vehicles in cancer therapy. In particular, among various nanoparticles, the mesoporous silica nanoparticles (MSN) exhibit numerous outstanding features, including mechanical thermal and chemical stability, huge surface area and ordered porous interior to store different anti-cancer therapeutics with high loading capacity and tunable release mechanisms. Furthermore, one can easily decorate the surface of MSN by attaching ligands for active targeting specifically to the cancer region exploiting overexpressed receptors. The controlled release of drugs to the disease site without any leakage to healthy tissues can be achieved by employing environment responsive gatekeepers for the end-capping of MSN. To achieve precise cancer chemotherapy, the most desired delivery system should possess high loading efficiency, site-specificity and capacity of controlled release. In this review we will focus on multimodal decorations of MSN, which is the most demanding ongoing approach related to MSN application in cancer therapy. Herein, we will report about the recently tried efforts for multimodal modifications of MSN, exploiting both the active targeting and stimuli responsive behavior simultaneously, along with individual targeted delivery and stimuli responsive cancer therapy using MSN.

Keywords: mesoporous silica nanoparticles; tumor targeting; stimuli responsive; multimodal decorations; targeted and controlled cargo release; cancer therapy and diagnosis

1. Introduction

Cancer is one of the most devastating diseases worldwide, characterized by unregulated cell division and cell growth, a fundamental aberration in cellular behaviors [1]. Consequently, the utmost ongoing challenge for the researchers is to restrain this dreadful disease. Even though, over the past decades, several therapeutic advances have been implemented in cancer treatment, including increases in survival rates [2], the metastasis and invasion associated with the malignant phenotype and heterogenic behavior of this disease still demands new therapeutic strategies [3]. Conventional methods for the treatment of cancer include chemotherapy, surgery and radiation therapy. Unfortunately, surgery and radiation therapy are limited for the treatment of cancers localized to one area of the body (solid cancers) [4]. On the other hand, although chemotherapy is widely used for the systemic treatment of advanced or malignant tumors, most of the chemotherapeutic agents are associated with severe side-effects of destroying the normal healthy cells and limited by cancer cell induced multidrug resistance (MDR) [5,6]. Therefore, developing efficient targeted cancer therapeutic strategies to reduce side-effects and overcome resistances is gaining increasing importance. Herein, researchers start to exploit the enhanced permeability and retention (EPR) effect of solid tumors [7]. Due to the presence of leaky vasculature and the lack of lymphatic drainage of small structures by solid tumors, nanoparticles can easily accrue in the tumor and represent promising delivery vehicles [8–10].

An ideal targeted nanoparticle delivery system should possess (i) the high loading capacity of multiple diverse chemotherapeutics, (ii) efficiency to protect the cargo until reaching the final destination, (iii) circulation stability in blood for prolonged periods without degradation and excretion, (iv) specificity toward target cancer cells to achieve off-target zero-delivery, (v) the ability of intracellular release and to facilitate controlled delivery of the cargo, and (vi) good biocompatibility and low toxicity [11–13]. Over the past decades, various types of organic and inorganic nanoparticles have been proposed as delivery vehicles to address those criteria [14–16]. Among the organic nanoparticles, liposomal-based drug delivery becomes one of the most promising approaches because of its high biocompatibility, flexibility in preparing various formulations, and easy synthesis to incorporate targeting moieties [17–19]. Furthermore, there are some already FDA-approved liposomal formulations; several polymeric and micelle based organic nanoparticles are also in clinical trials for use in cancer therapy [20,21]. However, the liposomal formulations and the polymer-based nanocarriers are limited, due to their invariant size and shape, inadequate loading efficiency, uncontrolled release of the cargo, and change in size and stability by changing loading parameters [22].

There are various inorganic materials developed so far as delivery systems trying to overcome the loading inefficiency, leakage and the uncontrolled release of the cargo, e.g., metal oxide nanoparticles, carbon nanotubes, and mesoporous silica nanoparticles (MSN) [23–27]. Few among the metal oxide nanoparticles are already in process for cancer therapy and diagnosis. A clinical (early phase I) study is also conducted with targeted MSN for image-guided operative sentinel lymph node mapping [28]. Particularly, in comparison to other nanoparticles, the MSN exhibit numerous outstanding features, including good biocompatibility, mechanical thermal and chemical stability, and most importantly, immense loading capacity of various cargos and their possible time-dependent release, thanks to the large surface area, high pore volume and narrow distribution of the tunable pore diameters of MSN [29,30]. For example, because of comprising large surface area one can load nearly a 1000-fold higher amount of doxorubicin in MSN compared to in the FDA-approved liposomal formulation Doxil® [31]. Moreover, silica is recognized by FDA as safe to be used in cosmetics and as a food-additive [32].

A comparative discussion about the pros and cons of MSN with other well-known nanomaterials for bio-applications was excellently provided by Chen et al. [33] and thus is discussed no further here.

In this review, we will discuss the efficacy of mesoporous silica-based systems for cancer therapy, the surface modification of MSN for passive and active targeting cancer therapy, and the modification of MSN for environment-responsive cancer therapy. Importantly, we will focus on multimodal decorations of MSN, which is the most demanding ongoing approach with respect to the present perspectives, and challenges related to MSN application in cancer therapy. Many reviews have summarized the synthesis of MSN, active targeting and environment-responsive drug delivery using MSN, whereas fewer involved in reporting the multimodal decorations of MSN for exploiting both the tumor targeting and stimuli responsive delivery of therapeutics simultaneously. Herein, we will review the multimodal approaches, including both the targeted delivery and stimuli responsive delivery simultaneously, along with individual targeted delivery and stimuli responsive delivery using MSN. As well, we will include the plausible applications of MSN in cancer diagnosis.

2. MSNs as Delivery Vehicles in Cancer Therapy

Despite the increasing numbers of anti-cancer drugs presented in the market and their ability to create potent and lethal interaction with cancer cells, their therapeutic efficacy remains affected by their low aqueous solubility and eventually not reaching a high enough concentration in the site of absorption, i.e., gastrointestinal (GI) lumen [34,35]. As for an example, camptothecin (CPT) is very effective at killing cancer cells in vitro, however, its clinical application has been limited due to poor water solubility. Additionally, researchers have tried to modify CPT as water-soluble salts to make intravenous injection possible, but this modification has altered its physicochemical characteristics and hampered its antitumor activity [36]. Another potent anti-cancer drug, paclitaxel, is also limited in vivo by its insolubility in aqueous systems, although it is very effective against various cancer cell lines [37].

With the aim to improve the drug solubility and oral bioavailability, a growing number of novel drug delivery systems, particularly nanostructures, have been developed [38,39]. The two foremost parameters determining the efficacy of a drug delivery system are the loading capacity and drug release profiles. To this end, with excellent features, including huge surface area and ordered porous interior, MSN can be used as reservoirs to store different anti-cancer drugs with high loading capacity and tunable release mechanisms [40,41]. As a promising drug delivery system, the pore size of MSN can be customized to selectively load either hydrophobic or hydrophilic anticancer agents, and their size and shape can be maintained to have the maximum cellular internalization [41,42]. There are mainly two ways that have been used to load the drug molecules into pores of MSN. One can load either in situ during synthesis or by the adsorption of cargo onto the pores of MSN (by physisorption or chemisorption). The adsorption method is the most widespread approach for the loading of therapeutic molecules, especially for poor water-soluble drugs [31,43]. During soaking of the MSN in a drug solution, the silanol groups present on the surface of MSN play the key role as adsorption sites. As the surface of MSN is negatively charged in the absence of any adsorbent under physiological conditions, the electrostatic adsorption method can be applied for the cargo having positive charge, as well as the lodging of water-soluble therapeutic agents into the pores of MSN. Moreover, the functionalization of MSN will increase the adsorbed amount of this group of cargo having additional interactions between adsorbate and adsorbent [44]. Pore size of MSN is another main controlling parameter to increase the extent of adsorption of hydrophobic molecules from organic solvents, if the molecular size of the cargo is in the range of the pore size of MSN [43,45]. Up until today, there have been various studies reported in favor of using MSN as efficient drug delivery nanosystem in cancer therapy. He et al. have reported the enhanced solubility of paclitaxel after loading into MSN [37]. Lu et al. have performed cytotoxicity assay with camptothecin (CPT)-loaded MSN and showed the clear growth inhibition of pancreatic cancer-cell lines (Capan-1, PANC-1, AsPC-1), stomach cancer-cell line (MKN45) and colon cancer-cell line (SW480) [36]. It was also reported that transplatin, a less potent anticancer drug (an inactive isomer of cisplatin), when loaded in MSN, became effective exhibiting enhanced cytotoxicity compared to that of cisplatin [46].

In this context we should also discuss about the protein adsorption and efficient protein delivery by MSN. The poor solubility and large sizes of the therapeutic proteins and their enzymatic and chemical degradation in the gastrointestinal tract commonly compromise their efficacy in cancer therapy. Additionally, the co-delivery of therapeutic proteins along with other therapeutic molecules is a big challenge for the conventional drug delivery systems, as the physicochemical properties of proteins, such as size, surface charge, stability, and susceptibility are very different than the other therapeutic molecules [47]. Herein, MSN are of special interest for protein delivery due to their possible easily tunable pore sizes, facile surface multi-functionalization, and enormous interior and exterior particle surface [48]. To expand the pore size of MSN depending on the sizes of the protein, generally two ways have been employed, exploiting polymers/surfactants with longer carbon chains/co-surfactants as templates, or the addition of suitable organic swelling agents to enlarge the sizes of surfactant templates [49]. There are variety of reported additives used as pore size expanding agents, such as *N*,*N*-dimethylhexadecylamine (DMHA), trimethylbenzene (TMB), aromatic hydrocarbons, auxiliary alkyl surfactant, and long-chain alkanes [50]. Moreover, positively charged amino silyl reagents or polymers have been widely used to compensate negative charges of the proteins, such as lysozyme, bovine serum albumin and myoglobin [51]. Protein loading amount in MSN can also be increased utilizing suitable surface functionalization, having strong electrostatic interaction between proteins and the pore channels. In this regards, Slowing et al. have first employed MSN for the intracellular delivery of native cytochrome c, a small protein, into human cervical cancer cells (Hela cells) [52]. There are several other reports about the cytochrome c delivery in cancer cells using MSN [53,54]. Zhang et al. have reported the high protein loading capacity of hollow silica vesicles and demonstrated cancer cell inhibition by the intracellular delivery of RNase A [55]. Besides, Niu et al. have modified MSN by employing hydrophobic C18-functionalization and Yang Y.N. et al. have utilized benzene bridged MSN for the effective intracellular delivery of RNase A [56,57]. Nonetheless, Yang and collaborators have reported multi-shell dendritic mesoporous organosilica nanoparticles to deliver protein antigens for cancer immunotherapy [58].

Along with efficient loading capacity, MSN have been used for controlled release of a variety of pharmaceutical drugs (e.g., DOX, TPT, and CPT) and therapeutic proteins/peptides [59,60]. It can be possible to release the cargo in a controlled manner, without any leakage before reaching the target destination, with the help of "gatekeeper" entities that can seal the pores of MSN. There are infinite gatekeepers reported for the end-capping of MSN to reside the drug molecules in the reservoir of MSNs, e.g., biomolecules, peptides, lipids, polymers, dendrimers, macrocyclic compounds, etc. [61–63] As reported below, we will discuss the gatekeeper systems to be used for controlled drug release.

3. Surface Modification of MSN for Passive and Active Targeting Cancer Therapy

Localizing MSN specifically into the cancer environment is one of the milestones to avoid side effects and damage to healthy cells. Several efforts have been executed to target the MSN to specific tissues, both through passive and/or active targeting [64]. At the beginning, MSN has been developed as anticancer drug delivery systems, mainly based on their efficacy to store high amount of chemotherapeutics into pores and exploit EPR effect for passive targeting to tumor tissues. In this part of the review, we will discuss the EPR effect and passive targeted cancer therapy using MSN. Later on, MSN surface modifications by conjugating targeting ligands have been introduced to enhance the uptake of MSN in targeted cells. Different targeting moieties have been employed to the surface of MSN, e.g., small molecules, aptamers, short peptides, antibodies and antibody fragments, etc. [31,65]. In the following part, we will review the targeted cancer therapy using MSN.

3.1. *Passive Targeting*

Since the beginning, the foremost important goal in chemotherapy is to achieve the tumor-specific delivery of chemotherapeutics. In this regard, most nanoparticles including MSN can passively target solid tumor tissue due to the EPR effect. In general, the body has its own pre-existing circulation network for the supply of food, nutrients and oxygen to the small primary tumor until the diameter exceeds 1–2 mm. Beyond this size, the tumor growth needs angiogenesis, i.e., the sprouting of new blood vessels from pre-existing vessels around the tumor, in order to supply food, nutrients, oxygen, survival factors etc. [66,67] Angiogenesis generates irregular blood vessels displaying a discontinuous and single thin layer of flattened endothelial cells with an absence of the basal membrane. Hence, nanoparticles having a diameter of at least 10 nm, which is the threshold of renal clearance, can leave the blood vessels and penetrate into the adjacent tumor tissue through the discontinuous leaky membrane. This effect is not applicable in normal tissue [68]. The penetrated nanoparticles remain longer in the tumor tissue without being cleared by the immune system, as the solid tumors commonly lack effective lymphatic drainage [69]. Moreover, particles having a diameter smaller than 4 nm can diffuse through the leaky endothelium back to the blood circulation and be reabsorbed, but the nanomaterials do not naturally return to the blood vessels, accumulating in the perivascular tumoral space [70]. In the nanomedicine field, this phenomenon is popularly known as the enhanced permeability and retention effect, or the "EPR" effect. To avail the efficient passive targeting particle size, the morphology and surface modifications of MSN have been considered. It is observed that the MSN should be at least 10 nm in diameter and have an optimal size of 100–200 nm to avoid the renal clearance of the particles [65]. To this end, Lee and co-workers have shown proficient cell death by the passive targeting of MSN loaded with doxorubicin (DOX) to the tumor site in a melanoma model [71]. Importantly, surface modifications of MSN also have a major influence to achieve efficient passive targeting by prolonging the circulation time of MSN in blood and subsequently reducing the renal clearance [72]. It has been reported by Zhu and colleagues that introducing PEGylation on hollow MSN improves cellular uptake in cervical cancer cells and mouse embryonic fibroblasts, compared to that of naked particles [73]. Huan and colleagues have demonstrated efficient biodistribution, accomplishing an 8% of the EPR effect at the tumor site in vivo of MSN functionalized with polyethyleneimine/polyethylene glycol (PEI/PEG), encapsulating doxorubicin together with P-glycoprotein siRNA [74]. With regard to passive targeting, another important factor is the 10 to 40 fold elevated interstitial fluid pressure (IFP) in solid tumors compared to normal tissue [75]. This pressure gradient may influence reduced nanoparticle distribution in tumor site. Actually, the necrotic tissues that are often present in the larger tumors and metastatic regions are highly hypovascularized, due to slower angiogenesis compared to tumor growth. As a result, the IFP becomes very high and the delivery of nanoparticles to this tumor region by passive targeting is hardly possible. Herein, the active targeting of nanoparticles including MSN is gaining increasing importance and we will discuss the advantage of active targeted drug delivery using MSN in the next part of the review.

3.2. *Active Targeting*

To deliver potent chemotherapeutics selectively to tumor environment, substantial progresses have been made by exploiting tumor cell-specific or tumor-associated cell-specific receptors [76]. A receptor highly expressed on tumor cells or tumor associated cells (compared to the normal cells) is a sensible target receptor for tumor specific drug delivery. If the surfaces of nanoparticles, including MSN, are decorated with ligands able to interact selectively with those overexpressed receptors, the specific retention and uptake of those nanoparticles by tumor cells will be enhanced. To design the targeting ligands grafted to MSN, various receptors over-expressed on the surface of tumor cells or tumor associated cells have been exploited (Figure 1) and we will discuss the decorated MSN mediated active targeted cancer therapy in this part of the review. Usually, the decorated MSN are taken up by the cancer cells via a receptor-mediated endocytosis process. Active targeting allows efficient particle uptake by the tumor cell and tumor microenvironment [77].

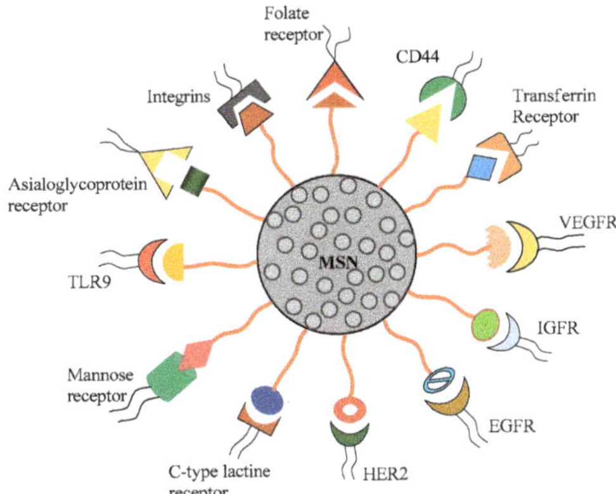

Figure 1. Plausible surface modifications of mesoporous silica nanoparticles (MSN) for active targeting to the over-expressed receptors in cancer microenvironment.

3.2.1. Targeting Folate Receptor

One of the most exploited targeting ligands, folic acid, has been employed to decorate MSN for targeting folate receptor, overexpressed in many tumors compared to healthy tissues [78,79]. The folate receptors are four glycopolypeptide members (FRα, FRβ, FRγ and FRδ), among which the alpha isoform, folate receptor α (FRα) is a glycosylphosphatidylinositol anchored cell surface receptor and has been reported to be overexpressed in solid tumors, such as ovarian, cervical, lung, breast, kidney, colorectal, and brain tumors [80]. In mostly 80–90% of epithelial ovarian cancers, other gynecological cancers, lung cancers and breast cancers, the FRα is highly overexpressed and gaining increasing importance to be exploited for targeted cancer therapy [81]. Considering this fact, several research groups have reported the enhanced specific cellular uptake of MSN in various cancer cells, having overexpressed folate receptors by modifying MSN surface with folic acid [82–86]. Nonetheless, using two different human pancreatic cancer xenografts on different mouse species, Lu et al. have also shown dramatic improvements in tumor-suppression effect by using folic acid functionalized camptothecin-loaded MSN in comparison with unfunctionalized MSN [87]. Moreover, along with using folic acid, López et al. have decorated MSN with triphenylphosphine (TPP), in order to target tumor cells, as well as the mitochondria of the tumor cells [88]. Conversely, instead of using folic acid, Rosenholm et al. have used methotrexate (MTX) as both a targeting ligand and a cytotoxic agent for cancer therapy, due to its high affinity for folate receptors and showed enhanced cancer-cell apoptosis by treating MTX incorporated MSN relative to free MTX [89].

3.2.2. Targeting Transferrin Receptor

There are two subtypes of transferrin receptors (TFRs), TFR1 and TFR2, which complexes with iron to facilitate iron metabolism in cells. Hence, the dysregulated expression of any subtype disorders can impair iron metabolism and eventually induce tumorigenesis and cancer progression [90]. It has been reported that TFR1 is abundantly expressed in many cancer types, e.g., liver, breast, lung, pancreatic, and colon cancer cells [90,91], and thus can be exploited as an important target for drug delivery. In order to improve the tumor specific delivery of MSN carrier, transferrin (Tf) which is a ligand of TFR1, has been widely exploited in surface modification of MSN [92]. As evidenced by the available studies targeting TFR1, Tf-modified MSN exhibit enhancement in nanoparticle uptake by Panc-1 cancer cells [93]. Additionally, Montalvo-Quiros et al. have used MSN as nanovehicles decorated with Tf to provide a nanoplatform for the nucleation and immobilization of silver nanoparticles (AgNPs) and demonstrated that only the nanosystem functionalized with Tf can transport the AgNPs inside the human hepatocarcinoma (HepG2) cells overexpressing Tf receptors [94]. Nevertheless, Tf-decorated MSN have been exploited for sorafenib delivery in thyroid cancer therapy [95]. Importantly, the overexpression of TFRs on the brain capillary endothelial cells (BCECs) of the blood-brain barrier (BBB) and glioblastoma multiforme (GBM) provides a route to allow effective chemotherapeutic penetration to the site of brain tumor [96]. Herein, few research groups have developed Tf-conjugated MSN to deliver the chemotherapeutics to glioma cells across the BBB [97,98].

3.2.3. Targeting Integrin Receptor and Nuclear Targeting

Integrin receptors, the α/β heterodimeric transmembrane glycoproteins, are overexpressed on angiogenetic endothelial cells and certain tumor cells, whereas they are absent (or present in basal levels) in pre-existing endothelial cells and normal tissues [19,99]. This makes integrins, especially $\alpha v\beta 3$ integrin receptors, a promising target in cancer therapy and RGD (arginine-glycine-aspartic acid) based peptides have found widespread exploitations for targeting chemotherapeutics to both tumor and tumor vasculatures via the overexpressed integrin receptors [100]. Therefore, peptides including the RGD motif have been widely used in surface decoration of MSN for targeted cancer therapy [101–106]. Moreover, Pan et al. have shown the in vivo efficacy of doxorubicin-loaded MSN grafted with RGD-motif. The same research group has further determined better tumor accumulation and reduced tumor size by coupling cell-penetrating and nuclear-targeting TAT peptide to the MSN along with RGD. Additionally, side effects of bare MSN to accumulate in liver and spleen have been distinctly minimized by treating RGD/TAT-MSN [107].

3.2.4. Targeting EGF Receptor and HER2 Receptor

Epidermal growth factor receptor (EGFR or ErbB1), a tyrosine kinase receptor, is a key factor in epithelial malignancies, in terms of enhancing tumor growth, invasion, and metastasis [108]. Overexpression of EGFR has been widely observed in many cancers including lung (especially non-small-cell lung carcinoma), colon, ovary, head and neck and breast cancers [109]. As EGFR has emerged as an attractive target for anti-lung cancer drug research, its ligand or antibody has been extensively employed in capping moiety for the active targeting of MSN in lung cancer cells. For example, She et al. have used amine functionalized MSN to conjugate with EGFs (epidermal growth factors) for targeting EGFR positive cells [110]. Sundarraj et al. have shown elevated accumulation of EGFR-MSN-cisplatin drug delivery system in EGFR overexpressed lung adenocarcinoma cells (A549) than that in normal lung cells (L-132). They have also used the non-small cell lung cancer nude mice model to determine the increased and prolonged cisplatin intratumoral distribution and enhanced tumor-cell apoptosis by treating EGFR-MSN-cisplatin [111]. On the other hand, Wang et al. have used cetuximab, a monoclonal antibody of EGFR as a capping agent of MSN loaded with anti-cancer drugs including doxorubicin and gefitinib, to specifically target lung cancer cells exploiting EGFR overexpression [112].

In addition to the EGFR, human epidermal growth factor receptor 2 (HER2)/ErbB2 is another member of the ErbB family of type-1 tyrosine kinases and a proto-oncogene, with a vast role of ErbB receptors in malignant transformation [113]. The overexpression of HER2 receptor in breast cancer alongside lungs, ovary and gastric/gastroesophageal cancers plays a major role in the angiogenic process and makes HER2 an important target in cancer therapy [114]. Furthermore, it has been reported that HER2 specific antibodies or antibody-fragments (e.g., trastuzumab) have been used in the surface modification of MSN for the selective targeting of breast cancer cells [115].

3.2.5. Targeting VEGF Receptor

The vascular endothelial growth factors (VEGFs) and their receptors (VEGFRs) play a critical role in tumor angiogenesis and metastasis. Among the three receptors (VEGFR1, VEGFR2, VEGFR3), VEGFR2 is widely explored as a direct stimulator of angiogenesis [116]. In addition to its constitutive expression on angiogenic endothelial cells, VEGFR2 is found to be overexpressed on several cancer cells such as breast cancer, lung cancer, pancreatic cancer, glioblastoma, gastrointestinal cancer, hepatocellular carcinoma, renal cell carcinoma, ovarian cancer, bladder cancer, and osteosarcoma cells [117]. To target VEGFR2, Weibo and co-workers have used $VEGF_{121}$, a natural VEGFR ligand which has a high binding affinity for VEGFR2 and observed a strong, specific binding of the MSN surface coated with $VEGF_{121}$ in HUVEC (VEGFR+), but not in 4T1 cells (VEGFR−) [118]. The same group has also demonstrated delivery of the MSN encapsulating the anti-cancer drug, sunitinib in a significantly higher amount to the U87MG tumor by targeting VEGFR exploiting $VEGF_{121}$ ligand in comparison with the non-targeted delivery [119]. Moreover, Zhang et al. have shown increased targeting ability and retention time of anti-VEGFR2 targeted MSN in anaplastic thyroid cancer tumor-bearing mouse [120]. Bevacizumab or related antibodies have been also exploited for targeting VEGF receptors.

3.2.6. Targeting Mannose Receptor and C-Type Lectin Receptor

Tumor-associated macrophages (TAMs) that exist in the tumor microenvironment promote tumor immunosuppression, angiogenesis, metastasis, and relapse. TAMs expressing the multi-ligand endocytic receptor mannose receptor (CD206/MRC1) have been suggested as a promising therapeutic target for cancer therapy [121]. It has been reported that MSN coupled with mannosylated polyethylenimine (MP) can target macrophage cells and enhance transfection efficiency through receptor-mediated endocytosis via mannose receptors [122]. Moreover, the C-type lectin receptor is also expressed exclusively by macrophages and exploited for cancer treatment. Lectin-functionalized MSN have recently been experimented in a mouse colon cancer model [123].

3.2.7. Other Active Targeted Delivery

There are several other receptors that have also been exploited for targeted delivery using surface-modified MSN. The overexpression of the insulin-like growth factor (IGF) receptor in ovarian cancer has been employed for the efficient targeted delivery of doxorubicin entrapped in surface modified MSN [124]. Quan et al. have developed lactosaminated MSN (Lac-MSN) for asialoglycoprotein receptor (ASGPR) targeted anticancer drug delivery and showed the effectively inhibited growth of HepG2 and SMMC7721 cells by treatment with docetaxel (DTX) loaded in Lac-MSN [125]. The surface of the MSN has also been functionalized with the ligands of somatostatin receptors [126] and also with hyaluronic acid to target CD44 receptors [127]. Furthermore, Chen et al. have shown the significantly larger tumor uptake of vasculature targeting anti-CD105 antibody (TRC105) conjugated MSN, compared to untargeted nanoparticles in a murine breast cancer model [128]. The same group has employed a TRC105 antibody fragment (Fab) for the surface modification of MSN to target tumor vasculature [129]. Besides, Sweeney et al. have attached a bladder-cancer specific peptide named Cyc6 to MSN for active targeting [130]. Apart from small molecules, peptides and antibodies, the synthetic single-stranded DNA or RNA oligonucleotides (aptamers) have been used to decorate MSN for targeting cancer cells [131,132]. Moreover, Nguyen et al. have shown the Toll-like receptor 9 mediated delivery of mesoporous silica cancer vaccine (antigen) to the dendritic cells (the body's most professional antigen presenting cells) [133].

4. Stimuli-Responsive Drug Delivery Using MSN

Although vast efforts have been devoted to active targeting therapy using MSN, the delivery efficacy still needs to be strengthened. During the blood circulation and penetration into the tumor matrix, anticancer drugs may leak from mesopores of MSN, leading to insufficient drug concentration at the tumor site. To overcome this obstacle, "smart" MSNs-modified with environment-responsive gatekeepers were designed. As the characteristics of tumor microenvironment differ from that of normal tissues (e.g., acidic pH, high concentration of glutathione, etc.), MSN can be modified introducing the moiety sensitive to the tumor microenvironment and release the cargo specifically at the tumor site [134,135]. There are internal and external stimuli that have been exploited for the controlled drug release (Figure 2). In this part of the review, we will discuss the pH, redox and enzyme internal stimuli responsive gatekeepers and also the magnetic, light and ultrasound external stimuli responsive gatekeepers frequently used to prepare stimuli responsive MSN.

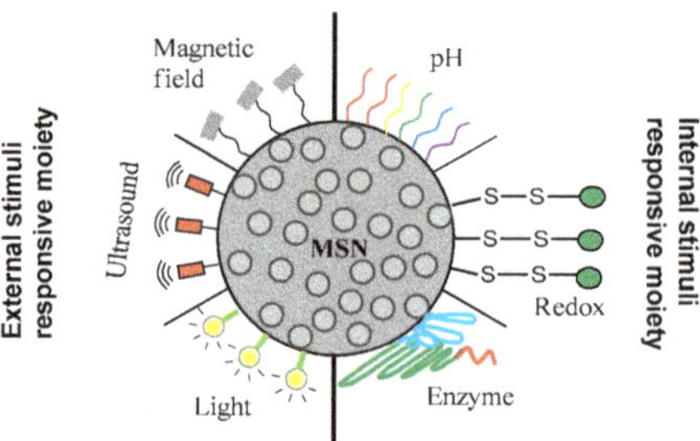

Figure 2. Most relevant stimuli responsive gatekeepers to decorate MSN for controlled cargo release in the cancer site.

4.1. PH-Responsive Gatekeepers

One of the most promising internal stimuli that has been employed for controlled drug release in cancer therapy is to exploit the lower pH values in most of the tumors in comparison with healthy tissues [136]. Actually, in cancer cells, because of high glycolysis rate, the production of lactic acid is high, thus eventually reducing the pH value in the tumor region. There are various reports in the literature regarding the pH-controlled delivery of chemotherapeutics by surface-engineered MSN in cancer therapy. Besides, there are mainly two ways in which they have been used to decorate the MSN for exploiting the pH sensitivity of tumor cells. One approach is to incorporate the pH responsive linkers in between MSN and the capping moiety usually used for blocking the pore entrances of MSN. There are several linkers that have been reported for the intracellular pH-responsive controlled delivery of anti-cancer drugs e.g., acetal linkers [137], boronate ester linkers [138], ferrocenyl linkers [139], aromatic amines [140], imine bonds [141] hydrazine linkers [142], acid labile amide bond [143], etc.

Another widely used approach is to modify the MSN surface with pH sensitive capping moiety, so that the MSN will only open up at acidic pH, release the cargo only in tumor environment and avoid any premature release of drugs on healthy tissues [144,145]. Yang and co-workers have reported that the MSN coated with pH-responsive chitosan/polymethacrylic acid polymer is more efficient to deliver doxorubicin in HeLa cells compared to the uncoated MSN [146]. The modification of the MSN surface using pH-sensitive self-immolative polymers, poly(acrylic acid), nanovalves, such as pseudorotaxane encircled by β-cyclodextrin, tannic acid, lipid coatings and many other nanoparticles have been reported [147–150]. Zhu and coworkers have used a pH-sensitive nanovector for the dissolution of ZnO nanoparticles functionalized onto the surface of MSN for the efficient delivery of doxorubicin in HeLa cells [151]. Moreover, pH degradable calcium phosphate coated MSN and gelatin capped MSN have also been described for intracellular acid-triggered drug delivery [152,153]. In a recent report, the MSN surface was modified with poly (styrene sulfonate) (PSS), which can act as a "nano-gate" for the pH responsive controlled release of curcumin [154].

4.2. Redox-Responsive Gatekeepers

Similar to the pH parameter, redox factor can also be exploited to achieve the controlled drug release from MSN specifically to the tumor environment. In general, glutathione (GSH) acts as a biological reducer and can cleave the redox-cleavable groups and trigger the bioactive agents. It has been observed that the GSH concentration in cancer cells is higher than that in normal cells [155]. Moreover, the intracellular concentration of GSH is in the range of 2–10 mM which is quite a bit higher than that in the extracellular part (2–20 nM); this concentration difference can allow the release of cargo from redox-responsive nanocarriers upon entering into the cytoplasm [156,157]. To take advantage of the high GSH concentration in cancer cells, the MSN surface has been decorated either with disulfide linkers or by incorporating any redox-cleavable group in capping moiety for the efficient release of cargo in cancer cells. As for an example, Kim et al. have used disulfide bonds as a linker in between MSN and the surface capping β-cyclodextrin moiety, and reported efficient doxorubicin toxicity in lung adenocarcinoma cells [158]. Moreover, Bräuchle and Bein research groups have reported cystein residues with disulfide linkers to modify the MSN surface [159]. Additionally, Wu et al. have used poly-(β-amino-esters) to seal the MSN pores and reported the intracellular reduction of disulfide linkers present between MSN and poly-(β-amino-esters) capping moiety [160]. The cargo release kinetics upon degradation of MSN can be further controlled by tuning the hindrance of disulfide or tetra-sulfide groups into the silica framework [161–163]. Besides, polymers cross-linked by cystamine, poly (propylene imine) dendrimer and polyethylenimine (PEI) via intermediate disulfide linkers are utilized to close the pores of MSN for a redox-responsive release of the chemotherapeutics by the degradation of polymeric networks in reducing the environment of the tumor site [164,165].

4.3. Enzyme-Responsive Gatekeepers

MSN drug release can also be modulated by the enzymatic cleavages of ester, peptide, urea, and oxamide bonds decorated on the MSN surface. Several enzymes such as esterase, protease, galactosidase, amylase, lipase, etc. have been exploited for enzyme responsive controlled drug release [166]. In this regard, Patel et al. have introduced ester bonds between MSN and the adamantine capping moiety, to employ the enzymatic role of porcine liver esterase for the controlled release of cargos [167]. Mondragón et al. have exploited protease cleavable ε-poly-L-lysine moiety to seal the camptothecin encapsulated MSN and reported the reduced viability of human cervix epitheloid carcinoma cells upon treatment of that nanosystem [168]. They have also reported some enzyme-responsive hydrolyzed starch products as saccharides to be used for controlled drug release [169]. There are various other protease-responsive moieties that have been used to cap the MSN pores and improve the drug release, e.g., protease-responsive biotin-avidin [170], arginine-rich protamine proteins [171], matrix metalloproteinase (MMP) degradable gelatin [172], avidin with MMP9-sensitive peptide linker (RSWMGLP) [173], poly (ethylene glycol) diacrylate moiety with protease-sensitive peptide linker (CGPQGIWGQGCR) [174]. Furthermore, cyclodextrin gatekeepers and HRP-polymer nanocapsules have also been employed on the MSN surface for enzyme-responsive drug release [175,176].

4.4. Magnetic Responsive Delivery System

One of the effective ways to exploit external stimuli is to exert the magnetic field on MSN, either to have magnetic guidance by applying the permanent magnetic field, or to increase the temperature by applying an alternating magnetic (AM) field [177,178]. In this regards, iron oxide has been widely exploited as the required magnetic component. There are mainly two ways that have been used to conjugate iron oxide with MSN, either using iron oxide core coated with mesoporous silica or MSN capped with iron oxide nanoparticles [179,180]. The most employed strategy consists on encapsulating superparamagnetic iron oxide nanoparticles (SPIONs) of ca. 5–10 nm within the MSN network during their synthesis [181,182]. These SPIONs are able to convert the magnetic energy into heat and can increase the local temperature of the system upon application of the AM field. If the surface of MSN has already been coated with temperature responsive moieties acting as gatekeepers, e.g., poly (N-isopropylacrylamide), pore opening and drug release from MSN can be triggered by applying an AM field [183]. Taken together, upon application of an AM field, SPIONs encapsulated in MSN can increase the local temperature up to a certain point, to change the conformation of the temperature responsive gatekeepers and open the pore entrances to release the anti-cancer drugs efficiently without having any premature leakage. There are several reports showing the controlled release of anti-cancer therapeutics by applying a magnetic stimulus [180,184,185]. Moreover, there are a few FDA-approved SPIONs for using as imaging agents and EU-approved iron oxide nanoparticles to use in glioblastoma therapy; these can be further exploited in magnetic responsive drug delivery [20].

Another strategy for the design of the magnetic responsive delivery system consists of the functionalization of drug-loaded MSN with a single DNA strand and then mixing this with SPIONs functionalized with the complementary DNA strand, to allow DNA hybridization that can act as a capping agent [186]. The reason behind selecting the DNA sequence is its melting temperature of 47 °C. Thus, upon application of an AM field, SPIONs encapsulated into the MSN network can increase the local temperature that subsequently trigger the double-stranded DNA melting and open the pores of MSN to release the drug. Interestingly, when the magnetic field is switched off, the DNA hybridization occurs again, thus closing the pores and stopping the drug release. This mechanism smartly provides the chance of exploiting the on-off drug release mechanism.

4.5. Light-Responsive Delivery System

The surface of MSN can be decorated introducing photo-cleavable linkers for triggering the cargo release from MSN, by applying lights with different wavelengths (ultraviolet, visible or near-infrared) [187,188]. Among all, as ultraviolet (UV) radiation has the highest power to easily break the bond, it has been the most commonly used light stimulus for the controlled drug release from MSN [187]. It has been reported that MSN coated with photo-responsive azobenzene-modified nucleic acid can trigger the drug release under UV light radiation [189]. However, the biomedical application of the UV light becomes restricted due to its toxicity and low tissue penetrability [190,191]. As an alternate, visible (Vis) light can be employed, as it is less harmful and has a higher tissue penetrability. Few Vis light-triggered MSN drug delivery systems have been reported [192,193]. For example, light responsive porphyrin nanocaps have been used to decorate the MSN. Porphyrin nanocaps are anchored via reactive oxygen species (ROS)-cleavable linkages, so that in response to the Vis light singlet oxygen molecules will be generated to break the sensitive linker and trigger the drug release by opening the pore of MSN [193].

Even though there are several advantages of using light (such as its easy application, non-invasiveness, low toxicity and precise focalization in the desired place), light-responsive delivery is restricted by its low tissue penetration capability (only a few millimeters). It has been observed that the best wavelengths for satisfactory tissue penetration are within the biological spectra, typically 800–1100 nm [134]. Likewise, Guardado-Alvarez et al. have exploited photolabile coumarine-molecules in the capping moiety of MSN surface to control the cargo release upon two-photon excitation at 800 nm [194]. Furthermore, Croissant and colleagues have shown that they can control drug release via a photo-transducer from mesoporous silica nanoimpellers in human cancer cells using two-photon light [195].

4.6. Ultrasound Based Delivery

Ultrasound (US) is an efficient stimulus to be used for controlled drug delivery, because of its advantage of being non-invasive, the absence of ionizing radiations in it and its capability to penetrate deep into living tissues by tuning the parameters, such as frequency, duty cycles and exposure times [171,196]. To exploit the US stimulus, the surface of the MSN has been decorated by employing US sensitive components in capping moiety to prevent the premature release of drugs in healthy tissues, e.g., 2-tetrahydropyranyl methacrylate. A hydrophobic monomer with a US-sensitive group can be transformed to hydrophilic methacrylic acid under US stimulus and this phase change can trigger the drug release from MSN pores [197,198]. Shi and co-workers have reported US responsive perfluorohexane encapsulated MSN to be exploited for drug delivery [199,200]. Moreover, Vallet-Regí and co-workers have decorated the MSN surface by using ultrasound-responsive copolymer (poly (2-(2methoxy-ethoxy) ethylmethacrylate-co-2-tetrahydropyranyl methacrylate) [201]. In fact, certain parts of the copolymer having chemical bonds that are cleavable under US radiation can change the hydrophobicity of the copolymer after their US-triggered cleavage, leading the conformational changes in polymer to open the pores of MSN and release the cargo at the target site [201].

5. Effective Combination of Active Targeting Therapy and Stimuli-Responsive Therapy Using MSN in Cancer Therapy

We have already discussed the various advantages of using MSN for drug delivery. Taken together, MSN exhibit large surface area, porous interior and tunable pore size to act as an excellent reservoir for different drug molecules and other materials of interest. Moreover, the various MSN syntheses approaches, mainly simple and adjustable, offer an ease optimization for sizes and shapes to maximize cellular uptake [202–204]. Importantly, one can easily decorate the surface of MSN by attaching small molecules, antibodies, aptamers, carrier proteins or peptide ligands for active targeting specifically to the cancer region, exploiting overexpressed receptors. Meanwhile, the controlled release of drugs to the disease site without any leakage to healthy tissues can be achieved by employing gatekeepers for the end-capping of MSN, triggered by various internal or external stimuli, such as pH, redox, enzyme activity, heat, light or magnetic field [205,206]. To achieve the precise chemotherapy of cancer, the most desired drug delivery system should possess high drug loading efficiency, site-specificity and the capacity of controlled drug release [207]. Hence, in this part of the review, we will report about the recently tried efforts for surface modification of MSN, exploiting both the active targeting and stimuli responsive behavior simultaneously (Figure 3), to obtain high efficacy with low dosage and minimize the off-target side effects of chemotherapy. Table 1 summarizes these simultaneously employed active targeting and stimuli responsive strategies developed up to date for MSN.

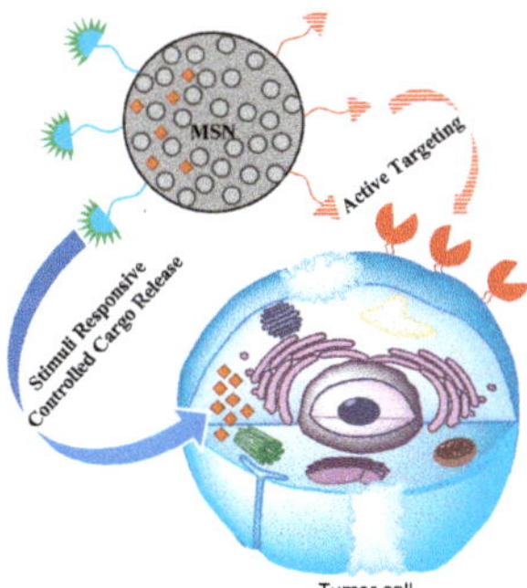

Figure 3. Multimodal decoration of MSN to achieve active targeting and stimuli responsive controlled release simultaneously.

Table 1. Simultaneously employed active targeting and stimuli responsive strategies using MSN in various cancer types.

Active Targeting		Stimuli Responsive Delivery		Cancer Therapeutics	Cancer Type	Outcome	Ref.
Ligand	**Receptor**	**Stimulus**	**Linker/Moiety**				
Folic acid	Folate	pH	poly(ethylene imine) (PEI)	-	cervical cancer	significantly higher number of particle internalization in cancer cells than normal cells	[208]
Folic acid	Folate	pH	poly(ethylene imine) (PEI)	Curcumin	colon cancer	suitable loading of fat-soluble antineoplastic drugs for sustained release	[209]
Folic acid	Folate	pH	polydopamine	Doxorubicin	cervical cancer	higher antitumor efficacy of MSNs@PDA-PEG-FA in vivo	[210]
Folic acid	Folate	Thermo/pH-coupling	poly[(*N*-isopropylacrylamide)-co-(methacrylic acid)]	Cisplatin	laryngeal carcinoma	higher cellular uptake, excellent drug release, greater cytotoxicity	[211],
Folic acid	Folate	Thermo/pH-coupling	poly[(*N*-isopropylacrylamide)-co-(methacrylic acid)]	siRNA against ABCG2 + cisplatin/5-fluorouracil (5-Fu)/paclitaxel	laryngeal carcinoma	down-regulation of ABCG2 significantly enhanced efficacy of chemotherapeutic drug-induced apoptosis of cancer cells	[212]
Folic acid	Folate	Redox	disulfide bonds	Curcumin	breast cancer	good biocompatibility, low toxicity, precise targeting and tumor growth inhibition	[213]
Folic acid	Folate	pH	chitosan-glycine	Colchicine (COL)	colon cancer	enhanced anticancer effects and reduced toxicity of free COL	[214]
Folic acid	Folate	pH	benzimidazole and β-cyclodextrin	valproic acid (VPA)	glioblastoma	enhanced effectiveness of radiotherapy	[215]
Folic acid	Folate	Magnetic field	iron oxide nanoparticles (IONPs)	Doxorubicin	breast cancer	effective active targeting and MRI-guided stimuli-responsive chemotherapy	[216]
Folic acid	Folate	pH and NIR light	polydopamine (PDA)	Doxorubicin	liver cancer	improved antitumor effect combining Dox-loaded MSN and NIR light	[217]
Folic acid	Folate	Enzyme (cathepsin B)	GFLG tetrapeptide linker	organotin-based cytotoxic compound	breast cancer	enhanced tumor growth inhibition with reduced hepatic and renal toxicity	[218]
Folic acid	Folate	Redox (Ascorbic acid)	cisplatin(IV) prodrug	cisplatin(IV) prodrug	cervical cancer	delivering cisplatin into cytosol, inducing DNA adducts and cell death	[219]
Transferrin	Transferrin	pH	chitosan or poly(d,l-lactide-co-glycolide) (PLGA)	Gemcitabine	pancreatic cancer	improved uptake of NPs by cancer cells, inhibition of cancer cell growth	[220]
Transferrin	Transferrin	pH and surface enhanced Raman scattering (SERS)	chitosan/poly(methacrylic acid) (CS-PMAA) and SERS reporter tagged Ag-NPs	Doxorubicin	cervical cancer	pH-responsive drug release, SERS-traceable characteristics and cancer cells targeting	[221]

Table 1. *Cont.*

Active Targeting		Stimuli Responsive Delivery		Cancer Therapeutics	Cancer Type	Outcome	Ref.
Ligand	Receptor	Stimulus	Linker/Moiety				
Transferrin	Transferrin	Redox	disulfide bonds	Doxorubicin	liver cancer	biocompatible system, potential in site-specific and controlled drug release	[222]
Transferrin	Transferrin	UV radiation (366 nm)	avidin, streptavidin and biotinylated photocleavable cross-linker	Doxorubicin	exposed tumors (skin, stomach and oesophagus)	efficient phototriggered drug delivery in accessible tumors and very high tumor cytotoxicity effect	[223]
Cetuximab	EGFR	photo	zinc phthalocyanine	ZnPcOBP	pancreatic Cancer	cell-line dependent photo-killing correlates well with EGFR expression levels	[224]
Trastuzumab	HER2	pH	poly(ethylene imine) (PEI)	siRNA against human HER2 oncogene	breast cancer	high batch-to-batch reproducibility, excellent safety profile, ready for clinical evaluation	[225]
HApt aptamer	HER2	pH	benzimidazole and β-cyclodextrin	Doxorubicin and biotherapeutic agent HApt	breast cancer	synergistic cytotoxic effects of chemotherapeutics in HER2-positive cancer cells	[226]
D-galactose	galactose receptor	pH	chitosan	5-fluorouracil (5-FU)	colon cancer	high drug loading capacity, possessed higher cytotoxicity on cancer cells	[227]
lectin concana-valin A (ConA)	glycans, sialic acids (SA)	pH	polyacrylic acid capping, acetal linker	Doxorubicin	bone cancer	increased antitumor effectiveness and decreased toxicity towards normal cell	[228]
cyclic RGDfC	$\alpha_v\beta_3$ integrin	photons	gold nanorods	-	breast cancer	enhanced radiosensitization of triple-negative breast cancer	[229]
cyclic RGDfC	$\alpha_v\beta_3$ integrin	Glutathione	thiol-functionalization	arsenic trioxide (ATO)	breast cancer	superior therapeutic ability of ATO-MSNs-RGD	[230]
RGD	$\alpha_v\beta_3$ integrin	Glutathione	β-Cyclodextrin, disulfide linker	Doxorubicin	MMP-rich tumor (colorectal and head and neck cancer)	tumor-triggered targeting drug delivery to cancerous cells	[231]
		matrix metallop-roteinase (MMP)	PLGVR peptide				
RGD and Tat_{48-60} peptide	Integrin and nuclear targeting	Glutathione	disulfide linker	Doxorubicin	cervical cancer	facilitated active targeting delivery and enhanced intracellular drug release	[232]
RGD	$\alpha_v\beta_3$ integrin	pH	α-amide-β-carboxyl group	Doxorubicin	glioblastoma	diversified multifunctional nanocomposites	[233]
$(RGDWWW)_2KC$	$\alpha_v\beta_3$ integrin	Glutathione	disulfide linker	Doxorubicin and therapeutic peptide	glioblastoma	tumor targeting and synergism of anticancer drug and therapeutic peptide	[234]

Table 1. *Cont.*

Active Targeting		Stimuli Responsive Delivery		Cancer Therapeutics	Cancer Type	Outcome	Ref.
Ligand	Receptor	Stimulus	Linker/Moiety				
K8(RGD)2	$\alpha_v\beta_3$ integrin	pH	acid-labile amides	Doxorubicin	glioblastoma	electrostatic repulsion induced nanovalve opening and drug release	[235]
RGD	$\alpha_v\beta_3$ integrin	pH	peptide-based amphiphile (P45)	Doxorubicin	Lung and breast cancer	targeted drug delivery and controlled drug release by the nanovalves	[236]
RGD	$\alpha_v\beta_3$ integrin	pH/NIR laser	gold nanostars (Au NSs)	Doxorubicin	glioblastoma	improved therapeutic efficacy combining chemotherapy and photothermal therapy (PTT)	[237]
cRGD and CREKA	$\alpha_v\beta_3$ integrin and fibronectin	radiofrequency (RF)	iron oxide core	Doxorubicin	brain tumor	remarkable increase in intratumoral drug levels	[238]
Asn-Gly-Arg (NGR)	cluster of differen-tiation 13 (CD13)	pH	polydopamine (PDA)	Doxorubicin	neovascular endothelial and glioma	greater BBB permeability, higher accumulation in intracranial tumor region	[239]
EpCAM aptamer	Epithelial cell adhesion molecule (EpCAM)	pH	citrate-capped gold nanoparticles	5-fluorouracil (5-FU)	hepatocellular carcinoma	preferential accumulation in tumor cells in vitro and in vivo	[240]
aptamer (Cy5.5-AS1411)	nucleolin (NCL)	laser irradiation (NIR light)	graphene oxide	Doxorubicin	breast cancer	synergism of chemotherapy and PTT	[241]
galactose (Gal) and TAT peptide	Asialoglycoprotein receptors and nuclear targeting	pH and Redox	poly(allylamine hydrochloride)-citraconic anhydride (PAH-Cit) and cysteine groups	Doxorubicinand VEGF-siRNA	hepato-carcinoma	effective and safe vector, sustained release, synergistic effect of chemodrugs and therapeutic genes	[242]
Phenylboronic acid (PBA)	sialic acid (SA)	MMP-2	PVGLIG peptide	Doxorubicin	liver cancer	tumor growth inhibition, minimal toxic side effects	[243]
YSA-BHQ1 and TAT-FITC	EphA2 receptor and nuclear targeting	pH	citraconic anhydride (Cit)	Doxorubicin	breast cancer	successfully developed anticancer drug delivery and imaging nanosystem	[244]
peptide CSNRDARRC	Targeting bladder cancer	pH	polydopamine (PDA)	Doxorubicin	bladder cancer	significantly superior antitumor effects of loaded nanocarriers than free drug	[245]
galactose (Gal) ligands and TAT peptide	Gal receptors and nuclear targeting	pH	poly(allylamine hydrochloride)-citraconic anhydride	Doxorubicin	hepato-carcinoma	improved tumorous distribution and potent therapeutic efficacy	[246]

Table 1. *Cont.*

Active Targeting		Stimuli Responsive Delivery		Cancer Therapeutics	Cancer Type	Outcome	Ref.
Ligand	Receptor	Stimulus	Linker/Moiety				
oligosaccharide of hyaluronic acid (oHA)	CD44	Glutathione	disulfide linker	6-mercaptopurine (6-MP)	colon cancer	increased stability and biocompatibility, efficient drug release in tumor cell	[247]
hyaluronic acid	CD44	Magnetic field	superparamagnetic Fe_3O_4 nanoparticles	Doxorubicin	breast cancer	active targeting to tumor cells and reduced off-target side effects	[248]
hyaluronic acid	CD44	NIR light	indocyanine green (ICG)	Doxorubicin	breast cancer	synergetic effect of chemotherapy and PTT	[249]
hyaluronic acid	CD44	Enzyme (MMP-2)	gelatin layer	Doxorubicin	breast cancer	successful bienzyme-responsive targeted and optimal drug delivery	[250]
hyaluronic acid	CD44	pH	DMMA (2,3-dimethylmaleic anhydride)	Doxorubicin	lung cancer	synergistic effect of active targeting and charge reversal in drug delivery	[251]

Besides, there are few reports that have used dual or multimodal response systems to improve the controlled release of the cargo. For example, Lu et al. have developed a pH/redox/near infrared (NIR) multi-stimuli responsive MSN to achieve efficient chemo-photothermal synergistic antitumor therapy [252]. Zhou et al. have also reported UV-light cross-linked and pH de-cross-linked coumarin-decorated cationic copolymer functionalized mesoporous silica nanoparticles for the improved co-delivery of anti-cancer drug and gene [253]. Moreover, Xu et al. have prepared a pH and redox dual-responsive (MSN)-sulfur (S)-S- chitosan (CS) controlled release drug delivery system [254]. Besides, a redox- and pH-sensitive dual response MSN system has been developed by Li and colleagues using ammonium salt to seal the pores [255]. Yan et al. have fabricated a pH/redox-triggered MSN nanosystem, for the codelivery of doxorubicin and paclitaxel in cancer cells [256]. Additionally, Anirudhan et al. have exploited both temperature and ultrasound sensitive gatekeepers for the surface modification of MSN [257].

6. MSN as Cancer Theranostics

Possible early detection and diagnosis is one of the most desired objectives to provide appropriate and extra real treatment for cancer. In order to overcome this hurdle along with the targeted and controlled delivery of chemotherapeutics, MSN have also been widely exploited for medical imaging and in situ diagnostics [258,259]. When both functions, i.e., therapy and diagnosis, are combined together, they are referred to as "theranostics" [260]. Herein, in this part of the review, we will discuss about various applications of MSN in cancer diagnosis such as exploiting MSN as imaging contrast agents, and utilizing MSN for proteomic analysis and fluorescent optical imaging.

Among the imaging technologies, magnetic resonance imaging (MRI) and ultrasound (US) have been mostly employed for cancer diagnosis due to their low-cost, low radioactivity and real-time monitoring properties [261]. There are various reports about the application of MSN decorated with specific targeting moiety as hyperpolarized, highly sensitive MRI agents having longer nuclear relaxation time [262,263]. As an example, Matsushita et al. have developed an MRI contrast agent comprising a core micelle with liquid perfluorocarbon inside the MSN for early cancer detection and diagnosis [264]. Additionally, a few research groups have systemically applied functionalized MSN to confer sufficient mean pixel intensity, to generate the higher quality US imaging of tumor bearing mice [265,266]. With imaging guidance from MRI or US, suspected cancerous tissues can be detected through biopsy. Furthermore, mesoporous silica-based chips with specific pore size provide a promising platform for proteomic analysis by mass spectrometry and chromatography, allowing the separation of low molecular weight proteins in serum from the higher weight proteins [267]. An analysis of mass spectrometry can identify unique protein signatures pertaining to various stages of cancer development, demonstrating plausible early cancer detection and therapy [268,269]. In addition, introducing metal ions or other functional groups enhances the selectivity and sensitivity of mesoporous silica chips to concentrate the low molecular weight proteins, analyze post-translational modifications in the human proteome and identify proteomic biomarkers in various cancers [270,271]. Importantly, fluorescent optical imaging exploiting MSN is gaining increasing attention in imaging-based therapy and cancer diagnosis [272,273]. The encapsulation of fluorescent dyes and bioluminescent proteins in MSN can overcome the associated limitations, such as rapid degradation, inadequate photo-stability and unpredictable toxicity of the fluorescent probes [274]. There are mainly two types of fluorescent MSN that have been reported for optical imaging, one is dye-doped MSN, prepared by incorporating fluorescent organic dye into pores of MSN and other one is combining QDs with MSN [275]. Yin et al. have synthesized folic acid-conjugated dye-entrapped MSN for in vivo cancer targeting and imaging [276]. Moreover, in contrast to the conventional organic dye, QDs appear more effective in optical imaging, due to possessing size-tunable wavelength absorption and emission, broad excitation wavelength, narrow emission bandwidth and a long fluorescent lifetime [277]. Functionalized QD-embedded MSN with high quantum yield have been largely exploited for selective tumor imaging in vivo, as well as for cancer cell imaging and detection in vitro by the intracellular internalization of QDs [278,279]. Recently, Zhao et al. have reported the synthesis of fluorescent Carbon Dot-MSN nanohybrids [86]. Nevertheless, Cheng et al. have reported tri-functionalized MSN, effectively decorated to be used in the field of theranostics coordinating the trio of target, imaging, and therapy in a discrete entity [280].

7. Challenges Regarding MSN Application in Cancer Therapy

Despite the recent advances of developing surface decorated MSN as an efficient carrier for the delivery of cancer chemotherapeutics, there are several challenges that need to be addressed for their further development. In particular, the scale up of MSN synthesis is one of the major issues limiting its commercial applications. On a small scale, the reproducibility on the synthesis of MSN can be maintained, but at the large scale, especially at an industrial level, it is very difficult to control batch to batch synthesis, as there are various different factors that need to be taken into account during the synthetic process. Hence, the clinical translation of MSN is taking a longer time than expected, as the therapeutic efficacy is not the only criteria for this [281].

In terms of the biological point of view, the clinical application of MSN is limited, because of the rapid clearance of nanoparticles by immune and excretory systems after administration [282,283]. Recent investigations have shown that MSN may be excreted, either in an intact or a degraded form, through hepatic or renal clearance [72,284]. However, the exact mechanism of the clearance is not known yet. Hence, the detailed in vivo analysis of pharmacokinetic and pharmacodynamic studies, possible immunogenicity and rigorous biodistribution of MSN-based systems should be employed before aiming to translate clinically [285,286]. A few reports highlighting half-life and biodistribution studies have demonstrated that in vivo biodegradation, systematic absorption and excretion, especially liver distribution and urinal excretion, are highly dependent on the physicochemical characteristics of MSN, such as geometries, porosities, surface chemistry, crystallinity, and different bio-nano interface interaction conditions [287–289]. For example, He et al. have evaluated the biodistribution and excretion of spherical MSN having various size ranges (80–360 nm) and pegylation (PEG-MSN) by fluorescence spectroscopy, and revealed accumulation of all the formulations in liver and spleen. They have also determined that, with a decrease in size and the pegylation of MSN, there is a reduction of the excretion rate from 45% to 15%, 30 min after administration [72]. In another study, Dogra et al. have shown that the increasing particle size of MSN from 32 to 142 nm results in a monotonic decrease in systemic bioavailability, along with accumulation in liver and spleen in healthy rats [290]. Furthermore, Sun et al. have completed a pharmacokinetic study of bevacizumab release from MSN-encapsulated bevacizumab nanoparticles in C57B/L mice and determined a significantly greater half-life, along with the sustained and slow release of MSN-encapsulated bevacizumab nanoparticles for a longer period of time than that of bevacizumab alone [291]. Additionally, Kong et al. have performed a biodistribution and pharmacokinetic study of Cy5-loaded hollow MSN in C57BL/6 mice and demonstrated gradual distribution in tumor and highest accumulation of MSN at 36 h after administration using fluorescence imaging. They have used the same MSN to deliver the cancer therapeutics (doxorubicin and interleukin-2) in the tumor microenvironment [292]. Regarding the limitation associated with bio-nano interface interactions, upon administration of MSN in the body and exposure to blood, proteins from blood serum and plasma adsorb onto the MSN surface and form a protein corona, which can eventually block the pores and decrease the release of cargo from the pores of MSN [293]. The protein corona formation is highly dependent upon the geometry of the MSN. Visalakshan et al. have shown a significantly lower amount of protein attaching from both plasma and serum on the spherical MSN, compared to the rod-like particles [294].

To address the biological limitations, a few research groups have started to introduce a lipid bilayer as gatekeeper and platform for surface modifications of MSN [295–298]. The advantages of using a lipid bilayer are its high biocompatibility, low immunogenicity, flexible formulation, and easy to incorporate targeting ligands and stimuli responsive moiety. For example, Brinker and co-workers have demonstrated MSN core for drug loading and a lipid bilayer as a gatekeeper to convey an EGFR-antibody for targeting leukemic cells efficiently in vitro and in vivo [299,300]. Samanta et al. have followed a similar approach of exploiting lipid bilayer around MSN to assist folate receptor targeted drug delivery in ovarian cancer [301]. Several other efforts have also been reported, exploiting organic/inorganic hybrid nanocarriers, L-tartaric acid, mucoadhesive delivery systems, organosilica-based drug delivery systems, to improve the biocompatibility of MSN [302–305]. Besides, cancer cell membranes have been utilized to coat MSN to improve immunocompatibility [306,307]. Moreover, an immunocompatible issue can be further resolved by replacing the commercially available lipids with the lipids derived from autologous extracellular vesicles (EVs) [308].

In conclusion, considering the various advantages of using MSN as a nanocarrier, along with the convincing preclinical results, it can be expected that, with the way out of related issues, MSN-based formulations may make exciting breakthroughs in cancer therapy.

Author Contributions: S.B. and V.C. wrote the paper. All authors have read and agreed to the published version of the manuscript.

Funding: S.B. receives funding from the European Commission under Grant Marie Skłodowska-Curie Actions (Standard European Individual Fellowships, H2020-MSCA-IF-2018, Grant Agreement No. 842964, Project Acronym 'MINT').

Conflicts of Interest: The authors declare no conflict of interest.

References

1. Anand, P.; Kunnumakara, A.B.; Sundaram, C.; Harikumar, K.B.; Tharakan, S.T.; Lai, O.S.; Sung, B.; Aggarwal, B.B. Cancer is a preventable disease that requires major lifestyle changes. *Pharm. Res.* **2008**, *25*, 2097–2116. [CrossRef] [PubMed]
2. Siegel, R.L.; Miller, K.D.; Jemal, A. Cancer statistics, 2020. *CA Cancer J. Clin.* **2020**, *70*, 7–30. [CrossRef]
3. Biankin, A.V.; Piantadosi, S.; Hollingsworth, S.J. Patient-centric trials for therapeutic development in precision oncology. *Nature* **2015**, *526*, 361–370. [CrossRef] [PubMed]
4. Baskar, R.; Itahana, K. Radiation therapy and cancer control in developing countries: Can we save more lives? *Int. J. Med. Sci.* **2017**, *14*, 13–17. [CrossRef]
5. Gillet, J.P.; Gottesman, M.M. Mechanisms of multidrug resistance in cancer. *Methods Mol. Biol.* **2010**, *596*, 47–76. [PubMed]
6. Vasan, N.; Baselga, J.; Hyman, D.M. A view on drug resistance in cancer. *Nature* **2019**, *575*, 299–309. [CrossRef] [PubMed]
7. Greish, K. Enhanced permeability and retention of macromolecular drugs in solid tumors: A royal gate for targeted anticancer nanomedicines. *J. Drug Target.* **2007**, *15*, 457–464. [CrossRef] [PubMed]
8. Davis, M.E.; Chen, Z.G.; Shin, D.M. Nanoparticle therapeutics: An emerging treatment modality for cancer. *Nat. Rev. Drug Discov.* **2008**, *7*, 771–782. [CrossRef]
9. Farokhzad, O.C.; Langer, R. Impact of nanotechnology on drug delivery. *ACS Nano* **2009**, *3*, 16–20. [CrossRef]
10. Maeda, H.; Wu, J.; Sawa, T.; Matsumura, Y.; Hori, K. Tumor vascular permeability and the EPR effect in macromolecular therapeutics: A review. *J. Control. Release* **2000**, *65*, 271–284. [CrossRef]
11. Pérez-Herrero, E.; Fernández-Medarde, A. Advanced targeted therapies in cancer: Drug nanocarriers, the future of chemotherapy. *Eur. J. Pharm. Biopharm.* **2015**, *93*, 52–79. [CrossRef] [PubMed]
12. Stylianopoulos, T.; Jain, R.K. Design considerations for nanotherapeutics in oncology. *Nanomedicine* **2015**, *11*, 1893–1907. [CrossRef]
13. Kydd, J.; Jadia, R.; Velpurisiva, P.; Gad, A.; Paliwal, S.; Rai, P. Targeting Strategies for the Combination Treatment of Cancer Using Drug Delivery Systems. *Pharmaceutics* **2017**, *9*, 46. [CrossRef]
14. Egusquiaguirre, S.P.; Igartua, M.; Hernández, R.M.; Pedraz, J.L. Nanoparticle delivery systems for cancer therapy: Advances in clinical and preclinical research. *Clin. Transl. Oncol.* **2012**, *14*, 83–93. [CrossRef] [PubMed]
15. Bhise, K.; Sau, S.; Alsaab, H.; Kashaw, S.K.; Tekade, R.K.; Iyer, A.K. Nanomedicine for cancer diagnosis and therapy: Advancement, success and structure-activity relationship. *Ther. Deliv.* **2017**, *8*, 1003–1018. [CrossRef] [PubMed]

16. Bayda, S.; Hadla, M.; Palazzolo, S.; Riello, P.; Corona, G.; Toffoli, G.; Rizzolio, F. Inorganic Nanoparticles for Cancer Therapy: A transition from lab to clinic. *Curr. Med. Chem.* **2018**, *25*, 4269–4303. [CrossRef]
17. Torchilin, V.P. Recent advances with liposomes as pharmaceutical carriers. *Nat. Rev. Drug Discov.* **2005**, *4*, 145–160. [CrossRef]
18. Mondal, G.; Barui, S.; Saha, S.; Chaudhuri, A. Tumor growth inhibition through targeting liposomally bound curcumin to tumor vasculature. *J. Control. Release* **2013**, *172*, 832–840. [CrossRef]
19. Barui, S.; Saha, S.; Mondal, G.; Haseena, S.; Chaudhuri, A. simultaneous delivery of doxorubicin and curcumin encapsulated in liposomes of pegylated RGDK-lipopeptide to tumor vasculature. *Biomaterials* **2014**, *35*, 1643–1656. [CrossRef]
20. Bobo, D.; Robinson, K.J.; Islam, J.; Thurecht, K.J.; Corrie, S.R. Nanoparticle-based medicines: A review of FDA-approved materials and clinical trials to date. *Pharm. Res.* **2016**, *33*, 2373–2387. [CrossRef]
21. García-Pinel, B.; Porras-Alcalá, C.; Ortega-Rodríguez, A.; Sarabia, F.; Prados, J.; Melguizo, C.; López-Romero, J.M. Lipid-based nanoparticles: Application and recent advances in cancer treatment. *Nanomaterials (Basel).* **2019**, *9*, 638. [CrossRef]
22. Noble, C.O.; Guo, Z.; Hayes, M.F.; Marks, J.D.; Park, J.W.; Benz, C.C.; Kirpotin, D.B.; Drummond, D.C. Characterization of highly stable liposomal and immunoliposomal formulations of vincristine and vinblastine. *Cancer Chemo. Pharm.* **2009**, *64*, 741–751. [CrossRef]
23. Li, W.; Cao, Z.; Liu, R.; Liu, L.; Li, H.; Li, X.; Chen, Y.; Lu, C.; Liu, Y. AuNPs as an important inorganic nanoparticle applied in drug carrier systems. *Artif. Cells Nanomed. Biotechnol.* **2019**, *47*, 4222–4233. [CrossRef]
24. Pinel, S.; Thomas, N.; Boura, C.; Barberi-Heyob, M. Approaches to physical stimulation of metallic nanoparticles for glioblastoma treatment. *Adv. Drug Deliv. Rev.* **2019**, *138*, 344–357. [CrossRef]
25. Negri, V.; Pacheco-Torres, J.; Calle, D.; López-Larrubia, P. Carbon nanotubes in biomedicine. *Top. Curr. Chem (Cham).* **2020**, *378*, 15. [CrossRef]
26. Li, T.; Shi, S.; Goel, S.; Shen, X.; Xie, X.; Chen, Z.; Zhang, H.; Li, S.; Qin, X.; Yang, H.; et al. Recent advancements in mesoporous silica nanoparticles towards therapeutic applications for cancer. *Acta Biomater.* **2019**, *89*, 1–13. [CrossRef]
27. Wang, Y.; Xie, Y.; Kilchrist, K.V.; Li, J.; Duvall, C.L.; Oupický, D. Endosomolytic and Tumor-Penetrating Mesoporous Silica Nanoparticles for siRNA/miRNA Combination Cancer Therapy. *ACS Appl. Mater. Interfaces* **2020**, *12*, 4308–4322. [CrossRef]
28. Bradbury, M.S.; Pauliah, M.; Zanzonico, P.; Wiesner, U.; Patel, S. Intraoperative mapping of SLN metastases using a clinically-translated ultrasmall silica nanoparticle. *Wiley Interdiscip. Rev. Nanomed. Nanobiotechnol.* **2016**, *8*, 535–553. [CrossRef]
29. Kumar, P.; Tambe, P.; Paknikar, K.M.; Gajbhiye, V. Mesoporous silica nanoparticles as cutting-edge theranostics: Advancement from merely a carrier to tailor-made smart delivery platform. *J. Control. Release* **2018**, *287*, 35–57. [CrossRef]
30. Iturrioz-Rodríguez, N.; Correa-Duarte, M.A.; Fanarraga, M.L. Controlled drug delivery systems for cancer based on mesoporous silica nanoparticles. *Int. J. Nanomed.* **2019**, *14*, 3389–3401. [CrossRef]
31. Watermann, A.; Brieger, J. Mesoporous Silica Nanoparticles as Drug Delivery Vehicles in Cancer. *Nanomaterials* **2017**, *7*, 189. [CrossRef]
32. US Food and Drug Administration GRAS Substances (SCOGS) Database-Select Committee on GRAS Substances (SCOGS) Opinion: Silicates. Available online: https://www.accessdata.fda.gov/scripts/fdcc/?set=SCOGS (accessed on 10 April 2020).
33. Chen, F.; Hableel, G.; Zhao, E.R.; Jokerst, J.V. Multifunctional Nanomedicine with silica: Role of silica in nanoparticles for theranostic, imaging, and drug monitoring. *J. Colloid Interface Sci.* **2018**, *521*, 261–279. [CrossRef]
34. Hauss, D.J. Oral lipid-based formulations. *Adv. Drug Deliv. Rev.* **2007**, *59*, 667–676. [CrossRef] [PubMed]
35. Kawabata, Y.; Wada, K.; Nakatani, M.; Yamada, S.; Onoue, S. Formulation design for poorly water-soluble drugs based on biopharmaceutics classification system: Basic approaches and practical applications. *Int. J. Pharm.* **2011**, *420*, 1–10. [CrossRef] [PubMed]
36. Lu, J.; Liong, M.; Zink, J.I.; Tamanoi, F. Mesoporous silica nanoparticles as a delivery system for hydrophobic anticancer drugs. *Small* **2007**, *3*, 1341–1346. [CrossRef] [PubMed]
37. He, Y.; Liang, S.; Long, M.; Xu, H. Mesoporous silica nanoparticles as potential carriers for enhanced drug solubility of paclitaxel. *Mater. Sci. Eng. C* **2017**, *78*, 12–17. [CrossRef]

38. Badruddoza, A.Z.M.; Gupta, A.; Myerson, A.S.; Trout, B.L.; Doyle, P.S. Low energy nanoemulsions as templates for the formulation of hydrophobic drugs. *Adv. Ther.* **2018**, *1*, 1700020. [CrossRef]
39. Wais, U.; Jackson, A.W.; He, T.; Zhang, H. Formation of hydrophobic drug nanoparticles via ambient solvent evaporation facilitated by branched diblock copolymers. *Int. J. Pharm.* **2017**, *533*, 245–253. [CrossRef]
40. Maleki, A.; Kettiger, H.; Schoubben, A.; Rosenholm, J.M.; Ambrogi, V.; Hamidi, M. Mesoporous silica materials: From physico-chemical properties to enhanced dissolution of poorly water-soluble drugs. *J. Control. Release* **2017**, *262*, 329–347. [CrossRef]
41. Zhou, Y.; Wu, B.; Quan, G.; Huang, Y.; Wu, Q.; Pan, X.; Zhang, X.; Wu, C. Mesoporous silica nanoparticles for drug and gene delivery. *Acta Pharm. Sin. B* **2018**, *8*, 165–177. [CrossRef]
42. Lu, J.; Liong, M.; Li, Z.; Zink, J.I.; Tamanoi, F. Biocompatibility, biodistribution, and drug-delivery efficiency of mesoporous silica nanoparticles for cancer therapy in animals. *Small* **2010**, *6*, 1794–1805. [CrossRef] [PubMed]
43. Jafari, S.; Derakhshankhah, H.; Alaei, L.; Varnamkhasti, B.S.; Saboury, A.A.; Fattahi, A. Mesoporous silica nanoparticles for therapeutic/diagnostic applications. *Biomed. Pharmacother.* **2019**, *109*, 1100–1111. [CrossRef] [PubMed]
44. Barkat, A.; Beg, S.; Panda, S.K.; Alharbi, S.K.; Rahman, M.; Ahmed, F.J. Functionalized mesoporous silica nanoparticles in anticancer therapeutics. *Semin. Cancer Biol.* **2019**, *S1044-579X*, 30104–X. [CrossRef] [PubMed]
45. Narayan, R.; Nayak, U.Y.; Raichur, A.M.; Garg, S. Mesoporous Silica Nanoparticles: A Comprehensive Review on Synthesis and Recent Advances. *Pharmaceutics* **2018**, *10*, 118. [CrossRef]
46. Tao, Z.; Toms, B.; Goodisman, J.; Asefa, T. Mesoporous silica microparticles enhance the cytotoxicity of anticancer platinum drugs. *ACS Nano* **2010**, *4*, 789–794. [CrossRef]
47. Castillo, R.R.; Lozano, D.; Vallet-Regí, M. Mesoporous silica nanoparticles as carriers for therapeutic biomolecules. *Pharmaceutics* **2020**, *12*, 432. [CrossRef]
48. Xu, C.; Lei, C.; Yu, C. Mesoporous silica nanoparticles for protein protection and delivery. *Front. Chem.* **2019**, *7*, 290. [CrossRef]
49. Knezevic, N.Z.; Durand, J.O. Large pore mesoporous silica nanomaterials for application in delivery of biomolecules. *Nanoscale* **2015**, *7*, 2199–2209. [CrossRef]
50. Liu, H.J.; Xu, P. Smart mesoporous silica nanoparticles for protein delivery. *Nanomaterials* **2019**, *9*, 511. [CrossRef]
51. Kim, S.I.; Pham, T.T.; Lee, J.W.; Roh, S.H. Releasing properties of proteins on SBA-15 spherical nanoparticles functionalized with aminosilanes. *J. Nanosci. Nanotechnol.* **2010**, *10*, 3467–3472. [CrossRef]
52. Slowing, I.I.; Trewyn, B.G.; Lin, V.S.Y. Mesoporous silica nanoparticles for intracellular delivery of membrane-impermeable proteins. *J. Am. Chem. Soc.* **2007**, *129*, 8845–8849. [CrossRef]
53. Méndez, J.; Morales Cruz, M.; Delgado, Y.; Figueroa, C.M.; Orellano, E.A.; Morales, M.; Monteagudo, A.; Griebenow, K. Delivery of chemically glycosylated cytochrome c immobilized in mesoporous silica nanoparticles induces apoptosis in HeLa cancer cells. *Mol. Pharm.* **2014**, *11*, 102–111. [CrossRef]
54. Choi, E.; Lim, D.-K.; Kim, S. Hydrolytic surface erosion of mesoporous silica nanoparticles for efficient intracellular delivery of cytochrome c. *J. Colloid Interface Sci.* **2020**, *560*, 416–425. [CrossRef]
55. Zhang, J.; Karmakar, S.; Yu, M.H.; Mitter, N.; Zou, J.; Yu, C.Z. Synthesis of silica vesicles with controlled entrance size for high loading, sustained release, and cellular delivery of therapeutical proteins. *Small* **2014**, *10*, 5068–5076. [CrossRef]
56. Niu, Y.; Yu, M.; Meka, A.; Liu, Y.; Zhang, J.; Yang, Y.; Yu, C. Understanding the contribution of surface roughness and hydrophobic modification of silica nanoparticles to enhanced therapeutic protein delivery. *J. Mater. Chem. B* **2016**, *4*, 212–219. [CrossRef]
57. Yang, Y.; Niu, Y.; Zhang, J.; Meka, A.K.; Zhang, H.; Xu, C.; Xiang, C.; Lin, C.; Yu, M.; Yu, C. Biphasic synthesis of large-pore and well-dispersed benzene bridged mesoporous organosilica nanoparticles for intracellular protein delivery. *Small* **2015**, *11*, 2743–2749. [CrossRef]
58. Yang, Y.; Lu, Y.; Abbaraju, P.L.; Zhang, J.; Zhang, M.; Xiang, G.; Yu, C. Multi-shelled dendritic mesoporous organosilica hollow spheres: Roles of composition and architecture in cancer immunotherapy. *Angew. Chem. Int. Ed.* **2017**, *56*, 8446–8450. [CrossRef]
59. Luo, G.F.; Chen, W.H.; Liu, Y.; Lei, Q.; Zhuo, R.X.; Zhanga, X.Z. Multifunctional enveloped mesoporous silica nanoparticles for subcellular co-delivery of drug and therapeutic peptide. *Sci. Rep.* **2014**, *4*, 6064. [CrossRef]

60. Shao, D.; Li, M.; Wang, Z.; Zheng, X.; Lao, Y.H.; Chang, Z.; Zhang, F.; Lu, M.; Yue, J.; Hu, H.; et al. Bioinspired diselenide-bridged mesoporous silica nanoparticles for dual-responsive protein delivery. *Adv. Mater.* **2018**, *30*, e1801198. [CrossRef]
61. Wen, J.; Yang, K.; Liu, F.; Li, H.; Xu, Y.; Sun, S. Diverse gatekeepers for mesoporous silica nanoparticle based drug delivery systems. *Chem. Soc. Rev.* **2017**, *46*, 6024–6045. [CrossRef]
62. Deodhar, G.V.; Adams, M.L.; Trewyn, B.G. Controlled release and intracellular protein delivery from mesoporous silica nanoparticles. *Biotechnol. J.* **2017**, *12*, 1600408. [CrossRef] [PubMed]
63. Argyo, C.; Weiss, V.; Bräuchle, C.; Bein, T. Multifunctional mesoporous silica nanoparticles as a universal platform for drug delivery. *Chem. Mater.* **2013**, *26*, 435–451. [CrossRef]
64. Vallet-Regí, M.; Colilla, M.; Izquierdo-Barba, I.; Manzano, M. Mesoporous silica nanoparticles for drug delivery: Current insights. *Molecules* **2018**, *23*, 47. [CrossRef] [PubMed]
65. Yang, Y.; Yu, C. Advances in silica based nanoparticles for targeted cancer therapy. *Nanomed. Nanotechnol. Biol. Med.* **2016**, *12*, 317–332. [CrossRef]
66. Folkman, J. Role of angiogenesis in tumor growth and metastasis. *Semin. Oncol.* **2002**, *29*, 15–18. [CrossRef]
67. Carmeliet, P.; Jain, R.K. Angiogenesis in cancer and other diseases. *Nature* **2000**, *407*, 249–257. [CrossRef]
68. Wilhelm, S.; Tavares, A.J.; Dai, Q.; Ohta, S.; Audet, J.; Dvorak, H.F.; Chan, W.C.W. Analysis of nanoparticle delivery to tumours. *Nat. Rev. Mater.* **2016**, *1*, 1–12. [CrossRef]
69. Fang, J.; Nakamura, H.; Maeda, H. The EPR effect: Unique features of tumor blood vessels for drug delivery, factors involved, and limitations and augmentation of the effect. *Adv. Drug Deliv. Rev.* **2011**, *63*, 136–151. [CrossRef]
70. Nakamura, H.; Fang, J.; Maeda, H. Development of next-generation macromolecular drugs based on the EPR effect: Challenges and pitfalls. *Expert Opin. Drug Del.* **2015**, *12*, 53–64. [CrossRef]
71. Lee, J.E.; Lee, N.; Kim, H.; Kim, J.; Choi, S.H.; Kim, J.H.; Kim, T.; Song, I.C.; Park, S.P.; Moon, W.K.; et al. Uniform mesoporous dye-doped silica nanoparticles decorated with multiple magnetite nanocrystals for simultaneous enhanced magnetic resonance imaging, fluorescence imaging, and drug delivery. *J. Am. Chem. Soc.* **2010**, *132*, 552–557. [CrossRef]
72. He, Q.; Zhang, Z.; Gao, F.; Li, Y.; Shi, J. *In vivo* biodistribution and urinary excretion of mesoporous silica nanoparticles: Effects of particle size and PEGylation. *Small* **2011**, *7*, 271–280. [CrossRef]
73. Zhu, Y.; Fang, Y.; Borchardt, L.; Kaskel, S. PEGylated hollow mesoporous silica nanoparticles as potential drug delivery vehicles. *Microporous Mesoporous Mater.* **2011**, *141*, 199–206. [CrossRef]
74. Meng, H.; Mai, W.X.; Zhang, H.; Xue, M.; Xia, T.; Lin, S.; Wang, X.; Zhao, Y.; Ji, Z.; Zink, J.I.; et al. Codelivery of an optimal Drug/siRNA combination using mesoporous silica nanoparticles to overcome drug resistance in breast cancer *in vitro* and *in vivo*. *ACS Nano* **2013**, *7*, 994–1005. [CrossRef]
75. Heldin, C.-H.; Rubin, K.; Pietras, K.; Östman, A. High interstitial fluid pressure-An obstacle in cancer therapy. *Nat. Rev. Cancer* **2004**, *4*, 806–813. [CrossRef]
76. Vyas, S.P.; Singh, A.; Sihorkar, V. Ligand-receptor-mediated drug delivery: An emerging paradigm in cellular drug targeting. *Crit. Rev. Ther. Drug Carrier Syst.* **2001**, *18*, 1–76. [CrossRef]
77. Ruoslahti, E.; Bhatia, S.N.; Sailor, M.J. Targeting of drugs and nanoparticles to tumors. *J. Cell Biol.* **2010**, *188*, 759–768. [CrossRef]
78. Parker, N.; Turk, M.J.; Westrick, E.; Lewis, J.D.; Low, P.S.; Leamon, C.P. Folate receptor expression in carcinomas and normal tissues determined by a quantitative radioligand binding assay. *Anal. Biochem.* **2005**, *338*, 284–293. [CrossRef]
79. Porta, F.; Lamers, G.E.M.; Morrhayim, J.; Chatzopoulou, A.; Schaaf, M.; den Dulk, H.; Backendorf, C.; Zink, J.I.; Kros, A. Folic acid-modified mesoporous silica nanoparticles for cellular and nuclear targeted drug delivery. *Adv. Healthc. Mater.* **2013**, *2*, 281–286. [CrossRef]
80. Zwicke, G.H.; Mansoori, G.A.; Jeffery, C.J. Utilizing the folate receptor for active targeting of cancer nanotherapeutics. *Nano Reviews* **2012**, *3*, 18496. [CrossRef] [PubMed]
81. Cheung, A.; Bax, H.J.; Josephs, D.H.; Ilieva, K.M.; Pellizzari, G.; Opzoomer, J.; Bloomfield, J.; Fittall, M.; Grigoriadis, A.; Figini, M.; et al. Targeting folate receptor alpha for cancer treatment. *Oncotarget* **2016**, *7*, 52553–52574. [CrossRef]
82. Slowing, I.; Trewyn, B.G.; Lin, V.S.Y. Effect of surface functionalization of MCM-41-type mesoporous silica nanoparticles on the endocytosis by human cancer cells. *J. Am. Chem. Soc.* **2006**, *128*, 14792–14793. [CrossRef]

83. Khosravian, P.; Ardestani, M.S.; Mehdi Khoobi, M.; Ostad, S.N.; Dorkoosh, F.A.; Javar, H.A.; Amanlou, M. Mesoporous silica nanoparticles functionalized with folic acid/methionine for active targeted delivery of docetaxel. *Onco.Targets Ther.* **2016**, *9*, 7315–7330. [CrossRef]
84. Yinxuea, S.; Binb, Z.; Xiangyang, D.; Yong, W.; Jie, Z.; Yanqiu, A.; Zongjiang, X.; Gaofenge, Z. Folic acid (FA)-conjugated mesoporous silica nanoparticles combined with MRP-1 siRNA improves the suppressive effects of myricetin on non-small cell lung cancer (NSCLC). *Biomed. Pharmacother.* **2020**, *125*, 109561. [CrossRef]
85. Zheng, G.; Shen, Y.; Zhao, R.; Chen, F.; Zhang, F.; Xu, A.; Shao, A. Dual-Targeting Multifuntional Mesoporous Silica Nanocarrier for Codelivery of siRNA and Ursolic Acid to Folate Receptor Overexpressing Cancer Cells. *J. Agric. Food Chem.* **2017**, *65*, 6904–6911. [CrossRef]
86. Zhao, S.; Sun, S.; Jiang, K.; Wang, Y.; Liu, Y.; Wu, S.; Li, Z.; Shu, Q.; Lin, H. In Situ Synthesis of Fluorescent Mesoporous Silica–Carbon Dot Nanohybrids Featuring Folate Receptor Overexpressing Cancer Cell Targeting and Drug Delivery. *Nano-Micro Lett.* **2019**, *11*, 32. [CrossRef]
87. Lu, J.; Li, Z.; Zink, J.I.; Tamanoi, F. *In vivo* tumor suppression efficacy of mesoporous silica nanoparticles-based drug-delivery system: Enhanced efficacy by folate modification. *Nanomedicine* **2012**, *8*, 212–220. [CrossRef]
88. López, V.; Villegas, M.R.; Rodríguez, V.; Villaverde, G.; Lozano, D.; Baeza, A.; Vallet-Regí, M. Janus mesoporous silica nanoparticles for dual targeting of tumor cells and mitochondria. *ACS Appl. Mater. Interfaces.* **2017**, *9*, 26697–26706. [CrossRef]
89. Rosenholm, J.M.; Peuhu, E.; Bate-Eya, L.T.; Eriksson, J.E.; Sahlgren, C.; Linden, M. Cancer-cell-specific induction of apoptosis using mesoporous silica nanoparticles as drug-delivery vectors. *Small* **2010**, *6*, 1234–1241. [CrossRef]
90. Daniels, T.R.; Bernabeu, E.; Rodríguez, J.A.; Patel, S.; Kozman, M.; Chiappetta, D.A.; Holler, E.; Ljubimova, J.Y.; Helguera, G.; Penicheta, M.L. Transferrin receptors and the targeted delivery of therapeutic agents against cancer. *Biochim. Biophys. Acta.* **2012**, *1820*, 291–317. [CrossRef]
91. Shen, Y.; Li, X.; Dong, D.; Zhang, B.; Xue, Y.; Shang, P. The review of TFR1 in cancer. *Am. J. Cancer Res.* **2018**, *8*, 916–931.
92. Jang, M.; Oh, I. Targeted drug delivery of Transferrin-Conjugated Mesoporous Silica Nanoparticles. *Yakhak Hoeji* **2017**, *61*, 241–247. [CrossRef]
93. Ferris, D.P.; Lu, J.; Gothard, C.; Yanes, R.; Thomas, C.R.; Olsen, J.C.; Stoddart, J.F.; Tamanoi, F.; Zink, J.I. Synthesis of Biomolecule-Modified Mesoporous Silica Nanoparticles for Targeted Hydrophobic Drug Delivery to Cancer Cells. *Small* **2011**, *7*, 1816–1826. [CrossRef]
94. Montalvo-Quiros, S.; Aragoneses-Cazorla, G.; Garcia-Alcalde, L.; Vallet-Regí, M.; González, B.; Luque-Garcia, J.L. Cancer cell targeting and therapeutic delivery of silver nanoparticles by mesoporous silica nanocarrirs: Insights into the action mechanisms using quantitative proteomics. *Nanoscale* **2019**, *11*, 4531–4545. [CrossRef]
95. Ke, Y.; Xiang, C. Transferrin receptor-targeted hMsN for sorafenib delivery in refractory differentiated thyroid cancer therapy. *Int. J. Nanomed.* **2018**, *13*, 8339–8354. [CrossRef]
96. Sun, T.; Wu, H.; Li, Y.; Huang, Y.; Yao, L.; Chen, X.; Han, X.; Zhou, Y.; Du, Z. Targeting transferrin receptor delivery of temozolomide for a potential glioma stem cell-mediated therapy. *Oncotarget* **2017**, *8*, 74451–74465. [CrossRef]
97. Luo, M.; Lewik, G.; Ratcliffe, J.C.; Choi, C.H.J.; Mäkilä, E.; Tong, W.Y.; Voelcker, N.H. Systematic evaluation of transferrin-modified porous silicon nanoparticles for targeted delivery of doxorubicin to glioblastoma. *ACS Appl. Mater. Interfaces* **2019**, *11*, 33637–33649. [CrossRef]
98. Sheykhzadeh, S.; Luo, M.; Peng, B.; White, J.; Abdalla, Y.; Tang, T.; Mäkilä, E.; Voelcker, N.H.; Tong, W.Y. Transferrin-targeted porous silicon nanoparticles reduce glioblastoma cell migration across tight extracellular space. *Sci. Rep.* **2020**, *10*, 2320. [CrossRef]
99. Barui, S.; Saha, S.; Yakati, V.; Chaudhuri, A. Systemic co-delivery of a homo-serine derived ceramide analog and curcumin to tumor vasculature inhibits mouse tumor growth. *Mol. Pharm.* **2016**, *13*, 404–419. [CrossRef]
100. Dal Corso, A.; Pignataro, L.; Belvisi, L.; Gennari, C. $\alpha v\beta 3$ Integrin-Targeted Peptide/Peptidomimetic-Drug Conjugates: In-Depth Analysis of the Linker Technology. *Curr. Top. Med. Chem.* **2016**, *16*, 314–329. [CrossRef]
101. Fang, I.J.; Slowing, I.I.; Wu, K.C.; Lin, V.S.; Trewyn, B.G. Ligand conformation dictates membrane and endosomal trafficking of arginine-glycine-aspartate (RGD)-functionalized mesoporous silica nanoparticles. *Chemistry* **2012**, *18*, 7787–7792. [CrossRef]

102. Hu, H.; You, Y.; He, L.; Chen, T. The rational design of NAMI-A-loaded mesoporous silica nanoparticles as antiangiogenic nanosystems. *J. Mater. Chem. B* **2015**, *3*, 6338–6346.
103. Hu, H.; Arena, F.; Gianolio, E.; Boffa, C.; di Gregorio, E.; Stefania, R.; Orio, L.; Baroni, S.; Aime, S. Mesoporous silica nanoparticles functionalized with fluorescent and MRI reporters for the visualization of murine tumors overexpressing $\alpha v \beta 3$ receptors. *Nanoscale* **2016**, *8*, 7094–7104. [CrossRef]
104. Sun, J.; Kim, D.H.; Guo, Y.; Teng, Z.; Li, Y.; Zheng, L.; Zhang, Z.; Larson, A.C.; Lu, G. A c(RGDfE) conjugated multi-functional nanomedicine delivery system for targeted pancreatic cancer therapy. *J. Mater. Chem. B* **2015**, *3*, 1049–1058. [CrossRef]
105. Chakravarty, R.; Goel, S.; Hong, H.; Chen, F.; Valdovinos, H.F.; Hernandez, R.; Barnhart, T.E.; Cai, W. Hollow mesoporous silica nanoparticles for tumor vasculature targeting and PET image-guided drug delivery. *Nanomedicine (Lond).* **2015**, *10*, 1233–1246. [CrossRef]
106. Mo, J.; He, L.; Ma, B.; Chen, T. Tailoring Particle Size of Mesoporous Silica Nanosystem To Antagonize Glioblastoma and Overcome Blood-Brain Barrier. *ACS Appl. Mater. Interfaces* **2016**, *8*, 6811–6825. [CrossRef]
107. Pan, L.; Liu, J.; He, Q.; Shi, J. MSN-mediated sequential vascular-to-cell nuclear-targeted drug delivery for efficient tumor regression. *Adv. Mater.* **2014**, *26*, 6742–6748. [CrossRef]
108. Kari, C.; Chan, T.O.; Rocha de Quadros, M.; Rodeck, U. Targeting the epidermal growth factor receptor in cancer: Apoptosis takes center stage. *Cancer Res.* **2003**, *63*, 1–5.
109. Sharma, S.V.; Bell, D.W.; Settleman, J.; Haber, D.A. Epidermal growth factor receptor mutations in lung cancer. *Nat. Rev. Cancer.* **2007**, *7*, 169–181. [CrossRef]
110. She, X.; Chen, L.; Velleman, L.; Li, C.; He, C.; Denman, J.; Wang, T.; Shigdar, S.; Duanc, W.; Kong, L. The control of epidermal growth factor grafted on mesoporous silica nanoparticles for targeted delivery. *J. Mater. Chem. B* **2015**, *3*, 6094. [CrossRef]
111. Sundarraj, S. Targeting efficiency and biodistribution of EGFR-conjugated mesoporous silica nanoparticles for cisplatin delivery in nude mice with lung cancer. *Ann. Oncol.* **2012**, *23*, ix70–ix71. [CrossRef]
112. Wang, Y.; Huang, H.; Yang, L.; Zhang, Z.; Ji, H. Cetuximab-modified mesoporous silica nano-medicine specifically targets EGFR-mutant lung cancer and overcomes drug resistance. *Sci. Rep.* **2016**, *6*, 25468. [CrossRef]
113. Iqbal, N.; Iqbal, N. Human epidermal growth factor receptor 2 (HER2) in cancers: Overexpression and therapeutic implications. *Mol. Biol. Int.* **2014**, *2014*, 852748. [CrossRef] [PubMed]
114. Orphanos, G.; Kountourakis, P. Targeting the HER2 receptor in metastatic breast cancer. *Hematol. Oncol. Stem Cell Ther.* **2012**, *5*, 127–137. [CrossRef] [PubMed]
115. Tsai, C.; Chen, C.; Hung, Y.; Changb, F.; Mou, C. Monoclonal antibody-functionalized mesoporous silicananoparticles (MSN) for selective targeting breast cancer cells. *J. Mater. Chem.* **2009**, *19*, 5737–5743. [CrossRef]
116. Ellis, L.M.; Hicklin, D.J. VEGF-targeted therapy: Mechanisms of anti-tumour activity. *Nat. Rev. Cancer* **2008**, *8*, 579–591. [CrossRef]
117. Costache, M.I.; Ioana, M.; Iordache, S.; Ene, D.; Costache, C.A.; Săftoiu, A. VEGF Expression in Pancreatic Cancer and Other Malignancies: A Review of the Literature. *Rom. J. Intern. Med.* **2015**, *53*, 199–208. [CrossRef]
118. Goel, S.; Chen, F.; Hong, H.; Valdovinos, H.F.; Barnhart, T.E.; Cai, W. VEGFR-targeted drug delivery in vivo with mesoporous silica nanoparticles. *J. Nucl. Med.* **2014**, *55* (Suppl. 1), 222.
119. Goel, S.; Chen, F.; Hong, H.; Valdovinos, H.F.; Hernandez, R.; Shi, S.; Barnhart, T.E.; Cai, W. $VEGF_{121}$-Conjugated Mesoporous Silica Nanoparticle: A Tumor Targeted Drug Delivery System. *ACS Appl. Mater. Interfaces* **2014**, *6*, 21677–21685. [CrossRef]
120. Zhang, R.; Zhang, Y.; Tan, J.; Wang, H.; Zhang, G.; Li, N.; Meng, Z.; Zhang, F.; Chang, J.; Wang, R. Antitumor effect of 131i-labeled anti-Vegfr2 targeted mesoporous silica nanoparticles in anaplastic thyroid cancer. *Nanoscale Res. Lett.* **2019**, *14*, 96. [CrossRef]
121. Scodeller, P.; Simón-Gracia, L.; Kopanchuk, S.; Tobi, A.; Kilk, K.; Säälik, P.; Kaarel Kurm, K.; Squadrito, M.L.; Kotamraju, V.R.; Rinken, A.; et al. Precision targeting of tumor macrophages with a CD206 binding peptide. *Sci. Rep.* **2017**, *7*, 14655. [CrossRef] [PubMed]
122. Park, I.Y.; Kim, I.Y.; Yoo, M.K.; Choi, Y.J.; Cho, M.H.; Cho, C.S. Mannosylated polyethylenimine coupled mesoporous silica nanoparticles for receptor-mediated gene delivery. *Int. J. Pharm.* **2008**, *359*, 280–287. [CrossRef] [PubMed]

123. Chen, N.-T.; Souris, J.S.; Cheng, S.-H.; Chu, C.-H.; Wang, Y.-C.; Konda, V.; Dougherty, U.; Bissonnette, M.; Mou, C.-Y.; Chen, C.-T.; et al. Lectin-functionalized mesoporous silica nanoparticles for endoscopic detection of premalignant colonic lesions. *Nanomed. Nanotechnol. Biol. Med.* **2017**, *13*, 1941–1952. [CrossRef]
124. Guo, X.; Guo, N.; Zhao, J.; Cai, Y. Active targeting co-delivery system based on hollow mesoporous silica nanoparticles for antitumor therapy in ovarian cancer stem-like cells. *Oncol. Rep.* **2017**, *38*, 1442–1450. [CrossRef] [PubMed]
125. Quan, G.; Pan, X.; Wang, Z.; Wu, Q.; Li, G.; Dian, L.; Chen, B.; Wu, C. Lactosaminated mesoporous silica nanoparticles for asialoglycoprotein receptor targeted anticancer drug delivery. *J. Nanobiotechnol.* **2015**, *13*. [CrossRef] [PubMed]
126. Paramonov, V.M.; Desai, D.; Kettiger, H.; Mamaeva, V.; Rosenholm, J.M.; Sahlgren, C.; Rivero-Müller, A. Targeting somatostatin receptors by functionalized mesoporous silica nanoparticles—are we striking home? *Nanotheranostics* **2018**, *2*, 320–346. [CrossRef] [PubMed]
127. Zhang, M.; Xu, C.; Wen, L.; Han, M.; Xiao, B.; Zhou, J.; Zhang, Y.; Zhang, Z.; Viennois, E.; Merlin, D. A hyaluronidase-responsive nanoparticle-based drug delivery system for targeting colon cancer cells. *Cancer Res.* **2016**, *76*, 7208–7218. [CrossRef] [PubMed]
128. Chen, F.; Hong, H.; Shi, S.; Goel, S.; Valdovinos, H.F.; Hernandez, R.; Theuer, C.P.; Barnhart, T.E.; Cai, W. Engineering of hollow mesoporous silica nanoparticles for remarkably enhanced tumor active targeting efficacy. *Sci. Rep.* **2014**, *4*, 5080. [CrossRef]
129. Chen, F.; Nayak, T.R.; Goel, S.; Valdovinos, H.F.; Hong, H.; Theuer, C.P.; Barnhart, T.E.; Cai, W. *In vivo* tumor vasculature targeted PET/NIRF imaging with TRC105(Fab)-conjugated, dual-labeled mesoporous silica nanoparticles. *Mol. Pharm.* **2014**, *11*, 4007–4014. [CrossRef]
130. Sweeney, S.K.; Luo, Y.; O'Donnell, M.A.; Assouline, J.G. Peptide-Mediated Targeting Mesoporous Silica Nanoparticles: A Novel Tool for Fighting Bladder Cancer. *J. Biomed. Nanotechnol.* **2017**, *13*, 232–242. [CrossRef]
131. Hicke, B.J.; Stephens, A.W.; Gould, T.; Chang, Y.-F.; Lynott, C.K.; Heil, J.; Borkowski, S.; Hilger, C.-S.; Cook, G.; Warren, S.; et al. Tumor targeting by an aptamer. *J. Nucl. Med.* **2006**, *47*, 668–678.
132. Yang, Y.; Zhao, W.; Tan, W.; Lai, Z.; Fang, D.; Jiang, L.; Zuo, C.; Yang, N.; Lai, Y. An efficient cell-targeting drug delivery system based on aptamer-modified mesoporous silica nanoparticles. *Nanoscale Res. Lett.* **2019**, *14*, 390. [CrossRef]
133. Nguyen, T.L.; Cha, B.G.; Choi, Y.; Im, J.; Kim, J. Injectable dual-scale mesoporous silica cancer vaccine enabling efficient delivery of antigen/adjuvant-loaded nanoparticles to dendritic cells recruited in local macroporous scaffold. *Biomaterials* **2020**, *239*, 119859. [CrossRef] [PubMed]
134. Mekaru, H.; Lu, J.; Tamanoi, F. Development of mesoporous silica-based nanoparticles with controlled release capability for cancer therapy. *Adv. Drug Deliv. Rev.* **2015**, *95*, 40–49. [CrossRef] [PubMed]
135. Vivero-Escoto, J.L.; Slowing, I.I.; Trewyn, B.G.; Lin, V.S.Y. Mesoporous silica nanoparticles for intracellular controlled drug delivery. *Small* **2010**, *6*, 1952–1967. [CrossRef] [PubMed]
136. Gatenby, R.A.; Gillies, R.J. Why do cancers have high aerobic glycolysis? *Nat. Rev. Cancer* **2004**, *4*, 891–899. [CrossRef]
137. Liu, R.; Zhang, Y.; Zhao, X.; Agarwal, A.; Mueller, L.J.; Feng, P. pH-responsive nanogated ensemble based on gold-capped mesoporous silica through an acid-labile acetal linker. *J. Am. Chem. Soc.* **2010**, *132*, 1500–1501. [CrossRef]
138. Gan, Q.; Lu, X.; Yuan, Y.; Qian, J.; Zhou, H.; Lu, X.; Shi, J.; Liu, C. A magnetic, reversible pH-responsive nanogated ensemble based on Fe_3O_4 nanoparticles-capped mesoporous silica. *Biomaterials* **2011**, *32*, 1932–1942. [CrossRef]
139. Xu, C.; Lin, Y.; Wang, J.; Wu, L.; Wei, W.; Ren, J.; Qu, X. Nanoceria-triggered synergetic drug release based on CeO_2-capped mesoporous silica host-guest interactions and switchable enzymatic activity and cellular effects of CeO_2. *Adv. Healthc. Mater.* **2013**, *2*, 1591–1599. [CrossRef]
140. Meng, H.; Xue, M.; Xia, T.; Zhao, Y.L.; Tamanoi, F.; Stoddart, J.F.; Zink, J.I.; Nel, A.E. Autonomous *in vitro* anti cancer drug release from mesoporous silica nanoparticles by pH-sensitive nanovalves. *J. Am. Chem. Soc.* **2010**, *132*, 12690–12697. [CrossRef]
141. Gao, Y.; Yang, C.; Liu, X.; Ma, R.; Kong, D.; Shi, L. A multifunctional nanocarrier based on nanogated mesoporous silica for enhanced tumor-specific uptake and intracellular delivery. *Macromol. Biosci.* **2012**, *12*, 251–259. [CrossRef]

142. Lee, C.H.; Cheng, S.H.; Huang, I.P.; Souris, J.S.; Yang, C.S.; Mou, C.Y.; Lo, L.W. Intracellular pH-responsive mesoporous silica nanoparticles for the controlled release of anticancer chemotherapeutics. *Angew. Chem.* **2010**, *122*, 8390–8395. [CrossRef]
143. Li, T.; Chen, X.; Liu, Y.; Fan, L.; Lin, L.; Xu, Y.; Chen, S.; Shao, J. pH-Sensitive mesoporous silica nanoparticles anticancer prodrugs for sustained release of ursolic acid and the enhanced anti-cancer efficacy for hepatocellular carcinoma cancer. *Eur. J. Pharm. Sci.* **2017**, *96*, 456–463. [CrossRef] [PubMed]
144. Feng, W.; Zhou, X.; He, C.; Qiu, K.; Nie, W.; Chen, L.; Wang, H.; Mo, X.; Zhang, Y. Polyelectrolyte multilayer functionalized mesoporous silica nanoparticles for pH-responsive drug delivery: Layer thickness dependent release profiles and biocompatibility. *J. Mater. Chem. B* **2013**, *1*, 5886–5898. [CrossRef] [PubMed]
145. Cauda, V.; Argyo, C.; Schlossbauera, A.; Bein, T. Controlling the delivery kinetics from colloidal mesoporous silicananoparticles with pH-sensitive gates. *J. Mater. Chem.* **2010**, *20*, 4305–4311. [CrossRef]
146. Tang, H.; Guo, J.; Sun, Y.; Chang, B.; Ren, Q.; Yang, W. Facile synthesis of pH sensitive polymer-coated mesoporous silica nanoparticles and their application in drug delivery. *Int. J. Pharm.* **2011**, *421*, 388–396. [CrossRef]
147. Popat, A.; Liu, J.; Lu, G.Q.M.; Qiao, S.Z. A pH-responsive drug delivery system based on chitosan coated mesoporous silica nanoparticles. *J. Mater. Chem.* **2012**, *22*, 11173–11178. [CrossRef]
148. Xiong, L.; Bi, J.; Tang, Y.; Qiao, S.Z. Magnetic Core-Shell Silica Nanoparticles with Large Radial Mesopores for siRNA Delivery. *Small* **2016**, *12*, 4735–4742. [CrossRef]
149. Gisbert-Garzarán, M.; Lozano, D.; Vallet-Regí, M.; Manzano, M. Self-immolative polymers as novel pH-responsive gatekeepers for drug delivery. *RSC Adv.* **2017**, *7*, 132–136. [CrossRef]
150. Yuan, L.; Tang, Q.; Yang, D.; Zhang, J.Z.; Zhang, F.; Hu, J. Preparation of pH-responsive mesoporous silica nanoparticles and their application in controlled drug delivery. *J. Phys. Chem. C* **2011**, *115*, 9926–9932. [CrossRef]
151. Muhammad, F.; Guo, M.; Qi, W.; Sun, F.; Wang, A.; Guo, Y.; Zhu, G. pH-triggered controlled drug release from mesoporous silica nanoparticles via intracelluar dissolution of ZnO nanolids. *J. Am. Chem. Soc.* **2011**, *133*, 8778–8781. [CrossRef]
152. Zou, Z.; He, D.; He, X.; Wang, K.; Yang, X.; Qing, Z.; Zhou, Q. Natural Gelatin Capped Mesoporous Silica Nanoparticles for Intracellular Acid-Triggered Drug Delivery. *Langmuir* **2013**, *29*, 12804–12810. [CrossRef] [PubMed]
153. Rim, H.P.; Min, K.H.; Lee, H.J.; Jeong, S.Y.; Lee, S.C. pH-tunable calcium phosphate covered mesoporous silica nanocontainers for intracellular controlled release of guest drugs. *Angew. Chem. Int. Ed.* **2011**, *50*, 8853–8857. [CrossRef]
154. Wibowo, F.R.; Saputra, O.A.; Lestari, W.W.; Koketsu, M.; Mukti, R.R.; Martien, R. pH-triggered drug release controlled by poly(styrene sulfonate) growth hollow mesoporous silica nanoparticles. *ACS Omega* **2020**, *5*, 4261–4269. [CrossRef]
155. Estrela, J.M.; Ortega, A.; Obrador, E. Glutathione in cancer biology and therapy. *Crit. Rev. Clin. Lab. Sci.* **2006**, *43*, 143–181. [CrossRef]
156. Croissant, J.; Cattoën, X.; Man, M.W.; Gallud, A.; Raehm, L.; Trens, P.; Maynadier, M.; Durand, J.O. Biodegradable ethylene-bis (Propyl) disulfide-based periodic mesoporous organosilica nanorods and nanospheres for efficient *in-vitro* drug delivery. *Adv. Mater.* **2014**, *26*, 6174–6180. [CrossRef]
157. Wang, D.; Xu, Z.; Chen, Z.; Liu, X.; Hou, C.; Zhang, X.; Zhang, H. Fabrication of single-hole glutathione-responsive degradable hollow silica nanoparticles for drug delivery. *ACS Appl. Mater. Interfaces* **2014**, *6*, 12600–12608. [CrossRef]
158. Kim, H.; Kim, S.; Park, C.; Lee, H.; Park, H.J.; Kim, C. Glutathione-induced intracellular release of guests from mesoporous silica nanocontainers with cyclodextrin gatekeepers. *Adv. Mater.* **2010**, *22*, 4280–4283. [CrossRef]
159. Sauer, A.M.; Schlossbauer, A.; Ruthardt, N.; Cauda, V.; Bein, T.; Bräuchle, C. Role of endosomal escape for disulfide-based drug delivery from colloidal mesoporous silica evaluated by live-cell imaging. *Nano Lett.* **2010**, *10*, 3684–3691. [CrossRef]
160. Wu, M.; Meng, Q.; Chen, Y.; Zhang, L.; Li, M.; Cai, X.; Li, Y.; Yu, P.; Zhang, L.; Shi, J. Large pore-sized hollow mesoporous organosilica for redox-responsive gene delivery and synergistic cancer chemotherapy. *Adv. Mater.* **2016**, *28*, 1963–1969. [CrossRef]

161. Nadrah, P.; Maver, U.; Jemec, A.; Tišler, T.; Bele, M.; Dražić, G.; Benčina, M.; Pintar, A.; Planinšek, O.; Gaberšček, M. Hindered disulfide bonds to regulate release rate of model drug from mesoporous silica. *ACS Appl. Mater. Interfaces* **2013**, *5*, 3908–3915. [CrossRef]
162. Prasetyanto, E.A.; Bertucci, A.; Septiadi, D.; Corradini, R.; Castro-Hartmann, P.; de Cola, L. Breakable hybrid organosilica nanocapsules for protein delivery. *Angew. Chem. Int. Ed.* **2016**, *55*, 3323–3327. [CrossRef]
163. Du, X.; Kleitz, F.; Li, X.; Huang, H.; Zhang, X.; Qiao, S.Z. Disulfide-bridged organosilica frameworks: Designed, synthesis, redox-triggered biodegradation, and nanobiomedical applications. *Adv. Funct. Mater.* **2018**, 1707325. [CrossRef]
164. Liu, R.; Zhao, X.; Wu, T.; Feng, P. Tunable Redox-Responsive Hybrid Nanogated Ensembles. *J. Am. Chem. Soc.* **2008**, *130*, 14418–14419. [CrossRef]
165. Sun, L.; Liu, Y.J.; Yang, Z.Z.; Qi, X.R. Tumor specific delivery with redox-triggered mesoporous silica nanoparticles inducing neovascularization suppression and vascular normalization. *RSC Adv.* **2015**, *5*, 55566–55578. [CrossRef]
166. Teng, Z.; Li, W.; Tang, Y.; Elzatahry, A.; Lu, G.; Zhao, D. Mesoporous organosilica hollow nanoparticles: Synthesis and applications. *Adv. Mater.* **2018**, 1707612. [CrossRef]
167. Patel, K.; Angelos, S.; Dichtel, W.R.; Coskun, A.; Yang, Y.-W.; Zink, J.I.; Stoddart, J.F. Enzyme responsive snap-top covered silica nanocontainers. *J. Am. Chem. Soc.* **2008**, *130*, 2382–2383. [CrossRef]
168. Mondragón, L.; Mas, N.; Ferragud, V.; de la Torre, C.; Agostini, A.; Martínez-Máñez, R.; Sancenón, F.; Amorós, P.; Pérez-Payá, E.; Orzáez, M. Enzyme-responsive intracellular-controlled release using silica mesoporous nanoparticles capped with ε-poly-L-lysine. *Chemistry* **2014**, *20*, 5271–5281. [CrossRef]
169. Bernardos, A.; Mondragón, L.; Aznar, E.; Marcos, M.D.; Martínez-Mánez, R.; Sancenón, F.; Soto, J.; Barat, J.M.; Pérez-Payá, E.; Guillem, C.; et al. Enzyme-responsive intracellular controlled release using nanometric silica mesoporous supports capped with "saccharides". *ACS Nano* **2010**, *4*, 6353–6368. [CrossRef]
170. Schlossbauer, A.; Kecht, J.; Bein, T. Biotin-Avidin as a protease-responsive cap system for controlled guest release from colloidal mesoporous silica. *Angew. Chem. Int. Ed.* **2009**, *48*, 3092–3095. [CrossRef]
171. Radhakrishnan, K.; Gupta, S.; Gnanadhas, D.P.; Ramamurthy, P.C.; Chakravortty, D.; Raichur, A.M. Protamine-capped mesoporous silica nanoparticles for biologically triggered drug release. *Part. Part. Syst. Charact.* **2013**, *31*, 449–458. [CrossRef]
172. Xua, J.-H.; Gao, F.-P.; Li, L.-L.; Ma, H.L.; Fan, Y.-S.; Liu, W.; Guo, S.-S.; Zhao, X.-Z.; Wang, H. Gelatin-mesoporous silica nanoparticles as matrix metalloproteinases-degradable drug delivery systems in vivo. *Microporous Mesoporous Mater.* **2015**, *204*, 226–234. [CrossRef]
173. Van Rijt, S.H.; Bölükbas, D.A.; Argyo, C.; Datz, S.; Lindner, M.; Eickelberg, O.; Königshoff, M.; Bein, T.; Meiners, S. Protease-mediated release of chemotherapeutics from mesoporous silica nanoparticles to ex vivo human and mouse lung tumors. *ACS Nano* **2015**, *9*, 2377–2389. [CrossRef]
174. Singh, N.; Karambelkar, A.; Gu, L.; Lin, K.; Miller, J.S.; Chen, C.S.; Sailor, M.J.; Bhatia, S.N. Bioresponsive mesoporous silica nanoparticles for triggered drug release. *J. Am. Chem. Soc.* **2011**, *133*, 19582–19585. [CrossRef]
175. Park, C.; Kim, H.; Kim, S.; Kim, C. Enzyme responsive nanocontainers with cyclodextrin gatekeepers and synergistic effects in release of guests. *J. Am. Chem. Soc.* **2009**, *131*, 16614–16615. [CrossRef]
176. Baeza, A.; Guisasola, E.; Torres-Pardo, A.; González-Calbet, J.M.; Melen, G.J.; Ramirez, M.; Vallet-Regí, M. Hybrid enzyme-polymeric capsules/mesoporous silica nanodevice for in situ cytotoxic agent generation. *Adv. Funct. Mater.* **2014**, *24*, 4625–4633. [CrossRef]
177. Mura, S.; Nicolas, J.; Couvreur, P. Stimuli-responsive nanocarriers for drug delivery. *Nat. Mater.* **2013**, *12*, 991–1003. [CrossRef]
178. Arcos, D.; Fal -Miyar, V.; Ruiz-Hernández, E.; García-Hernández, M.; Ruiz-González, M.L.; González-Calbet, J.; Vallet-Regí, M. Supramolecular mechanisms in the synthesis of mesoporous magnetic nanospheres for hyperthermia. *J. Mater. Chem.* **2012**, *22*, 64–72. [CrossRef]
179. Chen, P.-J.; Hu, S.-H.; Hsiao, C.-S.; Chen, Y.-Y.; Liu, D.-M.; Chen, S.-Y. Multifunctional magnetically removable nanogated lids of Fe3O4–capped mesoporous silica nanoparticles for intracellular controlled release and MR imaging. *J. Mater. Chem.* **2011**, *21*, 2535. [CrossRef]
180. Thomas, C.R.; Ferris, D.P.; Lee, J.H.; Choi, E.; Cho, M.H.; Kim, E.S.; Stoddart, J.F.; Shin, J.S.; Cheon, J.; Zink, J.I. Noninvasive remote-controlled release of drug molecules *in vitro* using magnetic actuation of mechanized nanoparticles. *J. Am. Chem. Soc.* **2010**, *132*, 10623–10625. [CrossRef] [PubMed]

181. Guisasola, E.; Baeza, A.; Talelli, M.; Arcos, D.; Vallet-Regí, M. Design of thermoresponsive polymeric gates with opposite controlled release behaviors. *RSC Adv.* **2016**, *6*, 42510–42516. [CrossRef]
182. Guisasola, E.; Baeza, A.; Talelli, M.; Arcos, D.; Moros, M.; de la Fuente, J.M.; Vallet-Regí, M. Magnetic Responsive Release Controlled by Hot Spot Effect. *Langmuir.* **2015**, *31*, 12777–12782. [CrossRef] [PubMed]
183. Baeza, A.; Guisasola, E.; Ruiz-Hernández, E.; Vallet-Regí, M. Magnetically triggered multidrug release by hybrid mesoporous silica nanoparticles. *Chem. Mater.* **2012**, *24*, 517–524. [CrossRef]
184. Cai, D.; Liu, L.; Han, C.; Ma, X.; Qian, J.; Zhou, J.; Zhu, W. Cancer cell membrane-coated mesoporous silica loaded with superparamagnetic ferroferric oxide and paclitaxel for the combination of chemo/Magnetocaloric therapy on MDA-MB-231 cells. *Sci. Rep.* **2019**, *9*, 14475. [CrossRef]
185. Chen, Y.; Wang, X.; Liu, T.; Zhang, D.S.; Wang, Y.; Gu, H.; Di, W. Highly effective antiangiogenesis via magnetic mesoporous silica-based siRNA vehicle targeting the VEGF gene for orthotopic ovarian cancer therapy. *Int. J. Nanomed.* **2015**, *10*, 2579–2594.
186. Ruiz-Hernández, E.; Baeza, A.; Vallet-Regí, M. Smart drug delivery through DNA/magnetic nanoparticle gates. *ACS Nano* **2011**, *5*, 1259–1266. [CrossRef]
187. Mal, N.K.; Fujiwara, M.; Tanaka, Y. Photocontrolled reversible release of guest molecules from coumarin-modified mesoporous silica. *Nature* **2003**, *421*, 350–353. [CrossRef]
188. Ferris, D.P.; Zhao, Y.-L.; Khashab, N.M.; Khatib, H.A.; Stoddart, J.F.; Zink, J.I. Light-operated mechanized nanoparticles. *J. Am. Chem. Soc.* **2009**, *131*, 1686–1688. [CrossRef]
189. Yuan, Q.; Zhang, Y.; Chen, T.; Lu, D.; Zhao, Z.; Zhang, X.; Li, Z.; Yan, C.-H.; Tan, W. Photon-manipulated drug release from a mesoporous nanocontainer controlled by azobenzene-modified nucleic acid. *ACS Nano* **2012**, *6*, 6337–6344. [CrossRef]
190. Wang, Z.; Boudjelal, M.; Kang, S. Ultraviolet irradiation of human skin causes functional vitamin A deficiency, preventable by all-trans retinoic acid pre-treatment. *Nat. Med.* **1999**, *5*, 418–422. [CrossRef]
191. Shindo, Y.; Witt, E.; Packer, L. Antioxidant defense mechanisms in murine epidermis and dermis and their responses to ultraviolet light. *J. Investig. Dermatol.* **1993**, *100*, 260–265. [CrossRef]
192. Olejniczak, J.; Carling, C.J.; Almutairi, A. Photocontrolled release using one-photon absorption of visible or NIR light. *J. Control. Release* **2015**, *219*, 18–30. [CrossRef]
193. Martínez-Carmona, M.; Lozano, D.; Baeza, A.; Colilla, M.; Vallet-Regí, M. A novel visible light responsive nanosystem for cancer treatment. *Nanoscale* **2017**, *9*, 15967–15973. [CrossRef] [PubMed]
194. Guardado-Alvarez, T.M.; Sudha Devi, L.; Russell, M.M.; Schwartz, B.J.; Zink, J.I. Activation of snap-top capped mesoporous silica nanocontainers using two near-infrared photons. *J. Am. Chem. Soc.* **2013**, *135*, 14000–14003. [CrossRef] [PubMed]
195. Croissant, J.; Maynadier, M.; Gallud, A.; Peindy N'Dongo, H.; Nyalosaso, J.L.; Derrien, G.; Charnay, C.; Durand, J.O.; Raehm, L.; Serein-Spirau, F.; et al. Two-photon-triggered drug delivery in cancer cells using nanoimpellers. *Angew. Chem. Int. Ed.* **2013**, *52*, 13813–13817. [CrossRef] [PubMed]
196. Sirsi, S.R.; Borden, M.A. State-of-the-art materials for ultrasound-triggered drug delivery. *Adv. Drug Deliv. Rev.* **2014**, *72*, 3–14. [CrossRef]
197. Wang, J.; Pelletier, M.; Zhang, H.J.; Xia, H.S.; Zhao, Y. High-frequency ultrasound-responsive block copolymer micelle. *Langmuir* **2009**, *25*, 13201–13205. [CrossRef]
198. Xuan, J.; Boissière, O.; Zhao, Y.; Yan, B.; Tremblay, L.; Lacelle, S.; Xia, H.; Zhao, Y. Ultrasound-responsive block copolymer micelles based on a new amplification mechanism. *Langmuir* **2012**, *28*, 16463–16468. [CrossRef]
199. Wang, X.; Chen, H.; Chen, Y.; Ma, M.; Zhang, K.; Li, F.; Zheng, Y.; Zeng, D.; Wang, Q.; Shi, J. Perfluorohexane-encapsulated mesoporous silica nanocapsules as enhancement agents for highly efficient High Intensity focused Ultrasound (HIFU). *Adv. Mater.* **2012**, *24*, 785–791. [CrossRef]
200. Wang, X.; Chen, H.; Zheng, Y.; Ma, M.; Chen, Y.; Zhang, K.; Zeng, D.; Shi, J. Au-nanoparticle coated mesoporous silica nanocapsule-based multifunctional platform for ultrasound mediated imaging, cytoclasis and tumor ablation. *Biomaterials* **2013**, *34*, 2057–2068. [CrossRef]
201. Paris, J.L.; Cabañas, M.V.; Manzano, M.; Vallet-Regí, M. Polymer-grafted mesoporous silica nanoparticles as ultrasound-responsive drug carriers. *ACS Nano* **2015**, *9*, 11023–11033. [CrossRef]
202. Cauda, V.; Schlossbauer, A.; Kecht, J.; Zürner, A.; Bein, T. Multiple core-shell functionalized colloidal mesoporous silica nanoparticles. *J. Am. Chem. Soc.* **2009**, *131*, 11361–11370. [CrossRef]
203. Cauda, V.; Argyo, C.; Piercey, D.G.; Bein, T. "Liquid-phase calcination" of colloidal mesoporous silica nanoparticles in high-boiling solvents. *J. Am. Chem. Soc.* **2011**, *133*, 6484–6486. [CrossRef]

204. Tang, F.; Li, L.; Chen, D. Mesoporous silica nanoparticles: Synthesis, biocompatibility and drug delivery. *Adv. Mater.* **2012**, *24*, 1504–1534. [CrossRef]
205. Martínez-Carmona, M.; Colilla, M.; Vallet-Regí, M. Smart mesoporous nanomaterials for antitumor therapy. *Nanomaterials* **2015**, *5*, 1906–1937. [CrossRef]
206. Castillo, R.R.; Colilla, M.; Vallet-Regí, M. Advances in mesoporous silica-based nanocarriers for co-delivery and combination therapy against cancer. *Expert Opin. Drug Deliv.* **2017**, *14*, 229–243. [CrossRef]
207. Aquib, M.; Farooq, M.A.; Banerjee, P.; Akhtar, F.; Filli, M.S.; Boakye-Yiadom, K.O.; Kesse, S.; Raza, F.; Maviah, M.B.J.; Mavlyanova, R.; et al. Targeted and stimuli-responsive mesoporous silica nanoparticles for drug delivery and theranostic use. *J. Biomed. Mater. Res.* **2019**, *107A*, 2643–2666. [CrossRef]
208. Rosenholm, J.M.; Meinander, A.; Peuhu, E.; Niemi, R.; Eriksson, J.E.; Sahlgren, C.; Lindén, M. Targeting of Porous Hybrid Silica Nanoparticles to Cancer Cells. *ACS Nano* **2009**, *3*, 197–206. [CrossRef]
209. Sun, X.; Wang, N.; Yang, L.Y.; Ouyang, X.K.; Huang, F. Folic acid and pei modified mesoporous silica for targeted delivery of curcumin. *Pharmaceutics* **2019**, *11*, 430. [CrossRef]
210. Cheng, W.; Nie, J.; Xu, L.; Liang, C.; Peng, Y.; Liu, G.; Wang, T.; Mei, L.; Huang, L.; Zeng, X. A pH-sensitive delivery vehicle based on folic acid-conjugated polydopamine-modified mesoporous silica nanoparticles for targeted cancer therapy. *ACS Appl. Mater. Interfaces* **2017**, *9*, 18462–18473. [CrossRef]
211. Liu, X.; Yu, D.; Jin, C.; Song, X.; Cheng, J.; Zhao, X.; Qi, X.; Zhang, G. A dual responsive targeted drug delivery system based on smart polymer coated mesoporous silica for laryngeal carcinoma treatment. *New J. Chem.* **2014**, *38*, 4830–4836. [CrossRef]
212. Qi, X.; Yu, D.; Jia, B.; Jin, C.; Liu, X.; Zhao, X.; Zhang, G. Targeting CD133+ laryngeal carcinoma cells with chemotherapeutic drugs and siRNA against ABCG2 mediated by thermo/pH-sensitive mesoporous silica nanoparticles. *Tumor Biol.* **2016**, *37*, 2209–2217. [CrossRef]
213. Li, N.; Wang, Z.; Zhang, Y.; Zhang, K.; Xie, J.; Liu, Y.; Li, W.; Feng, N. Curcumin-loaded redox-responsive mesoporous silica nanoparticles for targeted breast cancer therapy. *Artif. Cells Nanomed. Biotechnol.* **2018**, *46*, 921–935. [CrossRef]
214. AbouAitah, K.; Hassan, H.A.; Swiderska-Sroda, A.; Gohar, L.; Shaker, O.G.; Wojnarowicz, J.; Opalinska, A.; Smalc-Koziorowska, J.; Gierlotka, S.; Lojkowski, W. Targeted nano-drug delivery of colchicine against colon cancer cells by means of mesoporous silica nanoparticles. *Cancers* **2020**, *12*, 144. [CrossRef]
215. Zhang, H.; Zhang, W.; Zhou, Y.; Jiang, Y.; Li, S. Dual functional mesoporous silicon nanoparticles enhance the radiosensitivity of VPA in glioblastoma. *Transl. Oncol.* **2017**, *10*, 229–240. [CrossRef]
216. Gao, Q.; Xie, W.; Wang, Y.; Wang, D.; Guo, Z.; Gao, F.; Zhao, L.; Cai, Q. A theranostic nanocomposite system based on radial mesoporous silica hybridized with Fe_3O_4 nanoparticles for targeted magnetic field responsive chemotherapy of breast cancer. *RSC Adv.* **2018**, *8*, 4321. [CrossRef]
217. Cao, Y.; Wu, C.; Liu, Y.; Hu, L.; Shang, W.; Gao, Z.; Xia, N. Folate functionalized pH-sensitive photothermal therapy traceable hollow mesoporous silica nanoparticles as a targeted drug carrier to improve the antitumor effect of doxorubicin in the hepatoma cell line SMMC-7721. *Drug Delivery* **2020**, *27*, 258–268. [CrossRef]
218. Paredes, K.O.; Díaz-García, D.; García-Almodóvar, V.; Chamizo, L.L.; Marciello, M.; Díaz-Sánchez, M.; Prashar, S.; Gómez-Ruiz, S.; Filice, M. Multifunctional silica-based nanoparticles with controlled release of organotin metallodrug for targeted theranosis of breast cancer. *Cancers* **2020**, *12*, 187. [CrossRef]
219. Alvarez-Berríos, M.P.; Vivero-Escoto, J.L. *In vitro* evaluation of folic acid-conjugated redox-responsive mesoporous silica nanoparticles for the delivery of cisplatin. *Int. J. Nanomed.* **2016**, *11*, 6251–6265. [CrossRef]
220. Saini, K.; Bandyopadhyaya, R. Transferrin-Conjugated Polymer-Coated Mesoporous Silica Nanoparticles Loaded with Gemcitabine for Killing Pancreatic Cancer Cells. *ACS Appl. Nano Mater.* **2020**, *3*, 229–240. [CrossRef]
221. Fang, W.; Wang, Z.; Zong, S.; Chen, H.; Zhu, D.; Zhong, Y.; Cui, Y. pH-controllable drug carrier with SERS activity for targeting cancer cells. *Biosens. Bioelectron.* **2014**, *57*, 10–15. [CrossRef]
222. Chen, X.; Sun, H.; Hu, J.; Han, X.; Liu, H.; Hu, Y. Transferrin gated mesoporous silica nanoparticles for redox-responsive and targeted drug delivery. *Colloids Surf. B Biointerfaces* **2017**, *152*, 77–84. [CrossRef]
223. Martínez-Carmona, M.; Baeza, A.; Rodriguez-Milla, M.A.; García-Castro, J.; Vallet-Regí, M. Mesoporous silica nanoparticles grafted with a light-responsive protein shell for highly cytotoxic antitumoral therapy. *J. Mater. Chem. B* **2015**, *3*, 5746–5752. [CrossRef]

224. Er, Ö.; Colak, S.G.; Ocakoglu, K.; Ince, M.; Bresolí-Obach, R.; Mora, M.; Sagristá, M.L.; Yurt, F.; Nonell, S. Selective photokilling of human pancreatic cancer cells using cetuximab-targeted mesoporous silica nanoparticles for delivery of zinc phthalocyanine. *Molecules* **2018**, *23*, 2749. [CrossRef] [PubMed]
225. Ngamcherdtrakul, W.; Morry, J.; Gu, S.; Castro, D.J.; Goodyear, S.M.; Sangvanich, T.; Reda, M.M.; Lee, R.; Mihelic, S.A.; Beckman, B.L.; et al. Cationic polymer modified mesoporous silica nanoparticles for targeted siRNA delivery to HER2 breast cancer. *Adv. Funct. Mater.* **2015**, *25*, 2646–2659. [CrossRef] [PubMed]
226. Shen, Y.; Li, M.; Liu, T.; Liu, J.; Youhua Xie, Y.; Zhang, J.; Xu, S.; Liu, H. A dual-functional HER2 aptamer-conjugated, pH-activated mesoporous silica nanocarrier-based drug delivery system provides *in vitro* synergistic cytotoxicity in HER2-positive breast cancer cells. *Int. J. Nanomed.* **2019**, *14*, 4029–4044. [CrossRef]
227. Liu, W.; Zhu, Y.; Wang, F.; Li, X.; Liu, X.; Pang, J.; Pan, W. Galactosylated chitosan-functionalized mesoporous silica nanoparticles for efficient colon cancer cell-targeted drug delivery. *R. Soc. Open Sci.* **2018**, *5*, 181027. [CrossRef]
228. Martínez-Carmona, M.; Lozano, D.; Colilla, M.; Vallet-Regí, M. Lectin-Conjugated pH-Responsive Mesoporous Silica Nanoparticles for Targeted Bone Cancer Treatment. *Acta Biomater.* **2018**, *65*, 393–404. [CrossRef]
229. Zhao, N.; Yang, Z.; Li, B.; Meng, J.; Shi, Z.; Li, P.; Fu, S. RGD-conjugated mesoporous silica-encapsulated gold nanorods enhance the sensitization of triple-negative breast cancer to megavoltage radiation therapy. *Int. J. Nanomed.* **2016**, *11*, 5595–5610. [CrossRef]
230. Wu, X.; Han, Z.; Schur, R.M.; Lu, Z.R. Targeted mesoporous silica nanoparticles delivering arsenic trioxide with environment sensitive drug release for effective treatment of triple negative breast cancer. *ACS Biomater. Sci. Eng.* **2016**, *2*, 501–507. [CrossRef]
231. Zhang, J.; Yuan, Z.F.; Wang, Y.; Chen, W.H.; Luo, G.F.; Cheng, S.X.; Zhuo, R.X.; Zhang, X.Z. Multifunctional envelope-type mesoporous silica nanoparticles for tumor-triggered targeting drug delivery. *J. Am. Chem. Soc.* **2013**, *135*, 5068–5073. [CrossRef]
232. Cheng, Y.J.; Zhang, A.Q.; Hu, J.J.; He, F.; Zeng, X.; Zhang, X.Z. Multifunctional peptide-amphiphile end-capped mesoporous silica nanoparticles for tumor targeting drug delivery. *ACS Appl. Mater. Interfaces* **2017**, *9*, 2093–2103. [CrossRef]
233. Chen, G.; Xie, Y.; Peltier, R.; Lei, H.; Wang, P.; Chen, J.; Hu, Y.; Wang, F.; Yao, X.; Sun, H. Peptide-decorated gold nanoparticles as functional nano-capping agent of mesoporous silica container for targeting drug delivery. *ACS Appl. Mater. Interfaces* **2016**, *8*, 11204–11209. [CrossRef] [PubMed]
234. Xiao, D.; Hu, J.J.; Zhu, J.Y.; Wang, S.B.; Zhuo, R.X.; Zhang, X.Z. A redox-responsive mesoporous silica nanoparticle with a therapeutic peptide shell for tumor targeting synergistic therapy. *Nanoscale* **2016**, *8*, 16702–16709. [CrossRef] [PubMed]
235. Luo, G.F.; Chen, W.H.; Liu, Y.; Zhang, J.; Cheng, S.X.; Zhuo, R.X.; Zhang, X.Z. Charge-reversal plug gate nanovalves on peptide-functionalized mesoporous silica nanoparticles for targeted drug delivery. *J. Mater. Chem. B* **2013**, *1*, 5723–5732. [CrossRef] [PubMed]
236. Zhao, F.; Zhang, C.; Zhao, C.; Gao, W.; Fan, X.; Wu, G. A facile strategy to fabricate a pH-responsive mesoporous silica nanoparticle end-capped with amphiphilic peptides by self-assembly. *Colloids Surf. B Biointerfaces* **2019**, *179*, 352–362. [CrossRef]
237. Li, X.; Xing, L.; Hu, Y.; Xiong, Z.; Wang, R.; Xu, X.; Du, L.; Shen, M.; Shi, X. An RGD-modified hollow silica@Au core/shell nanoplatform for tumor combination therap. *Acta Biomater.* **2017**, *62*, 273–283. [CrossRef]
238. Turan, O.; Bielecki, P.; Tong, K.; Covarrubias, G.; Moon, T.; Rahmy, A.; Cooley, S.; Park, Y.; Peiris, P.M.; Ghaghada, K.B.; et al. Effect of dose and selection of two different ligands on the deposition and antitumor efficacy of targeted nanoparticles in brain tumors. *Mol. Pharmaceutics* **2019**, *16*, 4352–4360. [CrossRef]
239. Hu, J.; Zhang, X.; Wen, Z.; Tan, Y.; Huang, N.; Cheng, S.; Zheng, H.; Cheng, Y. Asn-Gly-Arg-modified polydopamine-coated nanoparticles for dual-targeting therapy of brain glioma in rats. *Oncotarget* **2016**, *7*, 73681–73696. [CrossRef]
240. Babaei, M.; Abnous, K.; Taghdisi, S.M.; Amel Farzad, S.; Peivandi, M.T.; Ramezani, M.; Alibolandi, M. Synthesis of theranostic epithelial cell adhesion molecule targeted mesoporous silica nanoparticle with gold gatekeeper for hepatocellular carcinoma. *Nanomedicine (Lond).* **2017**, *12*, 1261–1279. [CrossRef]

241. Tang, Y.; Hu, H.; Zhang, M.G.; Song, J.; Nie, L.; Wang, S.; Niu, G.; Huang, P.; Lu, G.; Chen, X. An aptamer-targeting photoresponsive drug delivery system using "off-on" graphene oxide wrapped mesoporous silica nanoparticles. *Nanoscale* **2015**, *7*, 6304–6310. [CrossRef]
242. Han, L.; Tang, C.; Yin, C. Dual-targeting and pH/redox-responsive multi-layered nanocomplexes for smart co-delivery of doxorubicin and siRNA. *Biomaterials* **2015**, *60*, 42–52. [CrossRef]
243. Liu, J.; Zhang, B.; Luo, Z.; Ding, X.; Li, J.; Dai, L.; Zhou, J.; Zhao, X.; Ye, J.; Cai, K. Enzyme responsive mesoporous silica nanoparticles for targeted tumor therapy *in vitro* and *in vivo*. *Nanoscale* **2015**, *7*, 3614–3626. [CrossRef] [PubMed]
244. Zhao, J.; Zhao, F.; Wang, X.; Fan, X.; Wu, G. Secondary nuclear targeting of mesoporous silica nano-particles for cancer-specific drug delivery based on charge inversion. *Oncotarget* **2016**, *7*, 70100–70112. [CrossRef] [PubMed]
245. Wei, Y.; Gao, L.; Wang, L.; Shi, L.; Wei, E.; Zhou, B.; Zhou, L.; Ge, B. Polydopamine and peptide decorated doxorubicinloaded mesoporous silica nanoparticles as a targeted drug delivery system for bladder cancer therapy. *Drug Deliv.* **2017**, *24*, 681–691. [CrossRef] [PubMed]
246. Han, L.; Tang, C.; Yin, C. pH-Responsive Core–Shell Structured Nanoparticles for Triple-Stage Targeted Delivery of Doxorubicin to Tumors. *ACS Appl. Mater. Interfaces* **2016**, *8*, 23498–23508. [CrossRef]
247. Zhao, Q.; Geng, H.; Wang, Y.; Gao, Y.; Huang, J.; Wang, Y.; Zhang, J.; Wang, S. Hyaluronic acid oligosaccharide modified redox-responsive mesoporous silica nanoparticles for targeted drug delivery. *ACS Appl. Mater. Interfaces* **2014**, *6*, 20290–20299. [CrossRef]
248. Fang, Z.; Li, X.; Xu, Z.; Du, F.; Wang, W.; Shi, R.; Gao, D. Hyaluronic acid-modified mesoporous silica-coated superparamagnetic Fe_3O_4 nanoparticles for targeted drug delivery. *Int. J. Nanomed.* **2019**, *14*, 5785–5797. [CrossRef]
249. Li, T.; Geng, Y.; Zhang, H.; Wang, J.; Feng, Y.; Chen, Z.; Xie, X.; Qin, X.; Li, S.; Wu, C.; et al. Versatile nanoplatform for synergistic chemo-photothermal therapy and multimodal imaging against breast cancer. *Expert Opin. Drug Del.* **2020**, *17*, 725–733. [CrossRef]
250. Zhang, Y.; Xu, J. Mesoporous silica nanoparticle-based intelligent drug delivery system for bienzyme-responsive tumour targeting and controlled release. *R. Soc. Open Sci.* **2018**, *5*, 170986. [CrossRef]
251. Liu, Y.; Dai, R.; Wei, Q.; Li, W.; Zhu, G.; Chi, H.; Guo, Z.; Wang, L.; Cui, C.; Xu, J.; et al. Dual-functionalized janus mesoporous silica nanoparticles with active targeting and charge reversal for synergistic tumor-targeting therapy. *ACS Appl. Mater. Interfaces* **2019**, *11*, 44582–44592. [CrossRef]
252. Lu, H.; Zhao, Q.; Wang, X.; Mao, Y.; Chen, C.; Gao, Y.; Sun, C. Siling Wang. Multi-stimuli responsive mesoporous silica-coated carbon nanoparticles for chemo-photothermal therapy of tumor. *Colloids Surf. B* **2020**, *190*, 110941. [CrossRef]
253. Zhou, S.; Ding, C.; Wang, C.; Fu, J. UV-light cross-linked and pH de-cross-linked coumarin-decorated cationic copolymer grafted mesoporous silica nanoparticles for drug and gene co-delivery *in vitro*. *Mater. Sci. Eng. C* **2019**. [CrossRef]
254. Xu, Y.; Xiao, L.; Chang, Y.; Cao, Y.; Chen, C.; Wang, D. pH and redox dual-responsive MSN-S-S-CS as a drug delivery system in cancer therapy. *Materials* **2020**, *13*, 1279. [CrossRef]
255. Li, Y.; Hei, M.; Xu, Y.; Qian, X.; Zhu, W. Ammonium salt modified mesoporous silica nanoparticles for dual intracellular-responsive gene delivery. *Int. J. Pharm.* **2016**, *511*, 689–702. [CrossRef]
256. Yan, J.; Xu, X.; Zhou, J.; Liu, C.; Zhang, L.; Wang, D.; Yang, F.; Zhang, H. Fabrication of a pH/redox-triggered mesoporous silica-based nanoparticle with microfluidics for anticancer drugs doxorubicin and paclitaxel codelivery. *ACS Appl. Bio Mater.* **2020**, *3*, 1216–1225. [CrossRef]
257. Anirudhan, T.S.; Nair, A.S. Temperature and ultrasound sensitive gatekeepers for the controlled release of chemotherapeutic drugs from mesoporous silica nanoparticles. *J. Mater. Chem. B* **2018**, *6*, 428–439. [CrossRef]
258. Lee, J.E.; Lee, N.; Kim, T.; Kim, J.; Hyeon, T. Multifunctional mesoporous silica nanocomposite nanoparticles for theranostic applications. *Acc. Chem. Res.* **2011**, *44*, 893–902. [CrossRef]
259. Nakamura, T.; Sugihara, F.; Matsushita, H.; Yoshioka, Y.; Mizukami, S.; Kikuchi, K. Mesoporous silica nanoparticles for (19)F magnetic resonance imaging, fluorescence imaging, and drug delivery. *Chem. Sci.* **2015**, *6*, 1986–1990. [CrossRef]
260. Chen, N.-T.; Cheng, S.-H.; Souris, J.S.; Chen, C.-T.; Mou, C.-Y.; Lo, L.-W. Theranostic applications of mesoporous silica nanoparticles and their organic/inorganic hybrids. *J. Mater. Chem. B* **2013**, *1*, 3128. [CrossRef]

261. Wu, X.; Wu, M.; Zhao, J.X. Recent development of silica nanoparticles as delivery vectors for cancer imaging and therapy. *Nanomedicine (Lond).* **2014**, *10*, 297–312. [CrossRef]
262. Cassidy, M.C.; Chan, H.R.; Ross, B.D.; Bhattacharya, P.K.; Marcus, C.M. *In vivo* magnetic resonance imaging of hyperpolarized silicon particles. *Nat. Nanotechnol.* **2013**, *8*, 363–368. [CrossRef]
263. Feng, Y.; Panwar, N.; Tng, D.J.H.; Tjin, S.C.; Wang, K.; Yong, K.-T. The application of mesoporous silica nanoparticle family in cancer theranostics. *Coord. Chem. Rev.* **2016**, *319*, 86–109. [CrossRef]
264. Matsushita, H.; Mizukami, S.; Sugihara, F.; Nakanishi, Y.; Yoshioka, Y.; Kikuchi, K. Multifunctional core-shell silica nanoparticles for highly sensitive (19)F magnetic resonance imaging. *Angew. Chem. Int. Ed. Engl.* **2014**, *53*, 1008–1011. [CrossRef] [PubMed]
265. Milgroom, A.; Intrator, M.; Madhavan, K.; Mazzaro, L.; Shandas, R.; Liu, B.L.; Park, D. Mesoporous Silica Nanoparticles as a Breast-Cancer Targeting Ultrasound Contrast Agent. *Colloids Surf. B Biointerfaces* **2014**, *116*, 652–657. [CrossRef] [PubMed]
266. Cha, B.G.; Kim, J. Functional mesoporous silica nanoparticles for bio-imaging applications. *Wiley Interdiscip. Rev. Nanomed. Nanobiotechnol.* **2019**, *11*, e1515. [CrossRef]
267. Bouamrani, A.; Hu, Y.; Tasciotti, E.; Li, L.; Chiappini, C.; Liu, X.; Ferrari, M. Mesoporous silica chips for selective enrichment and stabilization of low molecular weight proteome. *Proteomics* **2010**, *10*, 496–505. [CrossRef]
268. Jäger, T.; Szarvas, T.; Börgermann, C.; Schenck, M.; Schmid, K.; Rübben, H. Use of silicon chip technology to detect protein-based tumor markers in bladder cancer. *Der Urologe. Ausg. A* **2007**, *46*, 1152–1156. [CrossRef]
269. Liang, K.; Wu, H.; Hu, T.Y.; Li, Y. Mesoporous silica chip: Enabled peptide profiling as an effective platform for controlling bio-sample quality and optimizing handling procedure. *Clin. Proteom* **2016**, *13*, 34. [CrossRef]
270. Hu, Y.; Bouamrani, A.; Tasciotti, E.; Li, L.; Liu, X.W.; Ferrari, M. Tailoring of the nanotexture of mesoporous silica films and their functionalized derivatives for selectively harvesting low molecular weight protein. *ACS Nano* **2010**, *4*, 439–451. [CrossRef]
271. Hu, Y.; Peng, Y.; Lin, K.; Shen, H.; Brousseau, L.C., 3rd; Sakamoto, J.; Sun, T.; Ferrari, M. Surface engineering on mesoporous silica chips for enriching low molecular weight phosphorylated proteins. *Nanoscale* **2011**, *3*, 421–428. [CrossRef]
272. Wang, K.; He, X.; Yang, X.; Shi, H. Functionalized silica nanoparticles: A platform for fluorescence imaging at the cell and small animal levels. *Acc. Chem. Res.* **2013**, *46*, 1367–1376. [CrossRef]
273. Shi, S.; Chen, F.; Cai, W. Biomedical applications of functionalized hollow mesoporous silica nanoparticles: Focusing on molecular imaging. *Nanomedicine* **2013**, *8*, 2027–2039. [CrossRef] [PubMed]
274. Alford, R.; Simpson, H.M.; Duberman, J.; Hill, G.C.; Ogawa, M.; Regino, C.; Kobayashi, H.; Choyke, P.L. Toxicity of organic fluorophores used in molecular imaging: Literature review. *Mol. Imaging* **2009**, *8*, 341–354. [CrossRef] [PubMed]
275. Kesse, S.; Boakye-Yiadom, K.O.; Ochete, B.O.; Opoku-Damoah, Y.; Akhtar, F.; Filli, M.S.; Farooq, M.A.; Aquib, M.; Mily, B.J.M.; Murtaza, G.; et al. Mesoporous silica nanomaterials: Versatile nanocarriers for cancer theranostics and drug and gene delivery. *Pharmaceutics* **2019**, *11*, 77. [CrossRef] [PubMed]
276. Yin, F.; Zhang, B.; Zeng, S.; Lin, G.; Tian, J.; Yang, C.; Wang, K.; Xu, G.; Yong, K.-T. Folic acid-conjugated organically modified silica nanoparticles for enhanced targeted delivery in cancer cells and tumor in vivo. *J. Mater. Chem. B* **2015**, *3*, 6081–6093. [CrossRef] [PubMed]
277. Resch-Genger, U.; Grabolle, M.; Cavaliere-Jaricot, S.; Nitschke, R.; Nann, T. Quantum dots versus organic dyes as fluorescent labels. *Nat. Methods* **2008**, *5*, 763–775. [CrossRef]
278. Jun, B.H.; Hwang, D.W.; Jung, H.S.; Jang, J.; Kim, H.; Kang, H.; Kang, T.; Kyeong, S.; Lee, H.; Jeong, D.H.; et al. Ultrasensitive, biocompatible, quantum-dot-embedded silica nanoparticles for bioimaging. *Adv. Funct. Mater.* **2012**, *22*, 1843–1849. [CrossRef]
279. Zhou, S.; Huo, D.; Hou, C.; Yang, M.; Fa, H.; Xia, C.; Chen, M. Mesoporous silica-coated quantum dots functionalized with folic acid for lung cancer cell imaging. *Anal. Methods* **2015**, *7*, 9649–9654. [CrossRef]
280. Cheng, S.-H.; Lee, C.-H.; Chen, M.-C.; Souris, J.S.; Tseng, F.-G.; Yang, C.-S.; Mou, C.-Y.; Chen, C.-T.; Lo, L.-W. Tri-functionalization of mesoporous silica nanoparticles for comprehensive cancer theranostics—The trio of imaging, targeting and therapy. *J. Mater. Chem.* **2010**, *20*, 6149–6157. [CrossRef]
281. Ribeiro, T.; Rodrigues, A.S.; Calderon, S.; Fidalgo, A.; Gonçalves, J.L.M.; André, V.; Teresa Duarte, M.; Ferreira, P.J.; Farinha, J.P.S.; Baleizão, C. Silica nanocarriers with user-defined precise diameters by controlled template self-assembly. *J. Colloid Interface Sci.* **2020**, *561*, 609–619. [CrossRef]

282. He, Q.; Zhang, J.; Shi, J.; Zhu, Z.; Zhang, L.; Bu, W.; Guo, L.; Chen, Y. The effect of PEGylation of mesoporous silica nanoparticles on nonspecific binding of serum proteins and cellular responses. *Biomaterials* **2010**, *31*, 1085–1092. [CrossRef]
283. Manzano, M.; Vallet-Regí, M. Mesoporous silica nanoparticles for drug delivery. *Adv. Funct. Mater.* **2019**, 1902634. [CrossRef]
284. Farjadian, F.; Roointan, A.; Mohammadi-Samani, S.; Hosseini, M. Mesoporous silica nanoparticles: Synthesis, pharmaceutical applications, biodistribution, and biosafety assessment. *Chem. Eng. J.* **2019**, *359*, 684–705. [CrossRef]
285. Li, Z.; Zhang, Y.; Feng, N. Mesoporous silica nanoparticles: Synthesis, classification, drug loading, pharmacokinetics, biocompatibility, and application in drug delivery. *Expert Opin. Drug Deliv.* **2019**, *16*, 219–237. [CrossRef] [PubMed]
286. Rosenholm, J.M.; Mamaeva, V.; Sahlgren, C.; Linden, M. Nanoparticles in targeted cancer therapy: Mesoporous silica nanoparticles entering preclinical development Stage. *Nanomedicine* **2012**, *7*, 111–120. [CrossRef] [PubMed]
287. Huang, X.; Li, L.; Liu, T.; Hao, N.; Liu, H.; Chen, D.; Tang, F. The shape effect of mesoporous silica nanoparticles on biodistribution, clearance, and biocompatibility in vivo. *ACS Nano* **2011**, *5*, 5390–5399. [CrossRef]
288. Yu, T.; Hubbard, D.; Ray, A.; Ghandehari, H. *In vivo* biodistribution and pharmacokinetics of silica nanoparticles as a function of geometry, porosity and surface characteristics. *J. Control. Release* **2012**, *163*, 46–54. [CrossRef]
289. Li, L.; Liu, T.; Fu, C.; Tan, L.; Meng, X.; Liu, H. Biodistribution, excretion, and toxicity of mesoporous silica nanoparticles after oral administration depend on their shape. *Nanomed. Nanotechnol. Biol. Med.* **2015**, *11*, 1915–1924. [CrossRef]
290. Dogra, P.; Adolphi, N.L.; Wang, Z.; Lin, Y.-S.; Butler, K.S.; Durfee, P.N.; Croissant, J.G.; Noureddine, A.; Coker, E.N.; Bearer, E.L.; et al. Establishing the effects of mesoporous silica nanoparticle properties on *in vivo* disposition using imaging-based pharmacokinetics. *Nat. Commun.* **2018**, *9*, 4551. [CrossRef]
291. Sun, J.-G.; Jiang, Q.; Zhang, X.-P.; Shan, K.; Liu, B.-H.; Zhao, C.; Yan, B. Mesoporous silica nanoparticles as a delivery system for improving antiangiogenic therapy. *Int. J. Nanomed.* **2019**, *14*, 1489–1501. [CrossRef]
292. Kong, M.; Tang, J.; Qiao, Q.; Wu, T.; Qi, Y.; Tan, S.; Gao, X.; Zhang, Z. Biodegradable hollow mesoporous silica nanoparticles for regulating tumor microenvironment and enhancing antitumor efficiency. *Theranostics* **2017**, *7*, 3276–3292. [CrossRef]
293. Paula, A.J.; Araujo Júnior, R.T.; Martinez, D.S.T.; Paredes-Gamero, E.J.; Nader, H.B.; Durán, N.; Justo, G.Z.; Alves, O.L. Influence of protein corona on the transport of molecules into cells by mesoporous silica nanoparticles. *ACS Appl. Mater. Interfaces* **2013**, *5*, 8387–8393. [CrossRef] [PubMed]
294. Visalakshan, R.M.; García, L.E.G.; Benzigar, M.R.; Ghazaryan, A.; Simon, J.; Mierczynska-Vasilev, A.; Michl, T.D.; Vinu, A.; Mailänder, V.; Morsbach, S.; et al. The influence of nanoparticle shape on protein corona formation. *Small Nano Micro.* **2020**. [CrossRef]
295. Cauda, V.; Engelke, H.; Sauer, A.; Arcizet, D.; Bräuchle, C.; Rädler, J.; Bein, T. Colchicine-loaded lipid bilayer-coated 50 nm mesoporous nanoparticles efficiently induce microtubule depolymerization upon cell uptake. *Nano Lett.* **2010**, *10*, 2484–2492. [CrossRef]
296. Fei, W.; Zhang, Y.; Han, S.; Tao, J.; Zheng, H.; Wei, Y.; Zhu, J.; Li, F.; Wang, X. RGD conjugated liposome-hollow silica hybrid nanovehicles for targeted and controlled delivery of arsenic trioxide against hepatic carcinoma. *Int. J. Pharm.* **2017**, *519*, 250–262. [CrossRef]
297. Mackowiak, S.A.; Schmidt, A.; Weiss, V.; Argyo, C.; Constantin von Schirnding, C.; Bein, T.; Bräuchle, C. Targeted drug delivery in cancer cells with red-light photoactivated mesoporous silica nanoparticles. *Nano Lett.* **2013**, *13*, 2576–2583. [CrossRef]
298. Li, Y.; Miao, Y.; Chen, M.; Chen, X.; Li, F.; Zhang, X.; Gan, Y. Stepwise targeting and responsive lipid-coated nanoparticles for enhanced tumor cell sensitivity and hepatocellular carcinoma therapy. *Theranostics* **2020**, *10*, 3722–3736. [CrossRef]
299. Durfee, P.N.; Lin, Y.S.; Darren, R.; Dunphy, D.R.; Muñiz, A.J.; Butler, K.S.; Humphrey, K.R.; Lokke, A.J.; Agola, J.O.; Chou, S.S.; et al. Mesoporous silica nanoparticle-supported lipid bilayers (protocells) for active targeting and delivery to individual leukemia cells. *ACS Nano* **2016**, *10*, 8325–8345. [CrossRef]

300. Butler, K.S.; Durfee, P.N.; Theron, C.; Ashley, C.E.; Carnes, E.C.; Brinker, C.J. Protocells: Modular mesoporous silica nanoparticlesupported lipid bilayers for drug delivery. *Small* **2016**, *12*, 2173–2185. [CrossRef] [PubMed]
301. Samanta, S.; Pradhan, L.; Bahadur, D. Mesoporous lipid-silica nanohybrids for folate-targeted drug-resistant ovarian cancer. *New J. Chem.* **2018**, *42*, 2804–2814. [CrossRef]
302. Pan, G.; Jia, T.T.; Huang, Q.X.; Qiu, Y.Y.; Xu, J.; Yin, P.H.; Liu, T. Mesoporous silica nanoparticles (MSNs)-based organic/inorganic hybrid nanocarriers loading 5-Fluorouracil for the treatment of colon cancer with improved anticancer efficacy. *Colloids Surf. B Biointerfaces* **2017**, *159*, 375–385. [CrossRef]
303. Hu, B.; Wang, J.; Li, J.; Li, S.; Li, H. Superiority of L-tartaric acid modified chiral mesoporous silica nanoparticle as a drug carrier: Structure, wettability, degradation, bio-adhesion and biocompatibility. *Int. J. Nanomed.* **2020**, *15*, 601–618. [CrossRef] [PubMed]
304. Wang, B.; Zhang, K.; Wang, J.; Zhao, R.; Zhang, Q.; Kong, X. Poly(amidoamine)-modified mesoporous silica nanoparticles as a mucoadhesive drug delivery system for potential bladder cancer therapy. *Colloids Surf. B Biointerfaces* **2020**, *189*. [CrossRef] [PubMed]
305. Guimarães, R.S.; Rodrigues, C.F.; Moreira, A.F.; Correia, I.J. Overview of stimuli-responsive mesoporous organosilica nanocarriers for drug delivery. *Pharmacol. Res.* **2020**, *155*, 104742. [CrossRef]
306. Liu, C.M.; Chen, G.B.; Chen, H.H.; Zhang, J.B.; Li, H.Z.; Sheng, M.X.; Weng, W.B.; Guo, S.M. Cancer cell membrane-cloaked mesoporous silica nanoparticles with a pH-sensitive gatekeeper for cancer treatment. *Colloids Surf. B Biointerfaces* **2019**, *175*, 477–486. [CrossRef] [PubMed]
307. Yue, J.; Wang, Z.; Shao, D.; Chang, Z.; Hu, R.; Li, L.; Luo, S.; Dong, W. Cancer cell membrane-modified biodegradable mesoporous silica nanocarriers for berberine therapy of liver cancer. *RSC Adv.* **2018**, *8*, 40288–40297. [CrossRef]
308. Cauda, V.; Limongi, T.; Racca, L.; Canta, M.; Susa, F.; Piva, R.; Bergaggio, E.; Vitale, N.; Mereu, E. A biomimetic nanoporous carrier comprising an inhibitor directed towards the native form of IDH2 protein. Patent IB 2020/050401, 23 January 2019.

Article

Particle-Size-Dependent Delivery of Antitumoral miRNA Using Targeted Mesoporous Silica Nanoparticles

Lisa Haddick [1], Wei Zhang [2], Sören Reinhard [2], Karin Möller [1], Hanna Engelke [1], Ernst Wagner [2] and Thomas Bein [1,*]

1 Department of Chemistry and Center for NanoScience, University of Munich (LMU), Butenandtstrasse 5-13, 81377 Munich, Germany; lisa.haddick@cup.lmu.de (L.H.); K.Moeller@lmu.de (K.M.); hanna.engelke@cup.lmu.de (H.E.)

2 Pharmaceutical Biotechnology and Center for NanoScience, University of Munich (LMU), Butenandtstrasse 5-13, 81377 Munich, Germany; wei.zhang@cup.uni-muenchen.de (W.Z.); reinhard.soeren@cup.uni-muenchen.de (S.R.); Ernst.Wagner@cup.uni-muenchen.de (E.W.)

* Correspondence: bein@lmu.de

Received: 11 April 2020; Accepted: 27 May 2020; Published: 2 June 2020

Abstract: Multifunctional core-shell mesoporous silica nanoparticles (MSN) were tailored in size ranging from 60 to 160 nm as delivery agents for antitumoral microRNA (miRNA). The positively charged particle core with a pore diameter of about 5 nm and a stellate pore morphology allowed for an internal, protective adsorption of the fragile miRNA cargo. A negatively charged particle surface enabled the association of a deliberately designed block copolymer with the MSN shell by charge-matching, simultaneously acting as a capping as well as endosomal release agent. Furthermore, the copolymer was functionalized with the peptide ligand GE11 targeting the epidermal growth factor receptor, EGFR. These multifunctional nanoparticles showed an enhanced uptake into EGFR-overexpressing T24 bladder cancer cells through receptor-mediated cellular internalization. A luciferase gene knock-down of up to 65% and additional antitumoral effects such as a decreased cell migration as well as changes in cell cycle were observed. We demonstrate that nanoparticles with a diameter of 160 nm show the fastest cellular internalization after a very short incubation time of 45 min and produce the highest level of gene knock-down.

Keywords: mesoporous silica nanoparticles; core-shell; surface functionalization; cell targeting; size-dependent delivery; antitumoral microRNA (miRNA); confocal microscopy

1. Introduction

A novel modality in cancer therapy emerged over the past two decades in the form of gene therapy and in this context, small interfering RNA (siRNA) and micro RNA (miRNA) are promising alternatives to common anti-cancer medications [1,2]. Both nucleotide oligomers mediate the process of RNA interference (RNAi) in which cells use these short RNA strands of 19-24 nucleotides to recognize messenger RNA (mRNA) with a complementary sequence, induce their destruction and thus inhibit the translation into proteins. To provide cancer therapy, synthetically produced siRNA and miRNA mimics can be used that target and silence specific oncogenes. Recent studies have identified a number of cancer-related genes as potential targets for RNAi-based therapy [3–5]. One of these tumor suppressor miRNA is miR200c, which targets the proto-oncogene KRAS. It regulates cell differentiation, proliferation and survival [6], the epithelial to mesenchymal transition [7] and suppresses chemoresistance [8].

However, a major challenge for a widespread therapeutic application of RNAi is creating appropriate delivery vehicles to safely and effectively deliver and release siRNA and miRNA into the cytosol of disease-causing cells. Several excellent reviews summarize the requirements for siRNA delivery and comprehensively report on the efficiencies of gene silencing achieved with various delivery materials [9–11].

Some promising results were recently obtained from ongoing clinical trials concerning RNA delivery vehicles and in 2018 Alnylam's Onpattro (Patisiran) has been FDA-approved as the first ever RNAi drug against nerve damage caused by the rare hereditary disease, transthyretin amyloidosis [12]. However, Patisiran is a lipid-based nanocarrier system that does not actively target the cancer tissue, and most of the administered drug accumulates passively in the liver instead. Therefore, robust and specifically targeted carrier systems are needed for an efficient delivery of siRNA and miRNA to prevent a premature degradation of the unstable RNA. For this purpose, nanoparticles have attracted much attention because they provide a stable, nontoxic, and highly flexible platform.

Mesoporous silica nanoparticles (MSN) are emerging as potential nanocarriers for a variety of anti-cancer cargos including nucleic acids [13,14]. The properties that have rendered them specifically suitable for siRNA and miRNA delivery include a high surface area and pore volume, a controllable biodegradability [15,16], the tunability of particle size and pore size, their variable morphology and importantly, the possibility to create core-shell particles constructed of spatially separated regions with different surface functionalization [17,18].

This tunability of MSNs opens the possibility to systematically study the influence of the carrier properties on the gene silencing efficacy. Multiple parameters can affect successful gene silencing using carrier agents, e.g., the adsorption and release kinetics of the nucleic acid in the drug carrier, the dependence of cellular uptake or endosomal escape on nanoparticle size, surface characteristics such as charge and/or the attachments of functional residues, to name a few, but systematic studies addressing these issues are rare [19].

For the adsorption and release kinetics of nucleic acids using MSNs, the pore size, pore morphology and surface charge are important parameters. High loadings of nucleic acid were achieved by using large pore MSNs with a pore diameter of around 10–20 nm [20] and when pores were modified with cationic functional groups—aminopropyltriethoxysilane (APTES) or polylysine—to create a cationic layer for nucleic acid adsorption [14,21]. However, some of these systems suffer from poor release kinetics associated with low knockdown efficacies because of either the high affinity of polycations towards nucleic acid or because of a bottleneck-type pore morphology that features smaller pore openings than pore diameters, which both might decrease an efficient release of the highly charged nucleic acid molecules. Based on these findings, our group introduced novel medium-pore MSNs (pore diameter of around 5 nm) with stellate pore morphology, which were able to adsorb an exceptionally high amount of siRNA of 380 $\mu g\ mg^{-1}$ and which enabled a mainly electrostatically driven fast and efficient RNA desorption, resulting in a high silencing efficacy [21].

Cellular internalization of the nanocarrier is another factor that influences gene silencing efficacy. The relationship between shape and size of MSNs [22–25] or other nanocarriers [26,27] and cellular internalization is important as nanoparticle size may affect the uptake efficiency and kinetics and the internalization mechanism [28]. One of the great advantages of MSNs is that they can be designed to feature different sizes and shapes. For MSNs [22,29] as well as for Au [30], polystyrene [31] and iron oxide nanoparticles [32], a size-dependent uptake in cells was observed with a maximum uptake at a particle diameter of 30–50 nm. However, as size is only one among several parameters controlling cellular uptake of nanoparticles, optimal sizes may vary for different surfaces and different surface functionalizations. Additionally, targeting ligands were shown to improve selective cellular uptake and a better accumulation in tumor tissue [33,34]. The effect of particle size on gene transfection efficiency using silica-based nanoparticles with diameters from 125 to 570 nm as nanocarriers for plasmid DNA was studied by Yu et al. Here, the transfection efficiency was found to be a compromise between binding capacity of the nanocarriers and cellular uptake. Smaller particles showed higher cellular uptake but less binding capacity for plasmid DNA. Particles with a diameter of 330 nm showed the best gene transfection efficacy [35].

Gene silencing mediated by delivery of siRNA and miRNA with MSN nanocarriers is strongly dependent on the escape of siRNA or miRNA from the endosomes into the cytosol. To trigger this reaction, the surface of particles is often decorated with cationic polymers, which are known to support an endosomal escape. Polyethylenimine (PEI) is one well-studied polymer exhibiting good endosomal escape capability when used at high dosage, attributed to a 'proton sponge' effect, however, it has a poor toxicity profile [36]. For instance, the group of Gu showed very good knockdown efficacies when PEI was attached to the outside of their particles [37,38].

A systematic study of the correlation between gene knockdown efficacy and endosomal escape kinetics was performed by Wang et al. [39]. They used magnetic MSNs capped with PEI at low concentrations to avoid toxic effects, resulting in poor endosomal escape ability. When endosomal escape was induced by a chloroquine treatment at early transfection times they could still see a high gene silencing efficacy. In contrast, when applied at later stages, a low gene-silencing efficacy was found, presumably because the released RNA was already degraded within the endosome.

Although MSNs have shown great potential as an efficient carrier system for siRNA and miRNA delivery, systematic studies about particle properties and corresponding gene-silencing efficacy remain rare. Herein, we present a systematic investigation of the particle size-dependent delivery of miRNA using core-shell MSNs for gene silencing in T24 cells. A series of MSNs with uniform sizes ranging from 60 to 160 nm with an average pore diameter of around 5–6 nm was used to encapsulate an antitumoral microRNA mimic (miR200c) or a control RNA with a scrambled sequence (Ctrl). The core of these particles was functionalized with APTES to yield a positively charged inner surface to accommodate the negatively charged miRNA, while a thinner surface layer was enriched with mercaptosilane, forming a negatively charged shell to enable binding of the capping agent. This consisted of an amino-acid block-copolymer 454 (see Figure 1) to aid endosomal escape and which was further linked via polyethylene glycol (PEG) to the peptide GE11, targeting the epidermal growth factor receptor [40]. This vector is abbreviated as MSN-454-GE11. An alternative construct for the delivery of this miRNA was studied before by Müller et al. using RNA-encapsulating polyplexes consisting of the same copolymer 454 functionalized with GE11 [41]. Their particles, featuring a size of 120–150 nm, successfully showed antitumoral effects with two different therapeutic RNAs.

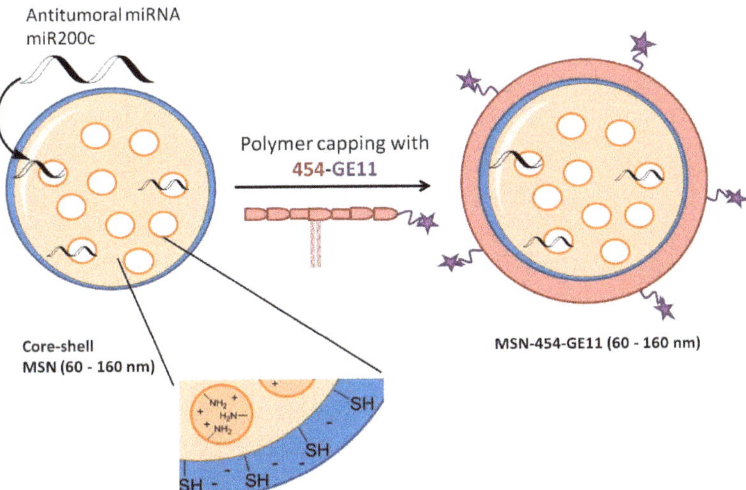

Figure 1. Schematic overview of the structure of core-shell mesoporous silica nanoparticles (MSN), the loading of miR200c and the polymer capping resulting in the MSN-454-GE11 vector. The positively

charged core in core-shell MSN enables a high loading capacity for miRNA. The mercapto-lined MSN shell associates with a positively charged block copolymer carrying the targeting ligand GE11, thus acting simultaneously as capping, endosomal release and targeting agent.

Our series of MSN-454-GE11 particles covers a broad size range but still provides comparable properties for each sample, including surface chemistry, surface charge, pore size and miRNA concentration to enable a conclusive study of the particle-size effect on miRNA delivery. Here, we show that good gene-silencing and antitumoral effects are obtained when miR200c is delivered by the largest MSN-454-GE11 particles in this sequence.

2. Materials and Methods

2.1. Reagents and Materials

2-(*N*-morpholino)ethanesulfonic acid (MES, Sigma-Adrich, Darmstadt, Germany), 3-aminopropyltriethoxysilane (APTES, Sigma-Adrich), 3-mercaptopropyl triethoxysilane (MPTES, >95%, Sigma-Adrich), ammonium fluoride (NH_4F, >98% Fluka, Darmstadt, Germany), Atto 633-carboxy dye (Atto-Tec, Siegen, Germany), 4′,6-diamidino-2-phenylindole (DAPI, Thermo Fisher, Schwerte, Germany), block copolymer surfactant (Pluronic F127 ($EO_{106}PO_{70}EO_{106}$), Sigma-Aldrich), cetyltrimethyl-ammonium chloride (CTAC, 25% in H_2O, Sigma-Adrich), *N*-(3-dimethylaminopropyl)-*N*′-ethylcarbodiimide hydrochloride (EDC, >97%, Sigma-Adrich), *N*-hydroxysulfosuccinimide (sulfo-NHS, 98%, Sigma-Adrich), octadecyltrimethylammonium bromide ($C_{18}Br$, Sigma-Adrich), propidium iodide, tetraethylorthosilicate (TEOS, >98%, Sigma-Aldrich), triethanolamine (TEA, 98%, Fluka), triisopropylbenzene (TiPB, 96%, Sigma-Adrich). Oligomer 454 [42] and Mal-PEG-GE11 [41] were synthesized as described before. siRNA and miRNA duplexes were purchased from Axolabs GmbH (Kulmbach, Germany): control RNA (sense: 5-AuGuAuuGG ccuGuAuuAGdTsdT-3′, antisense: 5′-CuAAuAcAGGCcAAuAcAUdTsdT-3), miR200c (sense: 5′-UCCAUCAUUACCCGGCAGUAUUA-3, antisense: 5′-UAAUACUGCCGGGUAAUGAUGGA-3).

2.2. Synthesis of MSN-NH2in-Shout

Two slightly different synthesis methods were used to prepare core-shell functionalized MSN containing 4.5% amino groups in the core and 2% thiol groups on the shell (MSN-NH_{2in}-SH_{out}).

Procedure A: Core-shell MSN particles were prepared by a co-condensation reaction according to our previous reports [17,43]. Early reports from our group showed that a consistent decrease in particle size can be obtained by decreasing the amount of TEA serving as organic base. This concept was applied for the synthesis of samples MSN160 nm, MSN130 nm and MSN100 nm [44]. A mixture of TEA (14 g, 93.8 mmol/10 g, 67 mmol/4 g, 26.8 mmol, respectively), tetraethylorthosilicate (TEOS, 1.74 mL, 7.84 mmol) and 3-aminopropyltriethoxysilane (APTES, 89 µL, 0.38 mmol) in a polypropylene reactor was heated in an oil bath at 90 °C for 20 min without stirring. A second solution was prepared consisting of water (2.41 mL, 1.21 mol), CTAC (2.71 mL, 1.83 mmol), TiPB (2.97 mL, 12 mmol) and ammonium fluoride (NH_4F, 100 mg, 2.7 mmol) and was heated to 60 °C under stirring. The second solution was subsequently added to the first solution under strong stirring. The whole mixture was further stirred at room temperature for 20 min. After this time, four portions of TEOS (4×51 µL, 0.92 mmol) were added in three minute intervals and the solution was stirred for another 30 min. To prepare the shell layer, a premixed solution of TEOS (42 µL, 0.190 mmol) and mercaptopropyl triethoxysilane (MPTES) (42 µL, 0.22 mmol) was added to the whole mixture. The condensation reaction was allowed to continue over night at room temperature.

Procedure B: MSN80 nm and MSN60 nm were prepared using the tri-block copolymer F127 as a particle growth inhibitor/dispersant. The procedure is based on an adapted recipe reported in the literature [45]. First, a mixture of TEA (7 g, 46.8 mmol/4 g, 26.7 mmol, respectively), TiPB (1.5 mL, 6.2 mmol), F127 (100 mg, 8 µmol) TEOS (2 mL, 9 mmol), and APTES (89 µL, 0.38 mmol) was prepared in a polypropylene reactor and heated at 90 °C for 20 min. Separately, NH_4F (100 mg, 2.7 mmol) and $C_{18}Br$

(0.35 g, 0.9 mmol) were dissolved in H_2O. This solution was heated to 90 °C under static conditions for 30 min and was then added under strong stirring to the first solution at once. The combined solutions were further heated at 60 °C for 30 min. The ingredients for the shell layer, TEOS (42 µL, 0.190 mmol) and MPTES (42 µL, 0.22 mmol), were mixed, preheated to 60 °C and added to the reaction mixture. Heating at 60 °C was continued for 30 min, and then the reaction mixture was allowed to cool down to room temperature under stirring and was continuously stirred overnight.

2.3. Template Extraction

Agglomerates were removed from the as- synthesized samples by centrifugation for a short time at low speed (7197 rcf, 5 min). The pellet was discarded and the remaining particles in the supernatant were collected via centrifugation (7197 rcf, 20 min) for template extraction. After centrifugation, the particles were resuspended in a solution containing 2 g ammonium nitrate in 100 mL ethanol and heated for 45 min under reflux (90 °C), cooled and collected by centrifugation (7197 rcf, 20 min). The first reflux treatment was followed by a second extraction step for 45 min under reflux (90 °C) using a solution of 10 mL concentrated hydrochloric acid in 90 mL ethanol. The MSNs were collected by centrifugation for 20 min (7197 rcf) and washed with ethanol (3 × 70 mL). Finally, MSNs were dispersed in 15 mL ethanol and used for further characterization.

2.4. Characterization

For transmission electron microscopy (TEM), samples were prepared by drying a diluted ethanolic suspension of MSN on a carbon-coated copper grid at room temperature for several hours. The measurements were performed on a Tecnai G2 20 S-Twin instrument operated at 200 kV with a TVIPS TemCam-F216 camera.

Dynamic light scattering (DLS) and zeta potential measurements were performed with a Malvern Zetasizer Nano instrument equipped with a 4 mW He-Ne-Laser (633 nm). For DLS data, a diluted colloidal suspension of the particles was measured in PMMA cuvettes at 25 °C. For zeta potential measurements, the additional Zetasizer titration system (MPT-2) was used based on diluted NaOH and HCl as titrants. For this purpose, a colloidal suspension of MSNs in water at a concentration of 0.1 mg mL^{-1} was prepared.

Raman spectroscopy was performed on a Bruker Equinox 55 with FRA-106 Raman attachment with a ND:YAG laser (1064 nm) and a laser power of 100 mW. Infrared spectra of dried sample powder were recorded on a ThermoScientific Nicolet iN10 IRmicroscope in reflection–absorption mode with a liquid-N_2 cooled MCT-A detector.

A Quantachrome Instrument NOVA 4000e was used for nitrogen sorption analysis at −196 °C. Sample outgassing was performed for 12 h at 120 °C and at a vacuum of 13.3×10^{-3} mbar. A quenched solid density functional theory (QSDFT) equilibrium model of N_2 on silica at a relative pressure $p/p_0 = 0.8$ was used to calculate the pore size and pore volume, based on the adsorption and desorption branch of the isotherm. A BET model in the range $p/p_0 = 0.05$–0.2 was used to determine the specific surface area.

Thermogravimetric analysis (TGA) of the powder samples (10 mg) was performed on a Netzsch STA 440 C TG/DSC using a heating rate of 10 °C/min up to 900 °C with a stream of synthetic air of about 25 mL·min^{-1}. siRNA concentrations were determined by UV measurements performed with the Nanodrop 2000c spectrometer (Thermo Scientific).

2.5. Loading of Ctrl and miR200c for Gene Silencing

In all experiments, 100 µL Ctrl or miR200c solution (c = 50 ng μL^{-1}) in MES buffer (pH = 5) was added to 100 µg MSN-NH_{2in}-SH_{out} samples, resulting in a final RNA concentration of 50 µg mg^{-1} of MSN carrier. Samples were vortexed and shaken at 37 °C for 30 min to complete adsorption. Particles were washed via centrifugation and were resuspended in 100 µL HEPES buffer. The supernatant was collected to verify complete adsorption of RNA.

2.6. Loading of Fluorescent Dye for Cellular Internalization Studies

A total of 1 mL MSN-NH_{2in}-SH_{out} samples (pH = 5; 1 mg mL^{-1} MES buffer) was mixed with 2 µL Atto-633 carboxy (2 mg mL^{-1} in anhydrous DMSO), 10 µL EDC and a catalytic amount of sulfo-NHS. The samples were vortexed and shaken for 4 h, were washed multiple times afterwards with MES and HEPES buffer (1 mL, respectively) (centrifugation steps: 10 min, 16,900 rcf) and were redispersed in 1 mL MES buffer.

2.7. Capping with 454-GE11

For capping of MSN-NH_{2in}-SH_{out} samples, 10 µL Mal-PEG-GE11 reagent (5 mg mL^{-1} in H_2O) was added to 100 µL 454 solution (5 mg mL^{-1} in H_2O) and shaken for 2 h at 37 °C. Then, 454-GE11 was mixed with 1 mL MSN suspension (1 mg mL^{-1} in HEPES buffer) and again shaken for 2 h at 37 °C. Resulting MSN-454-GE11 samples were centrifuged (16,900 rcf, 10 min) and washed with water and HEPES buffer. The samples were redispersed in 2 mL HEPES buffer.

2.8. miRNA Binding Assay

For the miRNA binding assay, a 2.5% agarose gel was prepared by dissolving agarose in TBE buffer (Trizma base 10.8 g, boric acid 5.5 g, disodium EDTA 0.75 g, and 1 L of water) and by heating to 100 °C. GelRed® (1:10,000) was added and the agarose gel was cast in the electrophoresis unit. MSN-454-GE11 samples loaded with Ctrl RNA were mixed with 4 µL loading buffer (6 mL glycerol, 1.2 mL of 0.5 M EDTA, 2.8 mL H_2O, 0.02 g bromophenol blue) and placed into the gel sample pockets. Electrophoresis was run at 120 V for 90 min.

2.9. Cell Culture

HeLa, T24 or T24/eGFPLuc-200cT (authenticated by DSMZ, Braunschweig, Germany) cells, stably expressing an eGFP-luciferase fusion gene under the control of the CMV promoter and with a target site for miR200c, were cultivated in Dulbecco's modified Eagle's medium (DMEM) supplemented with 10% fetal bovine serum (FBS), 100 U mL^{-1} penicillin and 100 U mL^{-1} streptomycin.

2.10. Cell Viability Determined by MTT Assay

MTT assay was performed in triplicate in 96-well plates. One day prior to transfection, T24 cells were seeded at 3500 cells/well. Before transfection, medium was replaced by 80 µL fresh medium. MSN samples (20 µL) loaded with or without RNA were added at different concentrations in HEPES buffer and incubated for 45 min at 37 °C followed by a medium exchange. Subsequently, cells were returned to the incubator and 48 h after transfection a viability assay was performed. For this, an MTT solution (100 µL/well; 0.5 mg mL^{-1} in DMEM) was added for 2 h. Then, the supernatant was removed and cells were lysed by freezing at −80 °C. DMSO (100 µL) was added and plates were incubated at 37 °C under shaking. Absorption at 590 nm against a reference wavelength of 630 nm was measured using a SpectraFluorTM Plus microplate reader S4 (Tecan, Groeding, Austria). Cell viability was calculated as percentage of absorption compared to wells treated with HEPES buffer.

2.11. Cellular Adhesion Determined by Flow Cytometry

T24/eGFPLuc-200cT cells were seeded 24 h before transfection on 24-well plates with a density of 5×10^5 cells per well in 1000 µL growth medium. After 24 h, the medium was replaced by 400 µL fresh medium. A total of 100 µL MSN, MSN-454-PEG and MSN-454-GE11 samples loaded with fluorescent dye (500 µg mL^{-1} MSN in HEPES buffer) were added and incubated for 45 min at 37 °C. After incubation time, the cells were washed three times with PBS, 500 I.U. heparin to remove particles non-specifically associated to the cell surface and were detached with trypsin/EDTA, taken up in growth medium, centrifuged and resuspended in PBS containing 10% FBS. Cellular internalization of the MSN samples was assayed by flow cytometry at an Atto-633 carboxy fluorescent dye excitation wavelength of 635 nm and detection of emission at 665 nm. Cells were gated by forward/sideward scatter and pulse width for exclusion of doublets. DAPI (4′,6-diamidino-2-phenylindole) was used to discriminate between viable and dead cells. Data were recorded by BD LSRFortessa™ (BD Biosciences, USA) and analyzed by FlowJo® 7.6.5 flow cytometric analysis software. All experiments were performed in triplicate.

2.12. Confocal Fluorescence Microscopy

Fluorescence microscopy was performed with a Zeiss Observer SD spinning disk confocal microscope using a Yokogawa CSU-X1 spinning disc unit and an oil objective (63× magnification) and BP 525/50 (WGA488) and LP 690/50 (Atto633) emission filters. A 488 nm and a 639 nm laser were used for excitation. At 24 h prior to transfection, 3500 T24 cells per well were seeded in 8-well plates in 280 μL growth medium. A total of 20 μL MSN-454-GE11 covalently labeled with Atto-633 carboxy (200 μg mL^{-1} MSN in HEPES buffer) were added to each well and incubated for 45 min or 6 h, respectively, at 37 °C followed by addition of WGA-488 to the medium for cell membrane staining. Cells were washed with PBS and after addition of 300 μL fresh medium, cells were directly imaged.

The cellular uptake of nanoparticles was quantified using the ImageJ macro "Particle_in_Cell-3D" developed by Adriano A. Torrano and Julia Blechinger, Department of Chemistry and Center for NanoScience (CeNS, University of Munich (LMU), Munich, Germany. http://imagejdocu.tudor.lu/doku.php?id=macro:particle_in_cell-3d)

2.13. Luciferase Gene Silencing

In all experiments, siRNA and miRNA delivery was performed in 96-well plates in triplicate. 3500 T24/eGFPLuc-200cT cells per well were seeded 24 h prior to transfection in 100 μL growth medium. Before transfection, medium was replaced with 80 μL fresh growth medium. 20 μL MSN, MSN-454-PEG and MSN-454-GE11 for RNA delivery (500 μg mL^{-1} MSN in HEPES buffer) were added to each well and incubated for 45 min at 37 °C followed by a medium exchange. Cells were then incubated with 100 μL fresh medium for an additional 48 h following transfection. After this time, cells were treated with 100 μL cell lysis buffer (Promega (Mannheim, Germany)). Luciferase activity in cell lysate (35 μL) was measured using a luciferin-LAR (1 M glycylglycine, 100 mM $MgCl_2$, 500 mM EDTA, DTT, ATP, coenzyme A) buffer solution on a luminometer for 10 s (Centro LB 960 plate reader luminometer, Berthold Technologies, Bad Wildbad, Germany).

2.14. Cell Cycle Analysis

At 24 h prior to transfection, 3.5×10^4 T24 cells per well were seeded in a 12-well plate in 1000 μL growth medium and then medium was replaced by 400 μL fresh medium. A total of 100 μL MSN-160 nm-454-GE11 (500 μg mL^{-1} in HEPES buffer) loaded with Ctrl RNA or miR200c was added and incubated at 37 °C for 4 h. After a selected incubation time, the medium was replaced with fresh medium and cells were incubated for an additional 72 h. Afterwards, cells were washed with PBS and detached with trypsin/EDTA. Cells were washed with PBS twice and suspended in 100 μL PBS. The cell suspension was added dropwise to 0.9 mL cold 70% ethanol and incubated at 4 °C for 2 h. Then, cell pellets were suspended in 1 mL PBS after centrifugation and incubated for 15 min at room temperature for counting. A total of 1×10^5 cells of each sample was incubated in 300 μL propidium iodide/TritonX-100 containing RNase solution for 15 min at 37 °C. For cell cycle analysis, the cells were analyzed by flow cytometry at an excitation wavelength of 488 nm and detection of emission with a 613/20 bandpass filter. Data were recorded by BD LSRFortessa™ (BD Biosciences, USA) and analyzed by FlowJo® 7.6.5 flow cytometric analysis software. All experiments were performed in triplicate.

2.15. Scratch Assay

At 24 h prior to transfection, 1×10^5 T24 and MDA-MB 231 cells per well were seeded in a 6-well plate in 2000 μL of growth medium. The medium was replaced by 800 μL fresh medium and 200 μL

MSN160 nm-454-GE11 (500 μg mL^{-1} in HEPES buffer) loaded with Ctrl RNA or miR200c was added. Cells were incubated with MSN160 nm-454-GE11 for 4 h, then the medium was changed. After 24 h additional incubation time, the cell layer was broken by a scratch using a 200 μL Eppendorf tip. Cells were washed with PBS and microscope (Axiovert 200, Zeiss, Oberkochen, Germany) pictures were taken after 24 h.

3. Results

3.1. Particle Synthesis

We synthesized a series of core-shell MSNs with uniform size, ranging from 60 to 160 nm via a delayed co-condensation method [17]. The core of all MSNs was functionalized with 4.5 mol% APTES (with respect to total silica) carrying amino groups in order to create a positively charged inner void volume capable of encapsulating the negatively charged RNA by electrostatic interaction. The outer surface was functionalized with 2 mol% mercaptopropyl triethoxysilane (MPTES) carrying a thiol functionalization to act as regiospecific linker for external binding of polymer and targeting ligand.

Two slightly different approaches were used to modulate the particle size of MSNs and to prepare a series of MSNs with uniform sizes over a range from 60 to 160 nm. Larger particles with average particle sizes of 160, 130 and 100 nm (MSN160 nm, MSN130 nm and MSN100 nm) were synthesized based on our previous reports using a decreasing amount of the base triethanolamine (TEA) in order to reduce the particle size (experimental details are described in the Materials and Methods section) [43,44,46]. Briefly, to prepare the core of our core-shell particles, a preheated solution containing cetyltrimethylammonium chloride (CTAC) as template and triisopropylbenzene (TiPB) as pore-expanding agent was mixed with a preheated solution containing TEA, tetraethoxysilane (TEOS) and APTES at an elevated temperature for 20 min. This was then followed by addition of the ingredients for the surface layer containing TEOS and MPTES. To prepare smaller particles with average particle sizes of 80 and 60 nm (MSN80 nm and MSN60 nm), we applied a slightly different procedure, additionally using F127 as growth inhibitor/pore expanding agent according to published reports [45]. The particle size and core-shell structure were then established as above, again by changing the molar ratio of TEOS:TEA. Hereby, a preheated solution of octadecyltrimethylammonium bromide ($C_{18}Br$) as template was mixed with a second preheated solution of TiPB, F127, TEA, TEOS and APTES, and stirred at 60 °C for 30 min. Afterwards, premixed TEOS and MPTES were added to form the negatively charged shell layer.

3.2. Characterization

The particles were characterized using a number of techniques including transmission electron microscopy (TEM), dynamic light scattering (DLS), thermogravimetric analysis (TGA), zeta potential measurements, Raman and IR spectroscopy. The particle size and the morphology were analyzed with TEM and DLS. All MSN samples show a spherical particle morphology with a disordered, wormlike pore structure, independent of size (Figure 2). Figure 2a,b shows a size distribution around 160 nm (sample MSN160 nm) for particles obtained with the CTAC synthesis using a molar ratio of TEOS:TEA = 1:10. As the molar ratio of TEOS:TEA was decreased to TEOS:TEA = 1:5 and 1:3, the average particle size decreased to 128 nm (sample MSN130 nm, Figure 2c,d) and 100 nm (sample MSN100 nm, Figure 2e,f), respectively. Particles obtained by the F127 synthesis show a mean particle size of 80 nm (sample MSN80 nm, TEOS:TEA = 1:5, Figure 2g,h). The smallest particles with a mean particle size of 60 nm (sample MSN60 nm, Figure 2i,j) were obtained with F127 and a TEOS:TEA ratio of 1:3. DLS measurements of suspended particles show slightly larger hydrodynamic diameters ranging from 90 to 250 nm (Figure S1) but clearly illustrate the same trend of a decreasing particle size with decreasing TEA concentration.

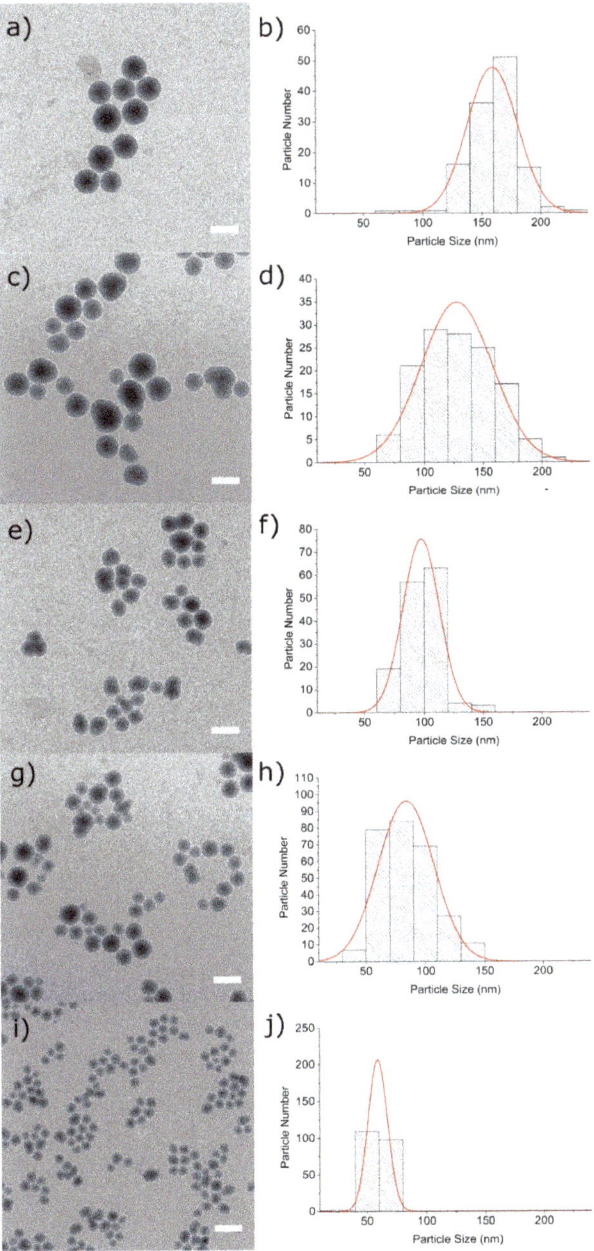

Figure 2. TEM micrographs of (**a**) MSN160 nm, (**c**) MSN130 nm, (**e**) MSN100 nm, (**g**) MSN80 nm, (**i**) MSN60 nm and corresponding particle size distribution histograms obtained from TEM images (**b**,**d**,**f**,**h**,**j**). Scale bar represents 150 nm.

Nitrogen sorption measurements resulted in typical type IV isotherms for all samples, as expected for MSNs (Figure 3a). Surface analysis indicates higher surface areas for particles prepared by the CTAC synthesis (samples MSN160 nm, MSN130 nm and MSN100 nm) ranging between 742–960 m^2/g, and shows a similar pore size of about 4.8 nm. Slightly smaller surface areas but larger pore sizes were obtained for particles prepared by the F127 synthesis (MSN80 nm, MSN60 nm; 660–685 m^2/g; pore size 6.0 nm, Figure 3b). Table 1 summarizes the properties of the particles obtained by the different synthesis methods.

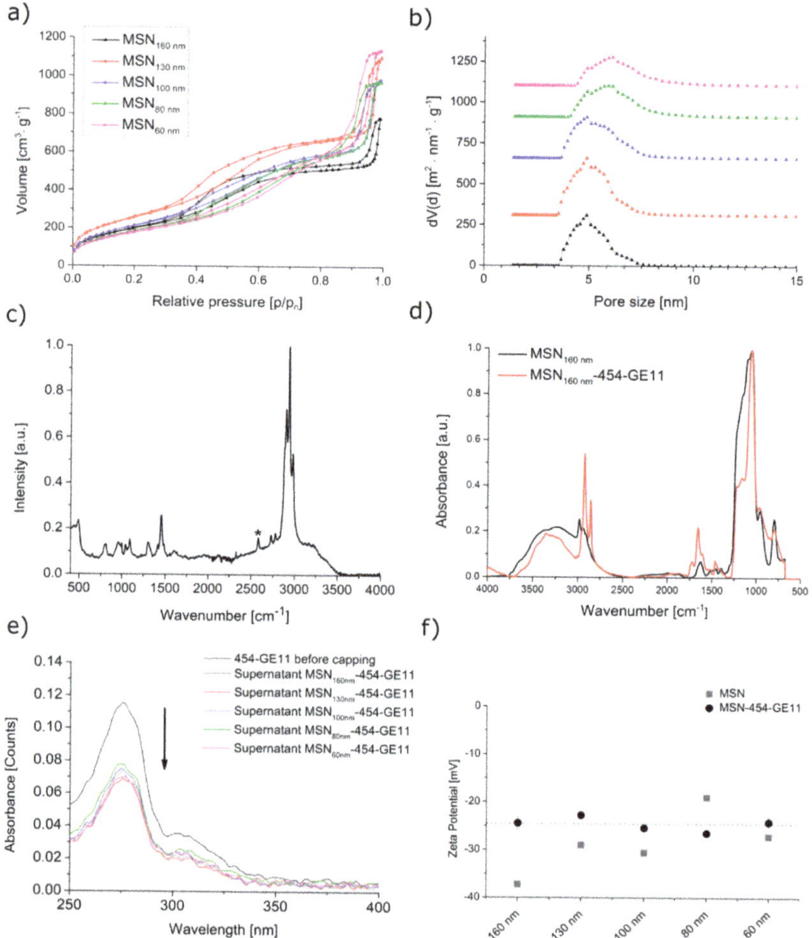

Figure 3. Characterization of MSN and MSN-454-GE11. (**a**) Nitrogen sorption isotherms and (**b**) corresponding pore size distributions. For clarity, the pore size distribution curves in panel b are shifted along the *y*-axis. (**c**) Raman spectrum of MSN160 nm. The signal at 2580 cm^{-1} (indicated by *) indicates the presence of thiol groups. (**d**) IR spectra, (**e**) UV VIS spectra of the capping solution before capping (black line) and the supernatants after capping, and (**f**) zeta potential measurements of MSN and MSN-454-GE11 at pH = 7.3.

Table 1. Synthesis and properties of MSN samples.

Name	Synthesis Method	Diameter TEM [a] [nm]	A_{BET} [m^2/g]	Pore Volume [b] [cc/g]	Pore Size [c] [nm]
MSN160 nm	CTAC Synthesis; TEOS:TEA = 1:10	159 ± 21	742	0.80	4.8
MSN130 nm	CTAC Synthesis; TEOS:TEA = 1:5	128 ± 30	961	1.06	4.8
MSN100 nm	CTAC Synthesis; TEOS:TEA = 1:3	100 ± 37	811	0.95	4.8
MSN80 nm	F127 Synthesis; TEOS:TEA = 1:5	84 ± 23	685	0.95	6.0
MSN60 nm	F127 Synthesis; TEOS:TEA = 1:3	59 ± 8	660	1.00	6.0

[a] The average particle size was obtained from TEM micrographs by measuring the diameter of around 200 particles of respective samples. [b] The total pore volume was determined at $p/p_0 = 0.9$ to exclude contributions of textural porosity. [c] Data were acquired from the adsorption branch of the nitrogen isotherm.

The presence of the amino and thiol groups originating from functionalization was verified with Raman and IR spectroscopy as well as with TGA. Representative results for sample MSN160 nm are shown in Figure 3c,d, respectively. Raman spectroscopy shows the S–H stretching mode of the thiol groups in the particle shell at 2580 cm^{-1} (Figure 3c), while the primary amines from core functionalization are seen at 1630 cm^{-1} with IR spectroscopy (MSN160 nm, black line, Figure 3d). TGA confirms the inclusion of organic functional groups by a weight loss of 20% (Figure S2). The decomposition of aminopropyl and successively, the mercaptopropyl groups starts at 290 °C.

3.3. Polymer Capping with 454-PEG and 454-GE11 and Targeting

The surface of the MSNs was capped with a modularly designed block copolymer 454 to prevent a premature release of the cargo and to enhance endosomal escape for intracellular delivery of miRNA. The structure of the copolymer 454 is shown in Figure S3. This T-shaped polymer 454 consists of a hydrophobic domain in the center made of two oleic acids attached to lysine units. Branched off at each side from the center are two succinyl-tetraethyl-pentamine (Stp) units [47], providing the polymer with a cationic charge. Each end contains a tyrosine trimer coupled to a terminal cysteine unit, potentially allowing additional functionalization. The combination of the central Stp and oleic acid units in 454 is assumed to facilitate interactions with the endosomal membrane, resulting in endosomal destabilization and thus enabling the nucleic acid cytosolic delivery. This block copolymer was successfully used for the formulation of INF-7-polyplex/siRNA vehicles [48] and was further essential for the highly efficient delivery of siRNA using MSNs, reported previously [21].

An anticancer therapy can potentially be improved by implementing cancer-cell-specific targeting molecules to the external surface of a carrier system. It is well known that EGF receptors (EGFR) are concentrated on many cancer cells. The small peptide GE11 has been used successfully before to specifically address EGFR-expressing cells [41,49,50]. Here, we exploit the mercapto residues of the cysteine groups in the capping polymer 454 for binding GE11 via a PEG linker to study the targeted, particle-size-dependent delivery of the miR200c.

For anchoring GE11 to the cysteine units in the block copolymer, we used a mercapto-reactive maleimide-PEG moiety consisting of 28 ethylene glycol monomer units that was previously linked to the GE11 ligand (Mal-PEG-GE11 reagent). The Mal-PEG-GE11 reagent was then mixed with the 454 block-copolymer (454-GE11). Only 0.1 equivalents (eq.; relative to 454-polymer) of the targeting ligand were used for 454-functionalization to maintain free thiol groups at the end of 454. The latter can potentially undergo a disulfide bridging with the terminal mercapto-groups on the outer surface of the MSNs, thus creating a stable copolymer capping. The attachment of the cationic polymer 454 to the surface is additionally electrostatically favored since zeta potential measurements of our pure MSNs reveal a negative surface charge for all samples at a pH higher than 5.5 (Figure S4). To perform the attachment of the premixed 454-GE11 to the surface of MSN, a suspension of MSN at pH 7.3 was mixed with 454-GE11 and the successful capping was confirmed using UV–VIS, IR, and zeta potential measurements. IR spectroscopy shows a significant increase of the CH stretching vibrations between 2850 and 2930 cm^{-1} and of the C-H bending vibrations of 1460 cm^{-1} as compared to the bare MSN particles (red graph in Figure 3d). Furthermore, the MSN-454-GE11 spectrum shows an increase of the N–H bending vibrations at 1650 cm^{-1} and a new signal at 1720 cm^{-1} related to a C=O stretching mode, indicating the multiple amide bonds of the copolymer. UV–VIS measurements were further used to quantify the amount of the 454-GE11 capping agent, which was similar in all samples and ranged from 17 to 20 wt% (Figure 3e, calibration curve Figure S5 and capping amount of different samples, Table S1).

Figure 3f shows the zeta potential measurements of all MSN samples before and after attachment of the 454-GE11 linker (MSN-454-GE11) at pH = 7.3. Besides the size of the nanoparticles, their surface charge is an important parameter for cellular internalization. Our pure MSN samples display a negative surface charge between −37 and −20 mV, as expected for mercapto-covered MSNs. The zeta potential of all capped MSN-454-GE11 samples is very comparable and has increased to about −24 mV when measured at pH 7.3. The observed difference before and after capping can be seen as an additional indication for a successful attachment of the targeted copolymer. The polymer capping did not affect the particle size distribution in water for most of the MSN-454-GE11 samples when measured by DLS. Only the sample with a mean particle size of 60 nm showed some moderate agglomeration (Figure S6).

3.4. Cell Internalization

As documented above, we have now established that our MSN carrier system shows physical properties such as surface area, pore size, zeta potential and capping concentration that are all very comparable, thus, leaving solely the particle size as a distinct variable. First, flow cytometry and confocal fluorescence microscopy data were used to investigate how the targeting ligand concentration as well as particle size might influence the cell internalization of the fully assembled MSN-454-GE11 samples.

Figure 4 shows the cellular internalization of the MSNs by T24 cells as examined via flow cytometry after a 45 min incubation time. MSN-454-GE11 samples were labeled with the Atto-633-carboxy fluorescent dye, which was covalently bound to the amino groups in the core of the particles. In order to investigate the best targeting concentration, we mixed different weight equivalents of targeting ligand GE11 with the 454 polymer in the range from 0.1 to 1 eq. before attaching the polymer to the MSN samples. Maleimide-PEG without ligand was used as negative control for polymer capping without targeting ligand (sample MSN-454-PEG). A significant targeting effect can be seen in Figure 4a when 0.1 eq. of the targeting ligand was used, while only a minor improvement is observed at higher concentrations.

Figure 4b documents the particle-size-dependent association of MSN-454-GE11 to cells as investigated via flow cytometry. Flow cytometry showed a similar degree of cell association for all samples, while a slight trend towards stronger association can be seen for smaller particles (quantification and statistical analysis can be found in SI, Figure S7). Since flow cytometry is not able to differentiate internalization from externally adhering particles, we followed this process also with confocal fluorescence microscopy as shown in Figure 5 (for enlarged images with additional orthogonal views of each image showing the particle internalization, see Figure S8, Supplementary Materials). Images were subsequently analyzed via the digital method 'Particle_in_Cell-3D' to quantify the cellular uptake of the differently sized MSN vectors. After imaging cells directly after an incubation time of 45 min, we found that only MSN160 nm-454-GE11 particles had penetrated through the cell membrane and were truly internalized. Other smaller MSN-454-GE11 particles were only attached to the outer cell membrane at this time. In contrast, all particle sizes were internalized after a 6 h incubation time. However, also after this time, MSN160 nm-454-GE11 particles showed the highest number of internalized particles of all samples (Figure S9).

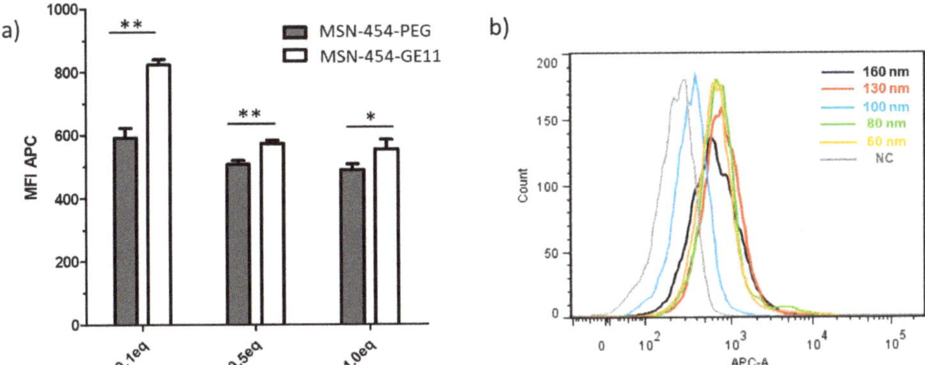

Figure 4. Cellular internalization determined by flow cytometry. Cells were incubated for 45 min, washed and analyzed. (**a**) MSN160 nm-454-PEG was used for passive and MSN160 nm-454-GE11 for receptor-mediated uptake. 454 was functionalized with different equivalents of Mal-PEG (454-PEG) and Mal-PEG-GE11 reagent (454-GE11), respectively, in the range from 0.1 to 1 eq. The mean fluorescence intensity of the Atto-633 signal (MFI APC) represents the amount of internalized nanoparticles. A high MFI corresponds to a large cell uptake of MSN. For statistical analysis, a two-tailed t-test was performed ($n = 3$, mean ± SD, * $p < 0.05$, ** $p < 0.01$) (**b**) Histograms of cellular internalization of Atto-633 labeled MSN-454-GE11 after 45 min incubation with particle sizes in the range of 160 nm to 60 nm and negative control (NC). The Atto-633 intensity (APC-A channel) is plotted against the number of events detected ('count'). 'Count' represents cumulative counts of cells with indicated Atto-633 fluorescence after appropriate gating by forward/sideward scatter and pulse width. For statistical analysis, see Figure S7.

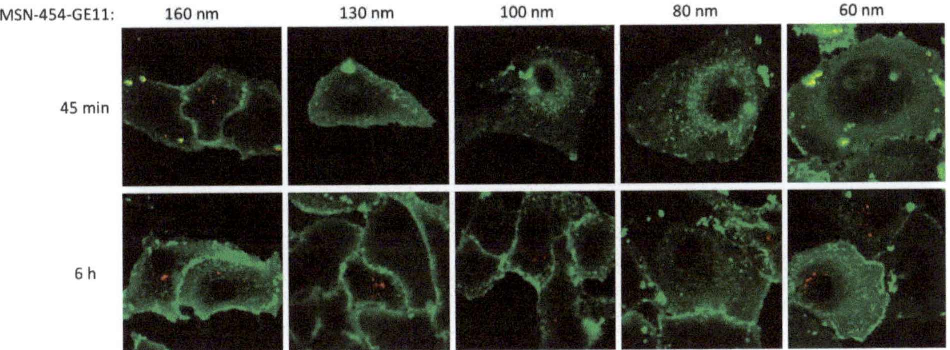

Figure 5. Representative confocal fluorescence microscopy images of Atto-633 labeled MSN-454-GE11 (red) with particle sizes in the range of 160 nm to 60 nm after 45 min (upper panel) and 6 h (lower panel) incubation on WGA488-stained T24 cells (green). For a statistical evaluation of particle-size-dependent uptake, see Figure S9, SI.

3.5. Loading of RNA

Loading of the genetic material into our MSN samples was performed with all samples prior to the attachment of the block copolymer. The adsorption of the miRNA is mainly driven by electrostatic interactions with the cationic, amino-functionalized core of the MSNs. The negatively charged thiol groups at the particle periphery are expected to minimize any external adsorption. For the best results, we performed the RNA adsorption in MES buffer at pH = 5. The MSN particles show a positive zeta potential at this pH due to protonation of the amino groups (zeta potential titration, see Figure S4).

Here, aliquots of 50 μg mg^{-1} RNA/MSN were used for loading and the actual uptake was calculated by difference measurements by determining the remaining RNA concentration in the supernatant. All samples were able to adsorb the amount offered, as no residual RNA could be detected in the supernatant after loading times as short as 30 min. The samples were subsequently capped with 454-GE11 to obtain the final MSN-454-GE11 samples used for subsequent experiments. The stable binding of RNA in these samples is reflected in gel shift results, as shown in Figure 6a). Only little RNA elution is visible upon applying a voltage of 100 V for 1 h to the gel.

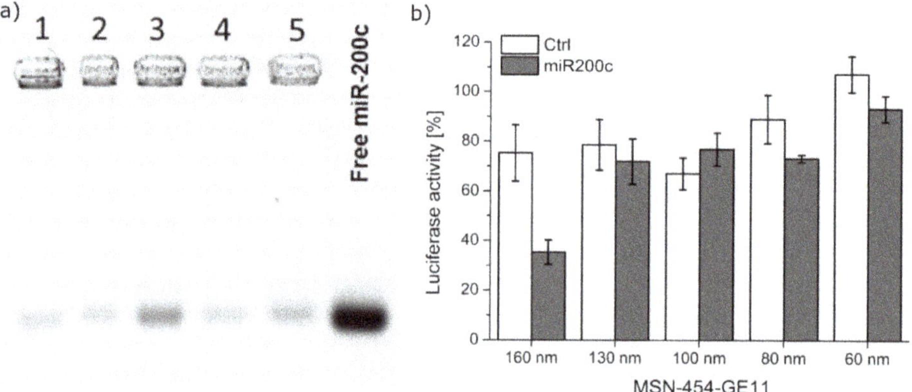

Figure 6. (**a**) Gel electrophoresis of samples MSN160 nm-454-GE11 (1), MSN130 nm-454-GE11 (2), MSN100 nm-454-GE11 (3), MSN80 nm-454-GE11 (4), and MSN60 nm-454-GE11 (5). (**b**) Gene silencing of T24/eGFPLuc-200cT cells transfected with MSN-454-GE11 containing either Ctrl (white) or miR200c (grey). After an incubation time of 45 min, cells were washed. At 48 h after transfection, gene-silencing effects were analyzed.

3.6. Gene Silencing

As a proof of concept, the gene silencing of the eGFP-luciferase reporter gene was performed using T24/eGFPLuc-200cT cells. This cell line stably expresses the eGFP-luciferase fusion protein, while simultaneously featuring a miR200c target site on the expressed mRNA. Therefore, the gene expression of eGFP-luciferase fusion protein can be blocked by the delivery of miR200c. The luciferase signal was thus used to evaluate the gene-silencing efficacy of miR200c. In order to simulate the dynamic conditions of drug delivery and to avoid a prolonged overexposure of the cells with particles suspended in the medium, we have used extremely short incubation times of 45 min. Cells were subsequently washed by exchanging the medium and the effect of this short exposure was then analyzed after time spans referred to as transfection time in the following. Cells were transfected with MSN and MSN-454-GE11 of different sizes loaded with either a synthetic miR200c mimic or a control siRNA without any target gene (Ctrl, Figure 6b). Gene silencing was never observed with the pure MSN samples (missing the 454 polymer construct) loaded with miR200c for silencing (Figure S10a). This might be caused by a premature release of the RNA or by endosomal trapping of the MSN samples missing the 454 polymer.

Similarly, none of the smaller sized MSN-454-GE11 samples showed any significant silencing efficacy. In contrast, with the larger 160 nm targeted polymer-capped MSN160 nm-454-GE11 sample, a luciferase gene knockdown of up to 65% was observed. These findings are in good agreement with the confocal fluorescence microscopy images, which showed the fastest internalization for particles with a size of 160 nm.

Flow cytometry data showed an improved cellular internalization of the targeted MSN160 nm-454-GE11 vector in comparison with the non-targeted control MSN-454-PEG when 0.1 eq. of the targeting ligand was used. This targeted vector showed a higher silencing efficacy compared to the non-targeted control, which is in good agreement with the increased uptake observed via flow cytometry.

As expected, the same sample loaded with just scrambled RNA (Ctrl) did not show a significant unspecific knockdown, thus excluding major toxic effects induced by the carrier system itself. Systematic cell viability studies using an MTT assay also show a good tolerance of the cells towards these samples (Figure S11).

These results show that a particle size of about 160 nm in combination with the copolymer 454 enables the fastest cell internalization and also the best transfection efficacy when short incubation times of only 45 min are used. Inhibition experiments addressing specific endocytic pathways were inconclusive with respect to a preferred mechanism for a certain particle size. However, endocytosis was substantially more blocked for the larger particles as compared to the 60 nm particles (see Figure S12, Supplementary Materials). Our data also confirm the endosomolytic activity of the 454 polymer since cell adherence/uptake was observed for all pure MSN samples (flow cytometry of pure MSN samples in Figure S13), but no transfection occurred without the 454 polymer (see Figure S10a). As described above, we further found that only MSN160 nm-454-GE11 showed gene-silencing activity, while all smaller MSN-454-GE11 samples were not active after the short incubation period applied here. Confocal fluorescence microscopy showed that MSN160 nm-454-GE11 particles were already internalized after 45 min, while all other MSN-454-GE11 samples needed much longer incubation times to achieve internalization. Thus, particle internalization of the smaller particles is likely too slow and prevents a larger efficacy.

Recently, miR200c was delivered by some of us using GE11 modified 454 polyplexes with a similar size of 120–150 nm [41]. The gene silencing efficiency achieved in the same T24/eGFPLuc-200cT cells with these polyplexes was around 60%, which is in the same range as the knockdown efficiency of the MSN-454-GE11 constructs presented here. This comparison suggests that the size of the targeted constructs is a determining feature controlling the knockdown efficiency. As MSN160 nm-454-GE11 constructs were shown to offer the best cellular-uptake behavior and gene-silencing efficiency, they were used for the following cell migration and cell cycle experiments.

3.7. Antitumoral Effects

The tumor suppressor miR200c inhibits the epithelial–mesenchymal transition, a process involved in metastasis by enhancing the motility and migration of tumor cells, by targeting ZEB1 and ZEB2. ZEB1 and ZEB2 are transcriptional repressors, which downregulate the marker E-cadherin [7,51]. Furthermore, Kopp et al. reported that one of the most prominent oncogenes KRAS is targeted by miR200c, which results in an altered cell cycle of the cancer cells [6]. Notably, they could show that miR200c inhibits cell cycle progression by decreasing the G1-population. We have performed direct investigations of these antitumoral effects through tumor cell migration and cell cycle analysis on two different EGFR overexpressing cell lines, T24 bladder cancer and HeLa cervical cancer cells using miR200c loaded MSN-454-GE11 samples.

Cell migration was studied using a scratch assay, as shown in Figure 7. T24 cells and HeLa cells were incubated with MSN-454-GE11 for 4 h (5 µg miR200c/well), then the medium was changed. After additional 24 h, the cell layer was broken by a scratch using a 200 µL Eppendorf pipette tip. The closure of the scratch, which is an indicator for cell migration, was measured at indicated time points. In both cell lines, the scratch was almost completely closed after 48 h when MSN-454-GE11 was loaded with Ctrl-RNA (88% scratch closure for T24 and 82% closure for HeLa cells). In contrast, with MSN-454-GE11 particles loaded with miR200c, a scratch closure of only 48% (for T24) and 53% (for HeLa cells) was observed after this time. Hence, miR200c delivered by MSN-454-GE11 significantly hinders cell migration.

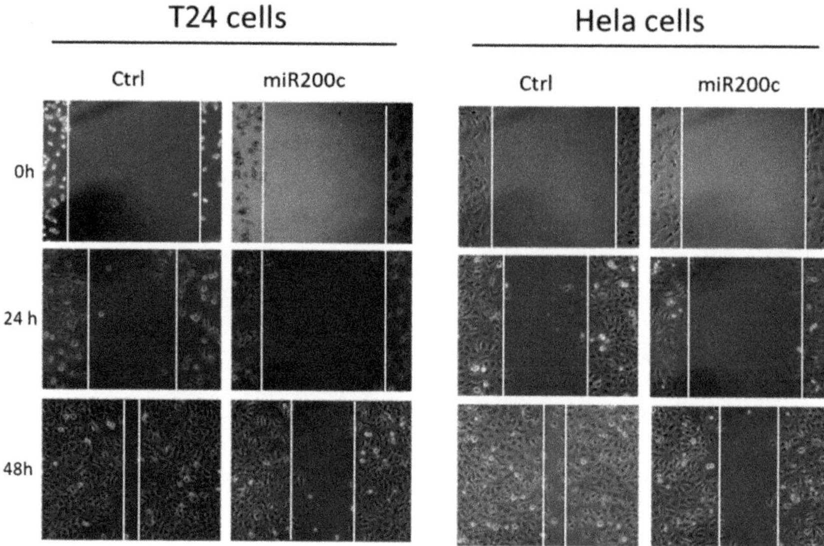

Figure 7. Inhibition of tumor cell migration. Images of a scratch assay of T24 and HeLa cells. Cells were treated with MSN160 nm-454-GE11 loaded with miR200c. The cell layer was broken after 24 h through a scratch and the closure was monitored for 48 h.

In addition, the effect of miR200c on the cell cycle was studied. Cells were transfected with MSN-454-GE11 loaded with miR200c or with the control RNA Ctrl and incubated for 4 h (Figure 8). A significant decrease in the number of cells in the G1 phase is observed when exposed to miR200C delivered by MSN, in combination with an increase in the number of cells in the S-phase. Thus, tumor cells transfected with MSN-454-GE11 loaded with miR200c showed the expected decreased migration and changes in the cell cycle.

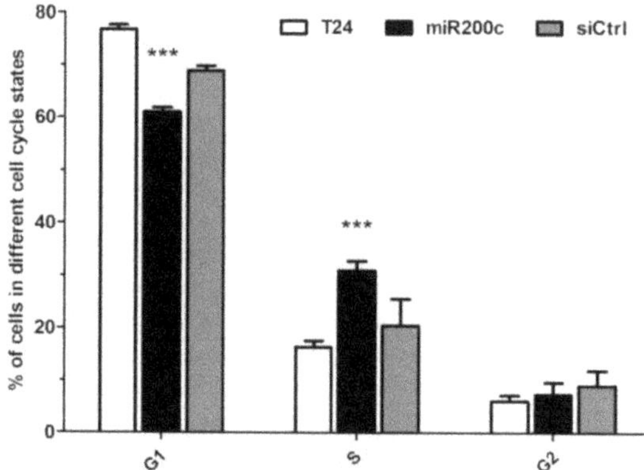

Figure 8. Cell cycle analysis via flow cytometry of cell stages G1, S and G2 of T24 cells at 72 h after treatment. For statistical analysis, a two-tailed t-test was performed (n = 3, mean ± SD, *** $p < 0.01$)

4. Conclusions

In this study, we exploited the tunability of MSNs to synthesize a series of core-shell MSNs with different particle sizes between 60 to 160 nm for studying the size effect of these carriers on antitumoral miRNA delivery. All other properties of the delivery vehicles, including surface area, pore size and zeta potential were kept comparable. The nanoparticles were capped with a positively charged block copolymer 454 equipped with the targeting agent Mal-PEG-GE11. Since gene silencing was only observed after capping the nanocarriers with this 454 polymer, we conclude that it is essential for the endosomal escape by destabilizing the endosomal membrane. It was shown that the targeting ligand GE11 enhances a receptor-mediated uptake. After capping, the MSN-454-GE11 vehicles were used for a systematic investigation of size-dependent gene silencing. While smaller particles did not lead to significant effects, MSN-454-GE11 with a size of 160 nm showed a remarkable gene knockdown efficacy and antitumoral effects such as a decreased migration and changes in cell cycle. Overall, we observed the fastest cellular internalization as well as the best knock-down efficacies with MSN-454-GE11 sized 160 nm. In contrast to FACS results that indicated a particle association with the cells independent of MSN particle size, we found with statistically evaluated image analysis that only the largest particles are truly internalized in the cells after short incubation times. Thus, cell studies as performed here, aiming to simulate dynamic conditions of in vivo drug delivery by washing the incubated cells after a short incubation time, might discriminate against all particles that are not well attached to the cell surface. Fast internalized particles are thus the winner. We hypothesize that due to their size, they expose a larger contact area to the cell membrane and simultaneously a larger number of targeting ligands, which, when spaced just right, allow for a maximum degree of endocytosis. Our study shows that fast cellular internalization is essential for a successful downregulation. In summary, the nanoscale MSN160 nm-454-GE11 vehicles show the most promising potential for future in vivo biomedical applications.

Supplementary Materials: The following are available online at http://www.mdpi.com/1999-4923/12/6/505/s1, Figure S1: DLS measurements of MSN samples, Figure S2: TGA, Figure S3: Structure of copolymer 454, Figure S4: Zeta potential titration curves of pure MSN samples, Figure S5: 454-GE11 calibration curve, Figure S6: DLS measurements of MSN-454-GE11 samples, Figure S7: Quantification of FACS data shown in Figure 4b, Figure S8: Confocal fluorescence microscopy images, Figure S9: Quantification of the cellular uptake of MSN-454-GE11, Figure S10: Gene-silencing assay, Figure S11: MTT cell viability study, Figure S12: Inhibition results, Figure S13: FACS of internalization of untargeted nanoparticles, Table S1: Estimated capping amount of 454-GE11 for different MSN samples.

Author Contributions: Conceptualization, T.B., E.W., K.M.; Measurements and Data Analysis, L.H., W.Z., S.R., H.E.; Writing—Original Draft Preparation, L.H.; Writing—Review & Editing, K.M., T.B. All authors have read and agreed to the published version of the manuscript.

Funding: Financial support from the DFG (SFB 1032), the Excellence Cluster Nanosystems Initiative Munich (NIM), the Center for NanoScience Munich (CeNS), as well as Ludwig-Maximilians-Universität (LMUexcellent funds) is gratefully acknowledged.

Acknowledgments: We thank Steffen Schmidt for electron microscopy.

Conflicts of Interest: Authors declare no conflict of interest. The funders had no role in the design of the study; in the collection, analyses, or interpretation of data; in the writing of the manuscript, or in the decision to publish the results.

References

1. Elbashir, S.M.; Harborth, J.; Lendeckel, W.; Yalcin, A.; Weber, K.; Tuschl, T. Duplexes of 21-Nucleotide Rnas Mediate RNA Interference in Cultured Mammalian Cells. *Nature* **2001**, *411*, 494. [CrossRef] [PubMed]
2. McCaffrey, A.P.; Meuse, L.; Pham, T.-T.T.; Conklin, D.S.; Hannon, G.J.; Kay, M.A. RNA Interference in Adult Mice. *Nature* **2002**, *418*, 38. [CrossRef] [PubMed]
3. Xin, Y.; Huang, M.; Guo, W.W.; Huang, Q.; Zhang, L.Z.; Jiang, G. Nano-Based Delivery of RNAi in Cancer Therapy. *Mol. Cancer* **2017**, *16*, 134. [CrossRef] [PubMed]
4. Friedman, R.C.; Farh, K.K.-H.; Burge, C.B.; Bartel, D.P. Most Mammalian mRNAs Are Conserved Targets of Micrornas. *Genome Res.* **2009**, *19*, 92–105. [CrossRef]
5. Lee, S.J.; Kim, M.J.; Kwon, I.C.; Roberts, T.M. Delivery Strategies and Potential Targets for siRNA in Major Cancer Types. *Adv. Drug Deliv. Rev.* **2016**, *104*, 2–15. [CrossRef]
6. Kopp, F.; Wagner, E.; Roidl, A. The Proto-Oncogene KRAS Is Targeted by Mir-200c. *Oncotarget* **2013**, *5*, 185–195. [CrossRef]

7. Hurteau, G.J.; Carlson, J.A.; Spivack, S.D.; Brock, G.J. Overexpression of the MicroRNA hsa-miR-200c Leads to Reduced Expression of Transcription Factor 8 and Increased Expression of E-Cadherin. *Cancer Res.* **2007**, *67*, 7972–7976. [CrossRef]
8. Kopp, F.; Oak, P.S.; Wagner, E.; Roidl, A. Mir-200c Sensitizes Breast Cancer Cells to Doxorubicin Treatment by Decreasing Trkb and Bmi1 Expression. *PLoS ONE* **2012**, *7*, e50469. [CrossRef]
9. Kim, H.J.; Kim, A.; Miyata, K.; Kataoka, K. Recent Progress in Development of siRNA Delivery Vehicles for Cancer Therapy. *Adv. Drug Deliv. Rev.* **2016**, *104*, 61–77. [CrossRef]
10. Kim, B.; Park, J.-H.; Sailor, M.J. Rekindling RNAi Therapy: Materials Design Requirements for in Vivo siRNA Delivery. *Adv. Mater.* **2019**, *31*, 1903637. [CrossRef]
11. Revia, R.A.; Stephen, Z.R.; Zhang, M. Theranostic Nanoparticles for Rna-Based Cancer Treatment. *Acc. Chem. Res.* **2019**, *52*, 1496–1506. [CrossRef] [PubMed]
12. Adams, D.; Gonzalez-Duarte, A.; O'Riordan, W.D.; Yang, C.C.; Ueda, M.; Kristen, A.V.; Tournev, I.; Schmidt, H.H.; Coelho, T.; Berk, J.L.; et al. Patisiran, an RNAi Therapeutic, for Hereditary Transthyretin Amyloidosis. *New Engl. J. Med.* **2018**, *379*, 11–21. [CrossRef] [PubMed]
13. Chen, Y.; Gu, H.; Zhang, D.S.-Z.; Li, F.; Liu, T.; Xia, W. Highly Effective Inhibition of Lung Cancer Growth and Metastasis by Systemic Delivery of siRNA Via Multimodal Mesoporous Silica-Based Nanocarrier. *Biomaterials* **2014**, *35*, 10058–10069. [CrossRef] [PubMed]
14. Hartono, S.B.; Gu, W.; Kleitz, F.; Liu, J.; He, L.; Middelberg, A.P.J.; Yu, C.; Lu, G.Q.; Qiao, S.Z. Poly-L-Lysine Functionalized Large Pore Cubic Mesostructured Silica Nanoparticles as Biocompatible Carriers for Gene Delivery. *ACS Nano* **2012**, *6*, 2104–2117. [CrossRef] [PubMed]
15. Möller, K.; Bein, T. Degradable Drug Carriers: Vanishing Mesoporous Silica Nanoparticles. *Chem. Mater.* **2019**, *31*, 4364–4378. [CrossRef]
16. Braun, K.; Pochert, A.; Beck, M.; Fiedler, R.; Gruber, J.; Linden, M. Dissolution Kinetics of Mesoporous Silica Nanoparticles in Different Simulated Body Fluids. *J. Sol.-Gel. Sci. Technol.* **2016**, *79*, 319–327. [CrossRef]
17. Kecht, J.; Schlossbauer, A.; Bein, T. Selective Functionalization of the Outer and Inner Surfaces in Mesoporous Silica Nanoparticles. *Chem. Mater.* **2008**, *20*, 7207–7214. [CrossRef]
18. Möller, K.; Bein, T. Talented Mesoporous Silica Nanoparticles. *Chem. Mater.* **2017**, *29*, 371–388.
19. Wang, J.; Lu, Z.; Wientjes, M.G.; Au, J.L.S. Delivery of siRNA Therapeutics: Barriers and Carriers. *AAPS J.* **2010**, *12*, 492–503. [CrossRef]
20. Gao, F.; Botella, P.; Corma, A.; Blesa, J.; Dong, L. Monodispersed Mesoporous Silica Nanoparticles with Very Large Pores for Enhanced Adsorption and Release of DNA. *J. Phys. Chem. B* **2009**, *113*, 1796–1804. [CrossRef]
21. Möller, K.; Müller, K.; Engelke, H.; Bräuchle, C.; Wagner, E.; Bein, T. Highly Efficient siRNA Delivery from Core–Shell Mesoporous Silica Nanoparticles with Multifunctional Polymer Caps. *Nanoscale* **2016**, *8*, 4007–4019.
22. Jin, Y.; Lohstreter, S.; Pierce, D.T.; Parisien, J.; Wu, M.; Hall, C.; Zhao, J.X. Silica Nanoparticles with Continuously Tunable Sizes: Synthesis and Size Effects on Cellular Contrast Imaging. *Chem. Mater.* **2008**, *20*, 4411–4419. [CrossRef]
23. Gan, Q.; Dai, D.; Yuan, Y.; Qian, J.; Sha, S.; Shi, J.; Liu, C. Effect of Size on the Cellular Endocytosis and Controlled Release of Mesoporous Silica Nanoparticles for Intracellular Delivery. *Biomed. Microdevices* **2012**, *14*, 259–270. [CrossRef] [PubMed]
24. Huang, X.; Teng, X.; Chen, D.; Tang, F.; He, J. The Effect of the Shape of Mesoporous Silica Nanoparticles on Cellular Uptake and Cell Function. *Biomaterials* **2010**, *31*, 438–448. [CrossRef] [PubMed]
25. Zhu, J.; Tang, J.; Zhao, L.; Zhou, X.; Wang, Y.; Yu, C. Ultrasmall, Well-Dispersed, Hollow Siliceous Spheres with Enhanced Endocytosis Properties. *Small* **2010**, *6*, 276–282. [CrossRef]
26. Chono, S.; Tanino, T.; Seki, T.; Morimoto, K. Uptake Characteristics of Liposomes by Rat Alveolar Macrophages: Influence of Particle Size and Surface Mannose Modification. *J. Pharm. Pharmacol.* **2007**, *59*, 75–80. [CrossRef]
27. Osaki, F.; Kanamori, T.; Sando, S.; Sera, T.; Aoyama, Y. A Quantum Dot Conjugated Sugar Ball and Its Cellular Uptake. On the Size Effects of Endocytosis in the Subviral Region. *J. Am. Chem. Soc.* **2004**, *126*, 6520–6521. [CrossRef]
28. Shang, L.; Nienhaus, K.; Nienhaus, G.U. Engineered Nanoparticles Interacting with Cells: Size Matters. *J. Nanobiotechnology* **2014**, *12*, 5. [CrossRef]
29. Lu, F.; Wu, S.-H.; Hung, Y.; Mou, C.-Y. Size Effect on Cell Uptake in Well-Suspended, Uniform Mesoporous Silica Nanoparticles. *Small* **2009**, *5*, 1408–1413. [CrossRef]

30. Chithrani, B.D.; Ghazani, A.A.; Chan, W.C.W. Determining the Size and Shape Dependence of Gold Nanoparticle Uptake into Mammalian Cells. *Nano Lett.* **2006**, *6*, 662–668. [CrossRef]
31. Varela, J.A.; Bexiga, M.G.; Åberg, C.; Simpson, J.C.; Dawson, K.A. Quantifying Size-Dependent Interactions between Fluorescently Labeled Polystyrene Nanoparticles and Mammalian Cells. *J. Nanobiotechnology* **2012**, *10*, 39. [CrossRef] [PubMed]
32. Huang, J.; Bu, L.; Xie, J.; Chen, K.; Cheng, Z.; Li, X.; Chen, X. Effects of Nanoparticle Size on Cellular Uptake and Liver MRI with Polyvinylpyrrolidone-Coated Iron Oxide Nanoparticles. *ACS Nano* **2010**, *4*, 7151–7160. [CrossRef] [PubMed]
33. Goel, S.; Chen, F.; Hong, H.; Valdovinos, H.; Hernandez, R.; Shi, S.; Barnhart, T.; Cai, W. Vegf121-Conjugated Mesoporous Silica Nanoparticle: A Tumor Targeted Drug Delivery System. *ACS Appl. Mater. Interfaces* **2014**, *6*, 21677–21685.
34. Chen, F.; Nayak, T.R.; Goel, S.; Valdovinos, H.F.; Hong, H.; Theuer, C.P.; Barnhart, T.E.; Cai, W. In Vivo Tumor Vasculature Targeted Pet/Nirf Imaging with Trc105(Fab)-Conjugated, Dual-Labeled Mesoporous Silica Nanoparticles. *Mol. Pharm.* **2014**, *11*, 4007–4014. [CrossRef]
35. Yu, M.; Niu, Y.; Zhang, J.; Zhang, H.; Yang, Y.; Taran, E.; Jambhrunkar, S.; Gu, W.; Thorn, P.; Yu, C. Size-Dependent Gene Delivery of Amine-Modified Silica Nanoparticles. *Nano Res.* **2016**, *9*, 291–305. [CrossRef]
36. Nayak, T.R.; Krasteva, L.K.; Cai, W. Multimodality Imaging of RNA Interference. *Curr. Med. chem.* **2013**, *20*, 3664–3675. [CrossRef]
37. Li, X.; Chen, Y.; Wang, M.; Ma, Y.; Xia, W.; Gu, H. A Mesoporous Silica Nanoparticle – Pei – Fusogenic Peptide System for siRNA Delivery in Cancer Therapy. *Biomaterials* **2013**, *34*, 1391–1401. [CrossRef]
38. Chen, Y.; Wang, X.; Liu, T.; Zhang, D.S.-Z.; Wang, Y.; Gu, H.; Di, W. Highly Effective Antiangiogenesis Via Magnetic Mesoporous Silica-Based siRNA Vehicle Targeting the Vegf Gene for Orthotopic Ovarian Cancer Therapy. *Int. J. Nanomed.* **2015**, *10*, 2579–2594.
39. Wang, M.; Li, X.; Ma, Y.; Gu, H. Endosomal Escape Kinetics of Mesoporous Silica-Based System for Efficient siRNA Delivery. *Int. J. Pharm.* **2013**, *448*, 51–57. [CrossRef]
40. Li, Z.; Zhao, R.; Wu, X.; Sun, Y.; Yao, M.; Li, J.; Xu, Y.; Gu, J. Identification and Characterization of a Novel Peptide Ligand of Epidermal Growth Factor Receptor for Targeted Delivery of Therapeutics. *Faseb. J.* **2005**, *19*, 1978–1985. [CrossRef]
41. Müller, K.; Klein, P.M.; Heissig, P.; Roidl, A.; Wagner, E. EGF Receptor Targeted Lipo-Oligocation Polyplexes for Antitumoral siRNA and Mirna Delivery. *Nanotechnology* **2016**, *27*, 464001. [CrossRef] [PubMed]
42. Troiber, C.; Edinger, D.; Kos, P.; Schreiner, L.; Kläger, R.; Herrmann, A.; Wagner, E. Stabilizing Effect of Tyrosine Trimers on PDNA and siRNA Polyplexes. *Biomaterials* **2013**, *34*, 1624–1633. [CrossRef]
43. Cauda, V.; Schlossbauer, A.; Kecht, J.; Zürner, A.; Bein, T. Multiple Core−Shell Functionalized Colloidal Mesoporous Silica Nanoparticles. *J. Am. Chem. Soc.* **2009**, *131*, 11361–11370. [CrossRef] [PubMed]
44. Möller, K.; Kobler, J.; Bein, T. Colloidal Suspensions of Nanometer-Sized Mesoporous Silica. *Adv. Funct. Mater.* **2007**, *17*, 605–612. [CrossRef]
45. Gu, J.; Huang, K.; Zhu, X.; Li, Y.; Wei, J.; Zhao, W.; Liu, C.; Shi, J. Sub-150nm Mesoporous Silica Nanoparticles with Tunable Pore Sizes and Well-Ordered Mesostructure for Protein Encapsulation. *J. Colloid Interface Sci.* **2013**, *407*, 236–242. [CrossRef] [PubMed]
46. Kobler, J.; Möller, K.; Bein, T. Colloidal Suspensions of Functionalized Mesoporous Silica Nanoparticles. *ACS Nano* **2008**, *2*, 791–799. [CrossRef]
47. Schaffert, D.; Badgujar, N.; Wagner, E. Novel Fmoc-Polyamino Acids for Solid-Phase Synthesis of Defined Polyamidoamines. *Org. Lett.* **2011**, *13*, 1586–1589. [CrossRef]
48. Zhang, W.; Muller, K.; Kessel, E.; Reinhard, S.; He, D.; Klein, P.M.; Hohn, M.; Rodl, W.; Kempter, S.; Wagner, E. Targeted siRNA Delivery Using a Lipo-Oligoaminoamide Nanocore with an Influenza Peptide and Transferrin Shell. *Adv. Healthc. Mater.* **2016**, *5*, 1493–1504. [CrossRef]
49. Genta, I.; Chiesa, E.; Colzani, B.; Modena, T.; Conti, B.; Dorati, R. Ge11 Peptide as an Active Targeting Agent in Antitumor Therapy: A Minireview. *Pharmaceutics* **2017**, *10*, 2. [CrossRef]

50. Biscaglia, F.; Rajendran, S.; Conflitti, P.; Benna, C.; Sommaggio, R.; Litti, L.; Mocellin, S.; Bocchinfuso, G.; Rosato, A.; Palleschi, A.; et al. Enhanced EGFR Targeting Activity of Plasmonic Nanostructures with Engineered Ge11 Peptide. *Adv. Healthc. Mater.* **2017**, *6*, 1700596. [CrossRef]
51. Korpal, M.; Lee, E.S.; Hu, G.; Kang, Y. The miR-200 Family Inhibits Epithelial-Mesenchymal Transition and Cancer Cell Migration by Direct Targeting of E-Cadherin Transcriptional Repressors Zeb1 and Zeb2. *J. Biol. Chem.* **2008**, *283*, 14910–14914. [CrossRef] [PubMed]

Article

Controlled Delivery of Insulin-like Growth Factor-1 from Bioactive Glass-Incorporated Alginate-Poloxamer/Silk Fibroin Hydrogels

Qing Min [1,†], Xiaofeng Yu [2,†], Jiaoyan Liu [2], Yuchen Zhang [1], Ying Wan [2,*] and Jiliang Wu [1,*]

1 School of Pharmacy, Hubei University of Science and Technology, Xianning 437100, China; baimin0628@hbust.edu.cn (Q.M.); zhangych@hbust.edu.cn (Y.Z.)

2 College of Life Science and Technology, Huazhong University of Science and Technology, Wuhan 430074, China; m201771729@hust.edu.cn (X.Y.); liujiaoyan@hust.edu.cn (J.L.)

* Correspondence: ying_wan@hust.edu.cn (Y.W.); jlwu@hbust.edu.cn (J.W.)

† These authors contributed equally to this work.

Received: 22 April 2020; Accepted: 11 June 2020; Published: 20 June 2020

Abstract: Thermosensitive alginate–poloxamer (ALG–POL) copolymer with an optimal POL content was synthesized, and it was used to combine with silk fibroin (SF) for building ALG–POL/SF hydrogels with dual network structure. Mesoporous bioactive glass (BG) nanoparticles (NPs) with a high level of mesoporosity and large pore size were prepared and they were employed as a vehicle for loading insulin-like growth factor-1 (IGF-1). IGF-1-loaded BG NPs were embedded into ALG–POL/SF hydrogels to achieve the controlled delivery of IGF-1. The resulting IGF-1-loaded BG/ALG–POL/SF gels were found to be injectable with their sol-gel transition near physiological temperature and pH. Rheological measurements showed that BG/ALG–POL/SF gels had their elastic modulus higher than 5kPa with large ratio of elastic modulus to viscous modulus, indicative of their mechanically strong features. The dry BG/ALG–POL/SF gels were seen to be highly porous with well-interconnected pore characteristics. The gels loaded with varied amounts of IGF-1 showed abilities to administer IGF-1 release in approximately linear manners for a few weeks while effectively preserving the bioactivity of encapsulated IGF-1. Results suggest that such constructed BG/ALG–POL/SF gels can function as a promising injectable biomaterial for bone tissue engineering applications.

Keywords: alginate–poloxamer copolymer; silk fibroin; dual network hydrogel; mesoporous bioactive glass; insulin-like growth factor-1

---

1. Introduction

The extracellular matrix (ECM) is the noncellular component presenting within all tissues and organs, and it usually has three-dimensional porous architecture. It provides not only essential physical scaffolding for the cellular constituents but also initiates crucial processes that are required for tissue morphogenesis, differentiation, and homeostasis [1]. Each tissue has its own ECM with a unique composition and topology, generated during tissue development [2]. Cells interact with biochemical and biophysical cues in their ECM in highly dynamic and reciprocal manners, and such cell–ECM interactions play a critical role in cell behavior mediation, cell function normality and even cell fate decisions, involving quite complicated processes from quiescence to activation and progenitor state to terminal differentiation [1,2]. Accordingly, it is pivotal to harness the interactions between ECM and the resident cells in developing strategies for effectively regenerating tissue or regulating disease [1,2]. To data, many kinds of hard biomaterials have been employed in tissue engineering for various purposes, but their applications in ECM biomimicry have been limited because these hard biomaterials are lack of ability to adequately mimic the structure and properties of ECM in body tissues [3]. Polymer hydrogels, which behave like soft and elastic objects, are usually constructed by physically or chemically crosslinked macromolecules. They contain large amounts of water while having highly porous architecture with tailorable physiochemical properties and easy diffusivity of small molecules. These features make them attractive candidates for mimicking the dynamic ECM [3,4].

Injectable polymer hydrogels having in situ gelling properties under physiological conditions have received a great deal of attention in tissue engineering owing to their two advantageous characteristics [5]. One is that they can be conveniently used to deliver cells, bioactive compounds, and therapeutic drugs alone or in combination by simply mixing these consignments with polymer solutions prior to gelation; and another is that they can be injected to the defect site via minimally invasive surgery followed by formation into solid-like fillers with discretional shapes [5,6]. Nowadays, varied kinds of natural polymers have been commonly used in the form of hydrogels. Among them, alginate has been extensively investigated for its hydrogel applications. Physically crosslinked alginate hydrogels can be easily built by using certain divalent cations, typically calcium ions (Ca^{2+}), as a crosslinker. Despite the convenience of preparation, so constructed alginate hydrogels often undergo progressive disintegration in vivo due to the ionic exchange between Ca^{2+} in the gels and monovalent cations (such as Na^+ and K^+) coming from the host tissue surrounding the applied gels, which often results in their unstable dimension and uncontrolled properties [7,8]. Another type of physically crosslinked alginate hydrogel was engineered by grafting a type of thermosensitive polymer, poloxamer (Pluronic F127), onto alginate backbone, and the sol-gel transition of alginate–poloxamer (ALG–POL) hydrogels can be trigged by thermosensitive action arisen from the poloxamer component [9]. Despite easy-handled and safe advantages, ALG–POL gels appear to be weak and brittle in nature, and are apt to disintegrate due to its high percentage of Pluronic F127 [10]. Thus, ALG–POL gels are incompetent for certain applications in cartilage or bone tissue engineering where sufficient strength and persistent dimension stability are concomitantly required.

Many studies have revealed that hydrogels with dual or multiple networks could be largely enhanced in their stability and mechanical performance when compared to the single network gel [11]. In spite of the mentioned advantages, multiple network gels are not all suitable for tissue engineering applications as they are usually fabricated via chemical crosslinking, and the involved crosslinking reactions could possibly impair the loaded cells or the host tissue surrounding the applied gel [5,12]. Silk fibroin (SF) is a kind of natural fibrous protein and it can be processed into hydrogels via enzyme-catalyzed crosslinking of amino acid phenolic groups by the aid of H_2O_2, and the obtained SF hydrogels show tunable strength and elasticity [13]. In the case of in vivo usage, the applied amount of H_2O_2 for crosslinking SF gels has to be controlled at a safe level since the resulting SF gels could be cytotoxic if the applied dose of H_2O_2 is higher than certain thresholds [14]. As a result, so prepared SF gels were usually weak [15]. Taking into account the gelable characteristics of ALG–POL and SF through their respectively independent gelling mechanisms, it is feasible to construct a new type of ALG–POL/SF gel with dual network structure while having enhanced performance.

Some growth factors have been proved to be highly effective for promoting bone repair, especially taking advantage of the controlled factor release by way of proper carriers [16]. Among various kinds of growth factors, insulin-like growth factor-1 (IGF-1) is considered to be crucial for longitudinal bone growth, skeletal maturation, and bone mass acquisition not only in the bone growth phase of young individuals but also in the maintenance of bone in adult life [17]. In the situation of bone repair, in addition to promoting cell proliferation and matrix synthesis, the applied IGF-1 at the defect site can also induce the chemotactic migration of osteoblasts to the repair site via local concentration gradients established by factor diffusion [18]. Like many other growth factors, IGF-1 has a short half-life when exposed to the circulatory system [19]. Therefore, when the need arises to maintain sustained release of IGF-1 at the local site in vivo, one of the practical strategies for its administration is to encapsulate IGF-1 into certain vehicles to preserve its activity while managing to modulate its dose and action duration. It is generally realized that directly encapsulating growth factors into hydrogels would result in their burst release because hydrogels commonly have rather porous structures with high water content [7,20]. It has been suggested that release kinetics of growth factors delivered by a hydrogel system could be mediated to varied degrees by encapsulating the growth factor into certain microcarriers such as microspheres (MPs) and nanoparticles (NPs) first, and then, embedding the factor-loaded microcarrier into the hydrogel system [20–22].

Bioactive glasses (BGs) have now been widely used as an attractive inorganic biomaterial for bone repair since they have the ability to strongly bond bone tissue via a hydroxycarbonate apatite interface layer with composition and function similar to naturally occurring bone hydroxyapatite [23]. Mesoporous BG microspheres or nanoparticles can also serve as a reservoir for delivering therapeutic drugs or bioactive molecules besides their functions for acting as bone repair material [24]. In addition to regular BG MPs or NPs, many of them have been doped with different kinds of compounds, and their ionic dissolution products are capable of inducing osteogenesis or angiogenesis at the bone defect site, depending on the variety of doped elements [25]. In this context, it would be rational to load IGF-1 into porous BG NPs, and then, to incorporate the IGF-1-loaded BG NPs into above mentioned dual network ALG–POL/SF gels. On the basis of such designed strategy, a multifunctional composite hydrogel system with mechanically strong features and capabilities for administering the release of IGF-1 could be obtained. In this study, an attempt was made to achieve this goal. Some formulated IGF-loaded BG/ALG–POL/SF gels were found to be injectable and mechanically strong, and to have affirmative abilities to control the release of IGF-1 while preserving its bioactivity.

2. Materials and Methods

2.1. Materials

Sodium alginate (ALG, M_n: 1.3×10^5 Da), Poloxamer 407 (POL, M_n: 12,600 Da), 1-ethyl-3-(3-dimethylaminopropyl)-carbodiimide (EDC), horseradish peroxidase (HRP), and *N*-hydroxyl succinimide (NHS) were procured from Aladdin Inc (Shanghai, China). Recombinant human IGF-1 and IGF-1 enzyme-linked immunosorbent assay (ELISA) Kit were purchased from PeproTech Inc (Cranbury, NJ, USA) and R&D systems (Minneapolis, MN, USA), respectively. Other reagents and chemicals were of analytical grade and purchased from Sinopharm Inc (Shanghai, China).

SF was produced using Bombyx Mori cocoons according to the reported method [15]. Cocoons were degummed in a Na_2CO_3 solution (0.02 M) at 100 °C for 30 min, and the retrieved silk fibers were rinsed with ultrapure water followed by drying in a ventilated hood. The obtained SF fibers were then dissolved in a LiBr solution (9.3 M) at 60 °C with stirring for 5 h, and the prepared solution was dialyzed against distilled water for 2 days using membrane tubes (MW cutoff: 3500) to remove impurities. The achieved dilute SF solution was further concentrated to varied concentrations by immersing the solution-loaded membrane tubes in a 50% PEG20000 solution, and the concentrated SF solutions were stored at 4 °C for further use.

2.2. Synthesis of Alginate-Poloxamer Copolymers

A two-step method was used to synthesize alginate–poloxamer (ALG–POL) copolymers. POL was first modified into monoamine-terminated POL (MATP) following reported methods [26,27], and the obtained MAPT was then grafted onto alginate at a fixed feed mass ratio of alginate to MATP at 1:30 to achieve alginate–poloxamer copolymers. Details for the synthesis of MATP and ALG–POL can be found in the Supplementary Materials.

2.3. Preparation of Bioactive Mesoporous Glasses

Two kinds of porous mesoporous BG NPs (named as MBG-1 and MBG-2, respectively) with different pore-sizes were prepared following reported methods. MBG-1 NPs were prepared as follows [28]. One gram of hexadecyl-trimethylammonium bromide was dissolved in an emulsion consisted of 150 mL of H_2O, 2 mL of aqueous ammonia, 40 mL of ethyl ether, 20 mL of ethanol, and 0.1125 g of calcium nitrate ($Ca(NO_3)_2{\cdot}4H_2O$). 600 μL of tetraethyl orthosilicate (TEOS) was then added to the mixture at a molar Ca/Si ratio of 15:85. After stirring at 30 °C for 4 h, the white sediment was collected by filtration, washed with distilled water, and dried in air at 60 °C, and finally, calcined at 550 °C for 5 h. The same method was used to prepare MBG-2 NPs with slight modification [29]. The above prepared emulsion was vigorously stirred at room temperature for 30 min, and then, 600 μL of TEOS was added with vigorous stirring at 30 °C for 4 h. The resulting precipitate was collected and processed in the same way as that applied to MBG-1 NPs. Parameters for these BG NPs are given in Table 1.

Table 1. Parameters for bioactive glass (BG) nanoparticles.

Sample Name	Surface Area (m^2/g)	Pore Volume (mL/g)	Pore Size (nm)	ζ-Potential (mV)	Particle Size (nm)	PDI
MBG-1	562.4 ± 10.2	0.49 ± 0.02	6.1 ± 0.13	−14.9 ± 0.5	227.2 ± 10.3	0.11
MBG-2	498.3 ± 9.4	0.61 ± 0.03	10.3 ± 0.18	−17.1 ± 0.7	264.7 ± 12.1	0.13

Several IGF-1 solutions in PBS (pH 7.4) with varied concentration of 50 ng/mL (low dose), 100 ng/mL (medium dose), and 150 ng/mL (high dose) were first prepared. In a typical preparation process, 1 mL of any IGF-1 solutions was introduced into a vial with inner protein-resistant coating, and 10 mg of blank MBG-1 or MBG-2 NPs was then added. The mixture was allowed to incubate overnight on an orbital shaker at 37 °C. The IGF-1-loaded BG NPs were collected by centrifugation and washed with PBS followed by freeze-drying. The amount of IGF-1 loaded in BG NPs was measured basing on the difference of IGF-1 concentrations in the loading medium before and after soaking BG NPs by using IGF-1 ELISA Kit. Loading efficiency (LE) of BG NPs was calculated by the following equation:

$$LE(\%) = (M_0/M_1) \times 100\% \quad (1)$$

where M_0 is the mass of IGF-1 encapsulated inside NPs, and M_1 is the feed mass of IGF-1. Parameters for the IGF-1oaded BG NPs are provided in Table 2.

Table 2. Parameters for insulin-like growth factor-1 (IGF-1)-loaded BG nanoparticles.

Sample Name	Feed Amount of IGF-1 (ng/mL)	Particle Size (nm)	PDI	ζ-Potential (mV)	LE (%)
BS1 [(a)]	50	229.1 ± 11.4	0.14	−10.3 ± 0.5	38.6 ± 1.7
BS2	100	230.4 ± 12.6	0.18	−9.1 ± 0.6	46.1 ± 2.1
BS3	150	232.7 ± 10.8	0.19	−8.6 ± 0.3	48.5 ± 2.5
BL1 [(b)]	50	265.3 ± 11.35	0.17	−10.5 ± 0.5	45.9 ± 2.4
BL2	100	268.2 ± 12.46	0.16	−8.1 ± 0.4	57.2 ± 1.9
BL3	150	267.9 ± 13.14	0.18	−7.2 ± 0.6	61.4 ± 2.3

(a) BSi (i = 1, 2 and 3) sample set was prepared using blank MBG-1 NPs. (b) BLj (j = 1, 2 and 3) sample set was prepared using blank MBG-2 NPs.

2.4. Preparation of Hydrogels

IGF-1-free composite solutions were prepared using ALG–POL, SF solution and blank BG NPs and they were used to construct blank BG/ALG–POL/SF gels for their compositional and structural optimization in order to save costly IGF-1. Some composite solutions containing IGF-1 were also prepared by directly adding IGF-1 or incorporating IGF-1-loaded BG NPs into the aqueous ALG–POL/SF mixture. These solutions were further processed into gels by incubating them in a water bath at 37 °C. The major parameters for them are summarized in Tables 3 and 4, respectively.

Table 3. Parameters for BG/alginate–poloxamer (ALG–POL)/silk fibroin (SF) hydrogels without factor loading (a).

Sample Name	SF (wt%)	ALG–POL (wt%)	Blank MBG-2 NPs (wt%) (b)	H_2O_2 (μL) (c)	HRP (μL) (d)	pH	Gelation Time (min) (d)	T_i (°C)
G-A	7.0	5.0	–	10	10	7.06 ± 0.09	11.75 ± 0.96	36.1 ± 0.42
G-B	7.0	5.0	1.0	10	10	7.13 ± 0.06	10.5 ± 1.29	35.2 ± 0.57
G-C	7.0	5.0	2.0	10	10	7.11 ± 0.07	9.75 ± 0.96	34.6 ± 0.49

(a) The full volume of solutions: 2mL. (b) See Table 1 for their parameters. (c) Concentration of H_2O_2: 500 mmol/L. (d) Concentration of HRP: 1000 U/mL. (e) Gelation time was determined by inverting vial every 1 min.

Table 4. Parameters for IGF-loaded BG/ALG–POL/SF hydrogels.

Sample Name	SF (wt%)	ALG–POL (wt%)	IGF-1-Loaded MBG-2 NPs (wt%) (b)	H_2O_2 (μL) (c)	HRP (μL) (d)	IGF-1 Content in Gel (ng/mL)
GEL-1 (a)	7.0	5.0	–	10	10	154.2 ± 11.27
GEL-2	7.0	5.0	–	10	10	512.9 ± 16.35
GEL-3	7.0	5.0	1.0	10	10	160.3 ± 10.84
GEL-4	7.2	5.0	2.0	10	10	520.7 ± 19.52

(a) GEL-1 and GEL-2 were directly loaded with prescribed amounts of IGF-1 for making comparisons. (b) IGF-1 load inside MBG-2 NPs was regulated by changing the IGF-1feed amount. (c) and (d) See Table 3 for their concentrations.

Gelation time was assessed using the inverted tube testing method. Typically, one of the IGF-1-free composite solutions (2.0 mL) was introduced into a glass vial and it was stirred in an ice/water bath for 5 min before being gelled. Fluidity of the composite solution was checked by regularly inverting the vial, and gelation time was recorded starting from the beginning of vial incubation in the water bath and ending at the point when the solution stopped flowing.

2.5. Characterization

Fourier transform infrared (FTIR) spectra of samples were performed on a spectrometer (Vertex 70, Bruker, Ettlingen, Germany). ^{1}H NMR spectra were recorded on a NMR spectrometer (Ascend TM 600 MHz, Bruker, Rheinstetten, Germany). The weight percentage of POL in ALG–POL copolymers was determined using an elemental analyzer (Vario EL III, Elementar, Hanau, Germany). The morphology of BG NPs was viewed with a transmission electron microscope (TEM, Tecnai G2, FEI, Hillsboro, OR, USA). Hydrodynamic size and zeta (ζ) potential of BG NPs were measured using a dynamic light scattering (DLS) instrument (Nano-ZS90, Malvern Instruments, Worcestershire, UK). A mass of BG NPs (100–150 mg) were dried for 12 h at 120 °C and degassed for 24 h at 200 °C under vacuum. The volume of nitrogen adsorbed and desorbed at different relative gas pressures was measured using a surface area and pore size analyzer (ASAP 2020 Plus, Micromeritics, Norcross, GA, USA), and it was then utilized to construct adsorption–desorption isotherms. The specific surface area was determined with the Brunauer–Emmett–Teller (BET) method. The pore volume and the mean pore size were derived from the adsorption branches of the isotherms using the Barrett–Joyner–Halanda (BJH) method. The porous cross-section structure of dry gels was examined by scanning electron microscopy (SEM, Quanta 200, FEI, Eindhoven, Netherlands). Dry gel samples with known weight (Wd, g) were immersed in PBS at 37 °C till they attained equilibrium, and their wet weight (Ws, g) was measured after removal of excess surface water with filter paper. Swelling index (SI) of gels was calculated using the following formula:

$$SI\ (g/g) = (W_s - W_d)/W_d \tag{2}$$

2.6. Rheological Analysis

Rheological measurements of fluids or hydrogels were carried out using a rheometer (Kinexus Pro KNX2100, Southborough, MA, USA) equipped with a parallel-plate sample holder. The temperature sweep curve was recorded in the range of 25 to 45 °C by heating liquid samples at a rate of 1 °C/min, and the incipient gelling temperature (T_i) of liquid samples was determined at the intersection point of storage modulus (G′) and loss modulus (G″). Isothermal frequency sweep experiments were conducted using hydrogel discs (10 mm in diameter and ca. 4 mm in height) prepared with a punch, and the obtained data were plotted as a function of modulus versus frequency at predefined strain amplitude of 1%. Shear viscosity of liquid samples was measured at 23 °C in a shear rate range between 0.1/s and 400/s.

2.7. Release of IGF-1

In vitro IGF-1 release profiles for IGF-1-loaded MBG-1 and MBG-2 NPs were first tested to find out their difference in release rates. Measurements were conducted using several sets of eppendorf tubes, each filled with 500 µL PBS containing 1% (*w/v*) bovine serum albumin. To each eppendorf tube, 5 mg of IGF-I-loaded MBG-1 or MBG-2 NPs was added, and all sets of tubes were incubated in a shaking water bath at 37 °C and 60 rpm for various periods up to 6 days. At predetermined time intervals, eppendorf tubes were withdrawn by group and they were centrifuged for 5 min at around 1000 g to collect supernatants. The release amount of IGF-I was determined using IGF-1 ELISA kit.

In the case of IGF-1 release from gel samples, some cylindrical gel samples were first produced. Each (0.5 mL) of IGF-1-loaded composite solutions (see Table 4) was filled into a cylindrical mold (diameter: 10 mm) and incubated at 37 °C for 20 min for gel formation. The gel samples were then introduced into different vials filled with 3 mL of PBS, and the vials were vortexed on a shaking table at 37 °C and 60 rpm. At prescribed time points, 1 mL of medium was withdrawn with replenishing the same volume of fresh buffer. The released amount of IGF-1 was measured using IGF-1 ELISA Kit.

2.8. Bioactivity of Released IGF-1

The IGF-1's ability to promote the proliferation of osteoblasts was tested to access the bioactivity of released IGF-1 [30–32]. MC3T3-E1 cells (Type Culture Collection of the Chinese Academy of Sciences, Shanghai, China) were used as testing cells. Cells were expanded in DMEM supplemented with 10% fetal bovine serum, 1% penicillin/streptomycin in a 5% CO_2 humidified atmosphere at 37 °C. The expanded cells were resuspended in PBS for further use.

Cells were cultured in 96-well plates at a density of 5×10^4 cells/well in complete culture medium at 37 °C for 24 h. The cells were serum-starved for 24 h, after which the media was replaced with either serum-free media (denoted as control, 0 ng/mL), serum-free media with free IGF-1 (5 or 50 ng/mL), or serum-free media with released IGF-1 (5 or 50 ng/mL). The cells were cultured for varied durations up to 72 h, and their proliferation was determined by 3-(4,5-dimethyl-2-thiazolyl)-2,5-diphenyl-2-H-tetrazolium bromide (MTT, Dojindo, Japan) essay. Briefly, the media was aspirated from the 96-well plates after the prescribed culture period, and a 20 µL aliquot of MTT stock solution and 200 µL of serum-free medium was added to each well. After being further cultured for 4 h, the media was removed from the wells, and 150 µL of DMSO was added to each well with shaking at 37 °C for 15 min. After that, the optical density (OD) was determined at 590 nm using a microplate reader (PerkinElmer Inc, USA).

Cells were also cultured on the surface of gels for making further comparison. Briefly, two kinds of IGF-1-contained composite solutions with their compositions respectively matching with GEL-2 and GEL-4 gels (see Table 4), 200 µL apiece, were pipetted into wells of 24-well plates, and cultured at 37 °C for gel formation. Subsequently, volume of MC3T3-E1 cell suspension (100 µL) was added to each well (5×10^4 cells/well), and cells were then cultured with serum-free media (500 µL) at 37 °C for varied periods up to 72 h. Cell proliferation was assessed by OD measurement using above described MTT assay. Cell cultured under monolayer condition in serum-free media (0 ng/mL) without exposing to gels were used as control.

2.9. Statistical Analysis

Data were presented as mean ± standard deviation. Analysis of the difference between groups was performed using one-way ANOVA. The statistical difference was considered to be significant when the p-value was less than 0.05.

3. Results and Discussions

3.1. Analysis of ALG–POL

ALG–POL copolymer was synthesized by grafting MAPT onto the ALG backbone through EDC/NHS chemistry. Figure 1 presents a schematic illustration to show the synthesis route for MAPT and ALG–POL copolymer. A representative NMR spectrum for MAPT is provided in Figure S1 in which the typical peaks respectively assigning to methyl and methylene confirm the successful synthesis of MAPT. The preliminary experimental results indicated that the composition of ALG–POL exerted significant effects on the thermoresponsivity, degradation tolerance, and strength of resulting ALG–POL/SF gels. Accordingly, the proportion of ALG and POL components in ALG–POL copolymer was optimized via orthogonal design method. By changing the ratio of MATP to ALG in a range between 20 and 35 at a step size of 5 during the synthesis of ALG–POL copolymers, one of ALG–POL copolymers was thus screened out by setting the MATP/ALG ratio at 30. The POL content in such optimized ALG–POL copolymer was thus measured to be around 66 wt% via elemental analysis. Figure S2 presents FTIR spectra for POL, ALG and ALG–POL. The spectrum of POL is characterized by three typical bands at 2891 (C–H stretch aliphatic), 1345 (in-plane O-H bend) and 1112 cm^{-1} (C–O stretching). The ALG spectrum shows specific absorbance bands of its COOH stretching at 1610 cm^{-1} and C–O–C stretching at 1305 cm^{-1} [33], respectively. In the spectrum for ALG–POL, the carbonyl absorption band for carboxylate sodium salt originally showing in the ALG spectrum disappeared while a new characteristic amide I band appeared at around 1637 cm^{-1}, suggesting that amide bonds have formed between ALG and POL [9,33]. FTIR spectra demonstrate that the ALG–POL copolymer has been successfully synthesized.

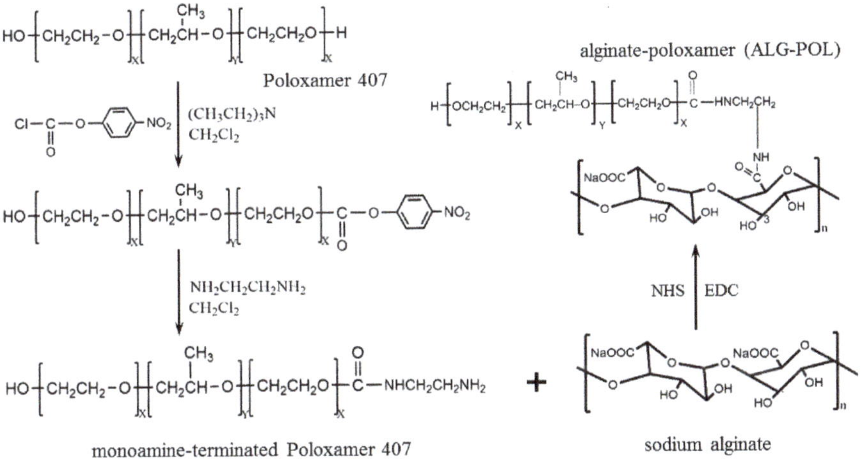

Figure 1. Schematic illustration for synthesis of alginate-poloxamer copolymer.

3.2. Parameters for Mesoporous BG Nanoparticles

Two kinds of mesoporous BG NPs were produced under slightly different processing conditions in order to attain certain BG NPs having proper pore-sizes and pore volumes for gaining high LE. Panels A and B in Figure 2 display two typical TEM micrographs for the prepared BG NPs, and these spherical BG NPs were seen to be porous with their size of about 200 nm. A mass of BG NPs were measured for their average particle size and ζ-potential, and relevant data are listed in Table 1. There were significant differences ($p < 0.05$) in the average size and ζ-potential of these BG NPs, signifying that the processing conditions significantly modulated their structure and property even though they were formulated with the same chemical composition [28,29]. Figure 2C displays the recorded N_2 adsorption–desorption isotherms for MBG-1 and MBG-2 NPs. Two isotherms exhibited their respective hysteresis loops of the desorption branch, indicative of the existence of large pores inside BG NPs [28,29]. In comparison to MBG-1 NPs, the hysteresis loop for MBG-2 NPs was shown to be steeper and its inception turning point was closer to the high pressure end of the N_2 isotherm, suggesting that MBG-2 NPs have more pores with larger size than MBG-1 NPs. Besides these differences, MBG-2 NPs also quite differed from MBG-1 NPs in pore volume and pore size distribution (Figure 2D). Two sets of BG NPs were measured for their major parameters and the obtained data are summarized in Table 1. It can be seen that the presently developed BG NPs had a high level of pore volume, large average pore size and negative ζ-potential, and thus, they could act as a practical vehicle for the IGF-1 delivery since IGF-1 is somewhat positively charged (isoelectric point, 8.6) at physiological pH and has its molecular weight of about 7.6 kD [34,35].

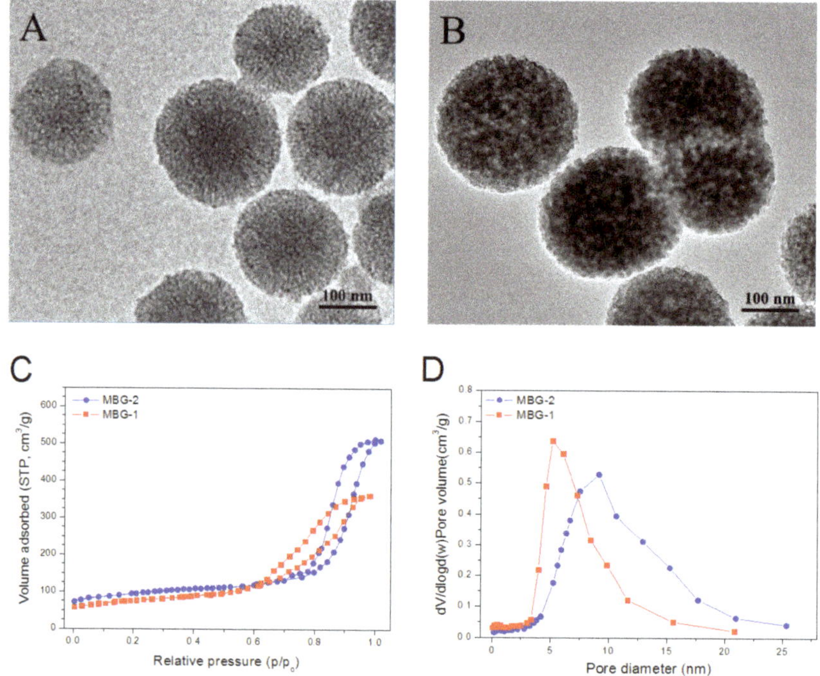

Figure 2. Images of MBG-1 (**A**) and MBG-2 (**B**) nanoparticles (NPs); N_2 adsorption isotherms (**C**) and pore size distribution (**D**) of NPs.

3.3. IGF-1 Release from BG Nanoparticles

Blank MBG-1 and MBG-2 NPs were loaded with IGF-1 under the condition of varying IGF-1 feed amounts, and two sets of IGF-1-loaded BG NPs were thus produced (Table 2). BS*i* (i = 1, 2 and 3) sample set was prepared by loading IGF-1 into MBG-1 NPs having smaller average pore size than that for MBG-2 NPs (see Table 1) whereas BL*j* (j = 1, 2 and 3) sample set was prepared by loading IGF-1 into MBG-2 NPs. Data in Table 2 reveal that these IGF-1-loaded NPs had similar average particle size ($p > 0.05$) but significantly higher ζ-potential ($p < 0.001$) as compared to their respective blank counterparts (see Table 1). The nearly unchanged average size for IGF-1-loaded NPs shown in Table 2 can be attributed to the very small IGF-1 mass when compared to NPs themselves, whereas the significantly increasing ζ-potential should be ascribed to the slightly positively charged nature of IGF-1 [34]. Table 2 indicates that the IGF-1-loaded NPs in BLj (j = 1, 2 and 3) set had significantly higher ($p < 0.05$) LE as compared to the counterpart in BSi (i = 1, 2 and 3) set. These differences are rational because the blank MBG-2 NPs used for preparing BLj (j = 1, 2 and 3) set have notably higher pore volume and larger pore size when compared to blank MBG-1 NPs used in BSi (i = 1, 2 and 3) set (see Table 1). Table 2 also shows that the IGF-1 feed amount exerted certain effects on LE, and this kind of effect would become insignificant once the IGF-1 feed amount reached 100 ng/mL or higher. It is worth mentioning that IGF-1 feed amounts were designated as such in order to test the LE for NPs. Actually, the IGF-1 load in these NPs can be effectively regulated by changing the feed amount of IGF-1.

The IGF-1 release patterns from IGF-1-loaded BG NPs are shown in Figure 3A,B. BS1, BS2, and BS3 NPs released around 50% of their initial IGF-1 load on the first day, and after 4-day release, the cumulative IGF-1 release amounted to about 70%. The IGF-1 load in these NPs did not impose any significant impacts on their release profiles. With respect to the cases associated with BL1, BL2, and BL3 NPs, their release profiles looked quite similar to that assigned for BS1, BS2, and BS3 NPs, respectively, with somewhat faster release rates (Figure 3B). Figure 3 verifies that these mesoporous BG NPs themselves are not able to effectively control the release kinetics of IGF-1 on account of their initial burst release features. LE is a key issue that is correlated to the rational use of IGF-1 because of high cost of IGF-1. In consideration of the similar release profiles illustrated in Figure 3 for both MBG-1 and MBG-2 NPs but significantly higher LE detected from MBG-2 NPs (see Table 2), MBG-2 NPs were thus chosen for the follow-up gel preparation.

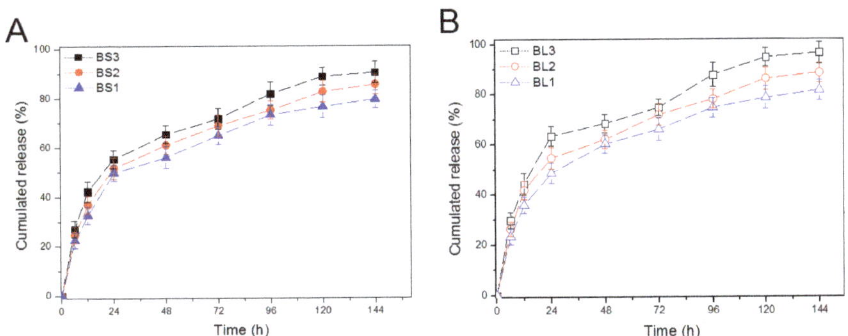

Figure 3. IGF-1 release profiles from 5 mg of BS1, BS2, and BS3 NPs ((**A**), initial IGF-1 load in BS1, BS2 and BS3 was 1.93, 4.61, and 7.27 ng/mg) and BL1, BL2, and BL3 NPs ((**B**), initial IGF-1 load in BL1, BL2, and BL3 was 2.29, 5.72, and 9.21 ng/mg) in 500 μL PBS (see Table 2 for their parameters).

3.4. *Rheological Properties of BG/ALG–POL/SF Gels without IGF-1 Load*

ALG–POL is a thermoresponsive copolymer and the thermal transition of ALG–POL solutions had strong concentration dependence. A previous study reported that ALG–POL was gelable when its solution concentration reached about 15 wt% or higher [9]. In the present study, the optimally synthesized ALG–POL copolymer was found to have clear sol-gel transition during a rational gelation period when its solution concentration reached 12 wt% or higher. Several optical images are presented in Figure S3 for showing changes of ALG–POL fluids after incubation. It can be noticed that the ALG–POL solution with its concentration of 9 wt% was remained as a fluid even though it was incubated at 37 °C for 60 min, and on the other hand, a 12 wt% ALG–POL solution was able to turn into gel via incubation at the same temperature during 14 min. Results in Figure S3 demonstrate that ALG–POL alone could be thermally gelable at 37 °C when its solution concentration is higher than a certain threshold.

In view of independent gelable mechanisms respectively belonging to ALG–POL and SF components, dual network gels with mechanically strong nature could be constructed by using ALG–POL and SF together. To achieve a ALG–POL/SF gel with required properties, a series of ALG–POL/SF composite solutions having their weight proportions of 4/8, 5/7, 6/6, and 7/5 was formulated for the preparation of blank ALG–POL/SF gels, and the optimal gel was sought out as 5 wt% for ALG–POL and 7 wt% for SF. Based on such designed composition for the ALG–POL/SF gel, blank MBG-2 NPs were embedded into the gel to fabricate three kinds of BG/ALG–POL/SF gels without IGF-1 load and the resulting gels were utilized to evaluate the rheological properties in order to save IGF-1. Major parameters for these blank BG/ALG–POL/SF gels are given in Table 3.

As seen from Table 3, G-A, G-B, and G-C gels had the same matrix and the difference in their composition was the percentage of blank MBG-2 NPs. Panels A, B, and C in Figure 4 elucidate the representative temperature sweep curves of G′ and G″ for different gels, and these gels were seen to respond to the thermal stimulus at different inception temperatures (T_i). G-A gel had its T_i at around 36 °C, whereas G-B and G-C gels showed their T_i near 35 °C, connoting that the introduction of BG NPs into the ALG–POL/SF gel has a very limited effect on their gelation temperature. It can be observed from Table 3 that the pH value and T_i of these gels were quite close to the physiological pH and temperature, and meanwhile, their gelation time was seen to be rational [5,6], suggesting their applicability under physiological conditions. Figure 4D presents the shear dependence of viscosity for different gels. Their viscosity was shown to be lower than 70 pa.s at 23 °C, and progressively decreased with rising shear rate, indicating their shear-thinning features. Given that the gel injection is usually performed at room temperature, curves graphed in Figure 4D validate that the presently formulated BG/ALG–POL/SF gels have well-defined injectability.

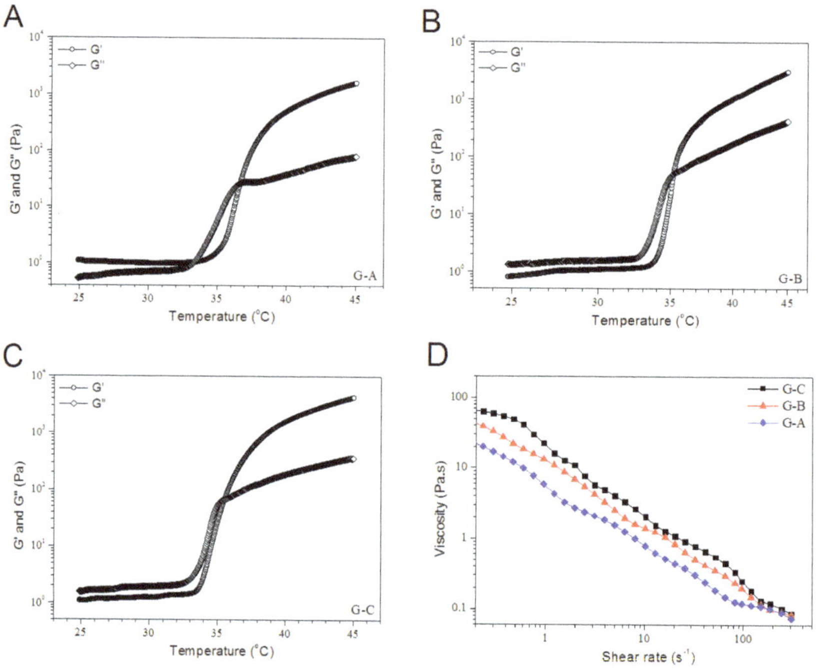

Figure 4. Temperature sweep curves (**A–C**) and shear viscosity ((**D**), 23 °C) for BG/ALG–POL/SF gels.

In principle, magnitude of G′ and G″ of hydrogels in the linear viscoelastic region (LVR) of their frequency sweep spectra together with G′/G″ ratio can be used to assess the gel strength [36]. In general, a strong hydrogel is characterized by high G′, and meanwhile, its G′ should be 1–2 orders of magnitude greater than its G″ [36]. Figure 5A,B show that at a fixed frequency in their respective LVR, for example, 1.0 HZ, three kinds of gels had their G′ of around 5 kPa or higher and their G″ greater than 300 Pa. The incorporation of BG NPs into the ALG–POL/SF gel seemed not to exert marked effects on G′ and G″ of the resulting gels. To make quantitative comparisons, G′ and G″ at 1.0 Hz for these gels were measured, and obtained average values are graphed in Figure 5C. The bar-graphs explicate that these gels had their G′ higher 5.5 kPa with the G′/G″ ratio greater than 15, verifying their mechanically strong features.

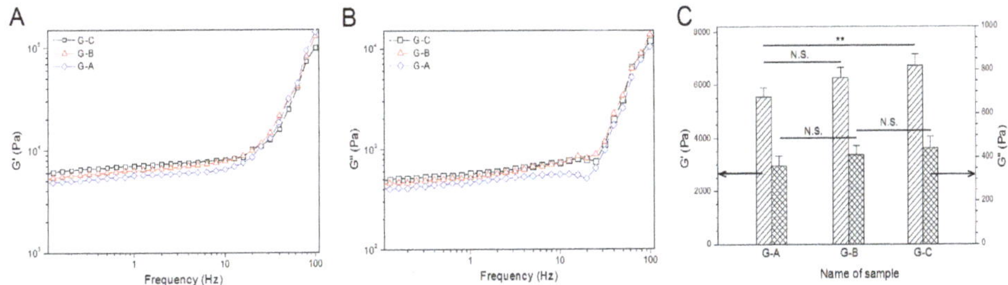

Figure 5. Frequency dependent functions (**A**,**B**) of modulus and average values (**C**) of modulus at 1.0 Hz and 37 °C for BG/ALG–POL/SF gels (**, $p < 0.001$; N.S., no significance).

3.5. Morphological Analysis of Dry Gels

The presently developed BG/ALG–POL/SF gels need to be porous because they are intended for use in bone repair where they will function as injectable materials for housing cells. G-A and G-C gels in Table 3 were selected and their lyophilized samples were examined to see their internal structures. A few SEM images for the dry gels are represented in Figure 6. These images show that dry gels were highly porous and their pore size changed from several tens of microns to more than two hundred microns with pore-interconnected characteristics. With respect to G-C gel, the incorporation of BG NPs did not significantly affect its pore structure when compared with G-A gel. The image with a larger magnification (Figure 6D) displays that the wall of pores in the gel was stuck or attached with many size-varied granules and these granules should be assigned to BG NPs or their aggregates. The average pore size for these dry gels is shown in Figure 6E. The dry gels had large average pore size without significant difference, which is advantageous for bone repair where large pore size and high porosity in the requisite gels are concurrently required.

SI of dry gels has been used as an approximate estimation for their porosity since the channels shaped inside the gels can regulate their swelling and deswelling behavior via water convection [37]. In general, dry gels with open-cell pores and high porosity have high SI and short swelling equilibrium time due to fast water convection [38]. The bar-graph in Figure 6F illustrates that these dry gels had their SI higher than 5, and meanwhile, the composition of the gels seemed not to exert significant impacts on their SI. In addition, it was found that these dry gels reached their respective swelling equilibrium in PBS less than 30 min. The similar SI together with their rapidly swollen features connotes that these gels have similar porosity.

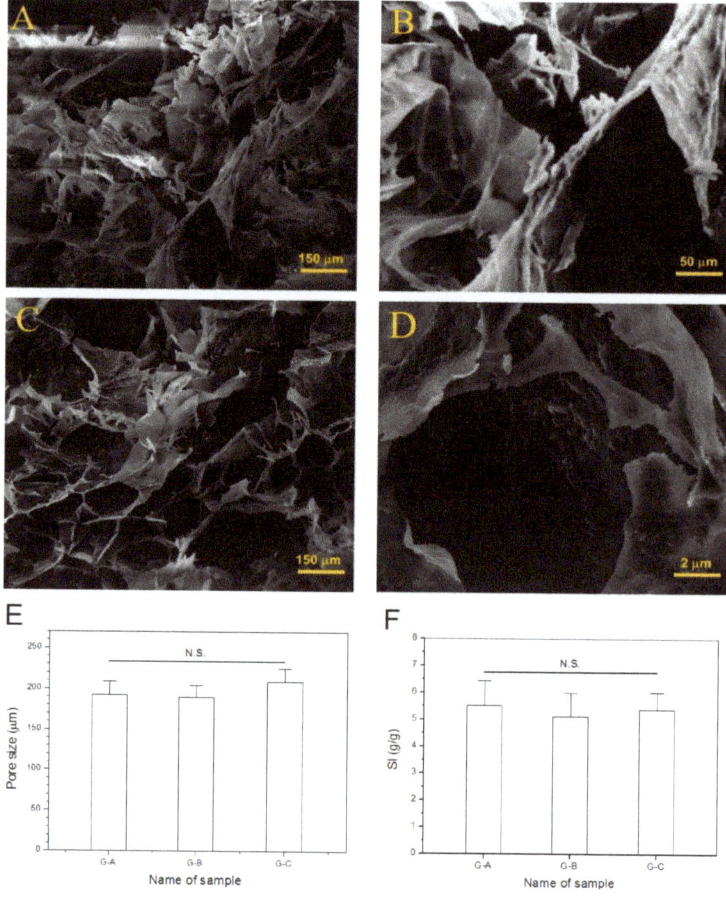

Figure 6. SEM micrographs ((**A**,**B**), G-A gel; (**C**,**D**), G-C gel), average pore-size (**E**) and swelling index (**F**) of BG/ALG–POL/SF dry gels (see Table 3 for parameters of gels; N.S., no significance).

3.6. IGF-1 Release of Gels

The gels were loaded with varied amounts IGF-1 in a designated way as illustrated in Table 4 and they were detected to access their capacity for administration of IGF-1 release. Release profiles for these gels are presented in Figure 7. Curves in Figure 7A exhibit that two kinds of gels directly loading with IGF-1 had fast IGF-1 release in the first few days at varied rates somewhat depending on their initial IGF-1 load, and their IGF-1 release became visibly slower after one-week release with similar release rate in the light of approximately constant distance between the two curves. In marked contrast to this observation, two gels embedded with IGF-1-loaded BG NPs behaved in quite different ways (Figure 7B). IGF-1 load of around 7% or less was released from the gels in the first day, and afterwards, the release patterns followed approximately linear behavior for a few weeks at various release rates. In comparison to the patterns shown in Figure 7A, the significantly reduced initial IGF-1 release and the subsequent release slowdown in Figure 7B can be attributed to the joint contribution of the gel matrix and BG NPs. As denoted in Table 4, GEL-3 and GEL-4 gels were prepared by embedding IGF-1-loaded BG NPs into ALG–POL/SF. In comparison to GEL-1 and GEL-2 gels, IGF-1 molecules in GEL-3 and GEL-4 gels will encounter increasing resistance derived from both BG NPs and gel matrix, which will certainly result in their release slowdown. When the release patterns in Figure 7B are compared each other, it can be observed that IGF-1 content in the gels remarkably affected the release rate. A possible reason could be ascribed to that the higher IGF-1 loading in a gel would form a larger IGF-1 concentration gradient inside the gel, which would force IGF-1 molecules to transport through the gel matrix faster and to reach the media earlier, leading to higher cumulative IGF-1 amount.

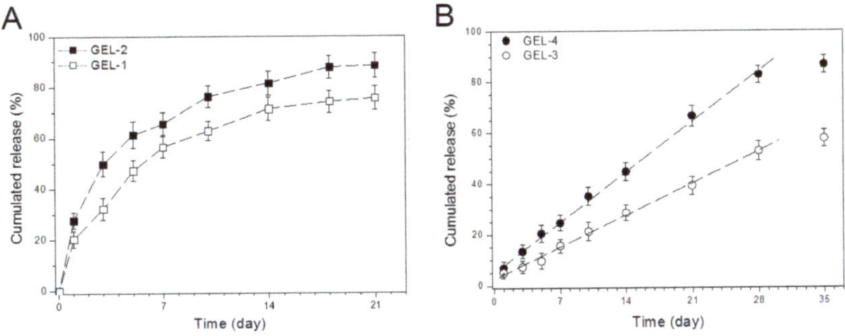

Figure 7. IGF-1 release patterns for gels directly loaded with IGF-1 (**A**) and for IGF-loaded BG/ALG–POL/SF gels (**B**).

3.7. Bioactivity Assessment of IGF-1

Bioactivity preservation of released IGF-1 is an important issue because it is closely correlated to the biological effects of IGF-1. In this study, MC3T3-E1 cells were used for assessment of IGF-1 bioactivity because IGF-1 is capable of promoting the proliferation of osteoid cells in dose-dependent manners [30–32]. To evaluate these gels on the same baseline, GEL-2 and GEL-4 gels were selected taking account of their similar and higher initial IGF-1 load. MC3T3-E1 cells were cultured with equivalent amount of released IGF-1 or free IGF-1 for varied durations up to 72 h and relevant results are elucidated in Figure 8. At a low level of IGF-1 (5 ng/mL), OD values matching with IGF-1-applied cell groups were remarkably higher than that of control group but no significant differences were detected among cell groups that were exposed to free IGF-1 or released IGF-1 as sampling time advanced (Figure 8A). By increasing the applied IGF-1 amount by 10 times (Figure 8B), the variation trend of OD values and their differences looked similar to that detected at the IGF-1 dosage of 5 ng/mL. In addition, by comparing each OD value in Figure 8B with the corresponding one in Figure 8A, IGF-1-dose dependent characteristics can be detected when the culture time reached 72h. These results support that the released IGF-1 was able to promote the proliferation of MC3T3-E1 cells in the way of dose-regulation, confirming that bioactivity of released IGF-1 can be well preserved.

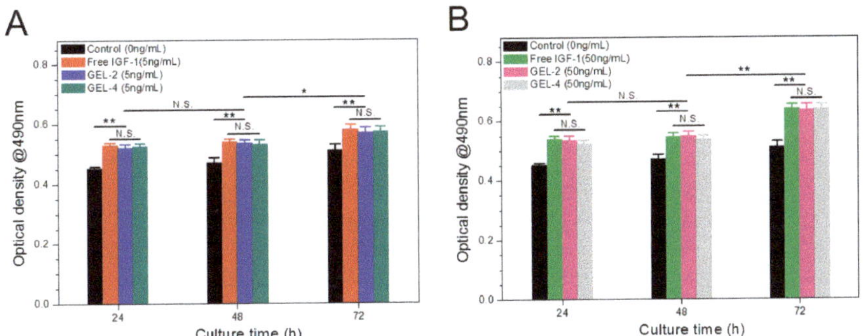

Figure 8. Response of MC3T3-E1 cells to 5 ng/mL (**A**) or 50 ng/mL (**B**) of free or released IGF-1 during various culture periods (*, $p < 0.05$; **, $p < 0.001$; N.S., no significance).

Besides these tests, MC3T3-E1 cells were also cultured on the surface of GEL-2 and GEL-4 gels for three days to further evaluate their growth, and the measured OD values are depicted in Figure 9. In these cases, the cumulated IGF-1 amount in the culture media would be dynamically altered because GEL-2 and GEL-4 gels would incessantly release IGF-1 at different rates despite their similar initial IGF-1 load. As shown in Figure 7, the cumulative amount of IGF-1 released from GEL-2 gel on the first, second and third days reached around 27, 39, and 50%, respectively; and the corresponding cumulative IGF-1 release from GEL-4 gel was about 7, 10, and 13%. Considering the patterns shown in Figure 7, it can be envisioned that in the current situation, the amount of available IGF-1 in GEL-2 group was significantly higher than that in GEL-4 group in the first three days. It can be seen that OD values measured from two gel groups were markedly higher than that detected from control group during the three sampling days, demonstrating that the released IGF-1 is bioactive and able to promote the growth of MC3T3-E1 cells (Figure 9). When GEL-2 and GEL-4 groups were compared each other, it shows that there was no significant difference in their OD value on the first day, but on the second and third days, OD value measured from GEL-2 group was notably larger than that detected from GEL-4 group. The possible reasons for these observations could be attributed to that (1) cells need a certain period of time to attach to the gels and to undergo recovery growth with low responsiveness to the released IGF-1 on the first day, resulting in insignificant difference in their OD value; and (2) after fully attaching and returning to their normal growth, cells seeded on GEL-2 gel would grow faster than those on GEL-4 gel because GEL-2 gel can release a notable higher IGF-1 amount than GEL-4 gel. These results further confirm that presently devised gels have ability to promote the proliferation of MC3T3-E1 cells.

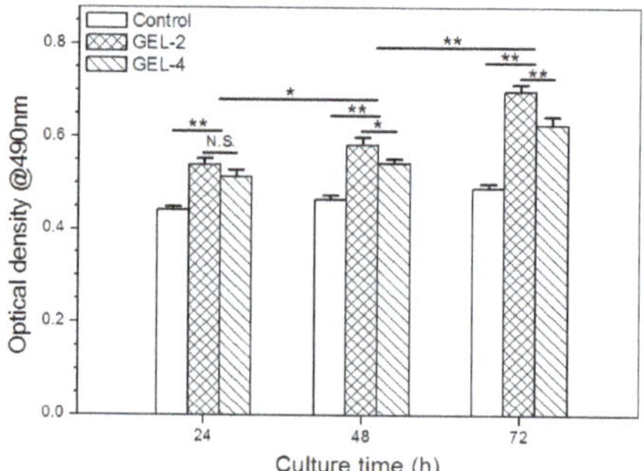

Figure 9. Optical density (OD) values of MC3T3-E1 cells cultured on the surface of gels during various periods (*, $p < 0.05$; **, $p < 0.001$; N.S., no significance).

4. Conclusions

Thermosensitive ALG–POL copolymer containing a necessitated percentage of POL was successfully synthesized. Such synthesized ALG–POL was found to be suitable for constructing hydrogels with dual network structure through combining with SF. The optimally obtained ALG–POL/SF gels were injectable at room temperature and mechanically strong with their sol-gel transition near physiological pH and temperature. Embedment of mesoporous BG nanoparticles into ALG–POL/SF gel did not significantly modify the gelation temperature, gelation time and pH of the resulting gels. The interior of the dry gels was seen to be highly porous with well-interconnected pore architecture. Direct incorporation of IGF-1 into ALG–POL/SF gels was inadvisable for administrating IGF-1 release. By embedding IGF-1-loaded BG nanoparticles into ALG–POL/SF gels, the resulting IGF-1-loaded BG/ALG–POL/SF gels showed a confirmative ability to administrate IGF-1 release in an approximately linear manner for a few weeks and their IGF-1 release rate could be effectively regulated by the IGF-1 load in BG nanoparticles. Cell tests confirmed that the released IGF-1 was bioactive as compared with the free IGF-1.

Supplementary Materials: The following are available online at http://www.mdpi.com/1999-4923/12/6/574/s1, Figure S1. ^{1}H NMR spectra for monoamine-terminated poloxamer (MATP). Figure S2. FTIR spectra for poloxamer (A), alginate (B) and alginate-poloxamer (C). Figure S3. ALG-POL fluids and formed ALG-POL gel after incubation. (Concentration of ALG-POL: 9 wt% (A); and 12 wt% (B)).

Author Contributions: Q.M., Y.W., and J.W. conceived and designed the experiments; X.Y., J.L., and Y.Z. performed the experiments; Q.M., Y.W., and J.W. wrote the paper. All authors have read and agreed to the published version of the manuscript.

Funding: This work was funded by the National Key R&D Program of China (Grant No. 2017YFC1103800) and the National Natural Science Foundation of China (Grant No. 81972065).

Conflicts of Interest: The authors declare no conflict of interest.

References

1. Jarvelainen, H.; Sainio, A.; Koulu, M.; Wight, T.N.; Penttinen, R. Extracellular matrix molecules: Potential targets in pharmacotherapy. *Pharmacol. Rev.* **2009**, *61*, 198–223. [CrossRef] [PubMed]
2. Frantz, C.; Stewart, K.M.; Weaver, V.M. The extracellular matrix at a glance. *J. Cell Sci.* **2010**, *123*, 4195–4200. [CrossRef] [PubMed]
3. Tibbitt, M.W.; Anseth, K.S. Hydrogels as extracellular matrix mimics for 3D cell. *Biotechnol. Bioeng.* **2009**, *103*, 655–663. [CrossRef] [PubMed]
4. Wang, C.; Varshney, R.R.; Wang, D.A. Therapeutic cell delivery and fate control in hydrogels and hydrogel hybrids. *Adv. Drug Delivery Rev.* **2010**, *62*, 699–710. [CrossRef]
5. Yang, J.A.; Yeom, J.; Hwang, B.W.; Hoffman, A.S.; Hahn, S.K. In situ-forming injectable hydrogels for regenerative medicine. *Prog. Polym. Sci.* **2014**, *39*, 1973–1986. [CrossRef]
6. Kretlow, J.D.; Klouda, L.; Mikos, A.G. Injectable matrices and scaffolds for drug delivery in tissue engineering. *Adv. Drug Deliv. Rev.* **2007**, *59*, 263–273. [CrossRef]
7. Tan, H.; Marra, K.G. Injectable, biodegradable hydrogels for tissue engineering applications. *Materials* **2010**, *3*, 1746–1767. [CrossRef]
8. Hermansson, E.; Schuster, E.; Lindgren, L.; Altskar, A.; Strom, A. Impact of solvent quality on the network strength and structure of alginate gels. *Carbohydr. Polym.* **2016**, *144*, 289–296. [CrossRef]
9. Chen, C.C.; Fang, C.L.; Al-Suwayeh, S.A.; Leu, Y.L.; Fang, J.Y. Transdermal delivery of selegiline from alginate-pluronic composite thermogels. *Int. J. Pharm.* **2011**, *415*, 119–128. [CrossRef]
10. Lin, H.R.; Sung, K.C.; Vong, W.J. In situ gelling of alginate/pluronic solutions for ophthalmic delivery of pilocarpine. *Biomacromolecules* **2004**, *5*, 2358–2365. [CrossRef]
11. Naseri, N.; Deepa, B.; Mathew, A.P.; Oksman, K.; Girandon, L. Nanocellulose-based interpenetrating polymer network (IPN) hydrogels for cartilage applications. *Biomacromolecules* **2016**, *17*, 3714–3723. [CrossRef] [PubMed]
12. Nie, J.; Pei, B.; Wang, Z.; Hu, Q. Construction of ordered structure in polysaccharide hydrogel: A review. *Carbohydr. Polym.* **2019**, *205*, 225–235. [CrossRef]
13. Partlow, B.P.; Hanna, C.W.; Rnjak-Kovacina, J.; Moreau, J.E.; Applegate, M.B.; Burke, K.A.; Marelli, B.; Mitropoulos, A.N.; Omenetto, F.G.; Kaplan, D.L.; et al. Highly tunable elastomeric silk biomaterials. *Adv. Funct. Mater.* **2014**, *24*, 4615–4624. [CrossRef] [PubMed]
14. Hopkins, A.M.; Laporte, L.D.; Tortelli, F.; Spedden, E.; Staii, C.; Atherton, T.J.; Hubbell, J.A.; Kaplan, D.L. Silk hydrogels as soft substrates for neural tissue engineering. *Adv. Funct. Mater.* **2013**, *23*, 5140–5149. [CrossRef]
15. Liu, J.; Yang, B.; Li, M.; Li, J.; Wan, Y. Enhanced dual network hydrogels consisting of thiolated chitosan and silk fibroin for cartilage tissue engineering. *Carbohydr. Polym.* **2020**, *227*, 115335. [CrossRef] [PubMed]
16. Santo, V.E.; Gomes, M.E.; Mano, J.F.; Reis, R.L. Controlled release strategies for bone, cartilage, and osteochondral engineering—Part II: Challenges on the evolution from single to multiple bioactive factor delivery. *Tissue Eng. Part B* **2013**, *19*, 327–352. [CrossRef] [PubMed]
17. Giustina, A.; Mazziotti, G.; Canalis, E. Growth hormone, insulin-like growth factors, and the skeleton. *Endocrine Rev.* **2008**, *29*, 535–559. [CrossRef]
18. Mantripragada, V.P.; Jayasuriya, A.C. IGF-1 release kinetics from chitosan microparticles fabricated using environmentally benign conditions. *Mater. Sci. Eng. C* **2014**, *42*, 506–516. [CrossRef]

19. Massicotte, F.; Fernandes, J.C.; Martel-Pelletier, J.; Pelletier, J.P.; Lajeunesse, D. Modulation of insulin-like growth factor 1 levels in human osteoarthritic subchondral bone osteoblasts. *Bone* **2006**, *38*, 333–341. [CrossRef]
20. Vo, T.N.; Kasper, F.K.; Mikos, A.G. Strategies for controlled delivery of growth factors and cells for bone regeneration. *Adv. Drug Deliv. Rev.* **2012**, *64*, 1292–1309. [CrossRef]
21. Wang, Z.; Wang, Z.; Lu, W.W.; Zhen, W.; Yang, D.; Peng, S. Novel biomaterial strategies for controlled growth factor delivery for biomedical applications. *NPG Asia Mater.* **2017**, *9*, e435. [CrossRef]
22. Min, Q.; Yu, X.; Liu, J.; Wu, J.; Wan, Y. Chitosan-based hydrogels embedded with hyaluronic acid complex nanoparticles for controlled delivery of bone morphogenetic protein-2. *Mar. Drugs* **2019**, *17*, 365. [CrossRef] [PubMed]
23. Hench, L.L. Bioceramics. *J. Am. Ceram. Soc.* **1998**, *81*, 1705–1728. [CrossRef]
24. Wu, C.; Chang, J. Multifunctional mesoporous bioactive glasses for effective delivery of therapeutic ions and drug/growth factors. *J. Control. Release* **2014**, *193*, 282–295. [CrossRef] [PubMed]
25. Boffito, M.; Pontremoli, C.; Fiorilli, S.; Laurano, R.; Ciardelli, G.; Vitale-Brovarone, C. Injectable thermosensitive formulation based on polyurethane hydrogel/mesoporous glasses for sustained co-delivery of functional ions and drugs. *Pharmaceutics* **2019**, *11*, 501. [CrossRef] [PubMed]
26. Cho, K.Y.; Chung, T.W.; Kim, B.C.; Kim, M.K.; Lee, J.H.; Wee, W.R.; Cho, C.S. Release of ciprofloxacin from poloxamer-graft-hyaluronic acid hydrogels in vitro. *Int. J. Pharm.* **2003**, *260*, 83–91. [CrossRef]
27. Hsu, S.H.; Leu, Y.L.; Hu, J.W.; Fang, J.Y. Physicochemical characterization and drug release of thermosensitive hydrogels composed of a hyaluronic acid/pluronic F127 graft. *Chem. Pharm. Bull.* **2009**, *57*, 453–458. [CrossRef]
28. Kim, T.H.; Singh, R.K.; Kang, M.S.; Kim, J.H.; Kim, H.W. Gene delivery nanocarriers of bioactive glass with unique potential to load BMP2 plasmid DNA and to internalize into mesenchymal stem cells for osteogenesis and bone regeneration. *Nanoscale* **2016**, *8*, 8300–8311. [CrossRef]
29. Mahapatra, C.; Singh, R.K.; Kim, J.J.; Patel, K.D.; Perez, R.A.; Jang, J.H.; Kim, H.W. Osteopromoting reservoir of stem cells: Bioactive mesoporous nanocarrier/collagen gel through slow-releasing FGF18 and the activated bmp signaling. *ACS Appl. Mater. Interfaces* **2016**, *8*, 27573–27584. [CrossRef]
30. Kim, S.K.; Kwon, J.Y.; Nam, T.J. Involvement of ligand occupancy in Insulin-like growth factor-I (IGF-I) induced cell growth in osteoblast like MC3T3-E1 cells. *BioFactors* **2007**, *29*, 187–202. [CrossRef]
31. Langdahl, B.L.; Kassem, M.; Moller, M.K.; Eriksen, E.F. The effects of IGF-I and IGF-II on proliferation and differentiation of human osteoblasts and interactions with growth hormone. *Eur. J. Clin. Invest.* **1998**, *28*, 176–183. [CrossRef] [PubMed]
32. Machwate, M.; Zerath, E.; Holy, X.; Pastoureau, P.; Marie, P.J. Insulin-like growth factor-I increases trabecular bone formation and osteoblastic cell proliferation in unloaded rats. *Endocrinology* **1994**, *134*, 1031–1038. [CrossRef] [PubMed]
33. Fang, J.Y.; Hsu, S.H.; Leu, Y.L.; Hu, J.W. Delivery of cisplatin from pluronic co-polymer systems: Liposome inclusion and alginate coupling. *J. Biomater. Sci. Polym. Ed.* **2009**, *20*, 1031–1047. [CrossRef] [PubMed]
34. Shavlakadze, T.; Winn, N.; Rosenthal, N.; Grounds, M.D. Reconciling data from transgenic mice that overexpress IGF-I specifically in skeletal muscle. *Growth Horm. IGF Res.* **2005**, *15*, 4–18. [CrossRef] [PubMed]
35. Rinderknecht, E.; Humbel, R.E. The amino acid sequence of human insulin like growth factor I and its structural homology, with proinsulin. *J. Biol. Chem.* **1978**, *253*, 2769–2776. [PubMed]
36. Clark, A.H.; Ross-Murphy, S.B. Structural and mechanical properties of biopolymer gel. *Adv. Polym. Sci.* **1987**, *83*, 60–192.
37. Yoshida, R.; Kaneko, Y.; Sakai, K.; Okano, T.; Sakurai, Y.; Bae, Y.H.; Kim, S.W. Positive thermo sensitive pulsatile drug release using negative thermosensitive hydrogels. *J. Control. Release* **1994**, *32*, 97–102. [CrossRef]
38. Dang, J.M.; Sun, D.D.N.; Shin-Ya, Y.; Sieber, A.N.; Kostuik, J.P.; Leong, K.W. Temperature-responsive hydroxybutyl chitosan for the culture of mesenchymal stem cells and intervertebral disk cells. *Biomaterials* **2006**, *27*, 406–418. [CrossRef]

Article

Injectable Thermosensitive Formulation Based on Polyurethane Hydrogel/Mesoporous Glasses for Sustained Co-Delivery of Functional Ions and Drugs

Monica Boffito [1,†], Carlotta Pontremoli [2,†], Sonia Fiorilli [2,*], Rossella Laurano [1], Gianluca Ciardelli [1,‡] and Chiara Vitale-Brovarone [2,‡]

1 Department of Mechanical and Aerospace Engineering, Politecnico di Torino, Corso Duca degli Abruzzi 24, 10129 Torino, Italy; monica.boffito@polito.it (M.B.); rossella.laurano@polito.it (R.L.); gianluca.ciardelli@polito.it (G.C.)

2 Department of Applied Science and Technology, Politecnico di Torino, Corso Duca degli Abruzzi 24, 10129 Torino, Italy; carlotta.pontremoli@polito.it (C.P.); chiara.vitale@polito.it (C.V.-B.)

* Correspondence: sonia.fiorilli@polito.it; Tel.: +39-011-090-4683

† These authors contributed equally.

‡ Joint last authorship.

Received: 7 August 2019; Accepted: 19 September 2019; Published: 1 October 2019

Abstract: Mini-invasively injectable hydrogels are widely attracting interest as smart tools for the co-delivery of therapeutic agents targeting different aspects of tissue/organ healing (e.g., neo-angiogenesis, inflammation). In this work, copper-substituted bioactive mesoporous glasses (Cu-MBGs) were prepared as nano- and micro-particles and successfully loaded with ibuprofen through an incipient wetness method (loaded ibuprofen approx. 10% *w*/*w*). Injectable hybrid formulations were then developed by dispersing ibuprofen-loaded Cu-MBGs within thermosensitive hydrogels based on a custom-made amphiphilic polyurethane. This procedure showed almost no effects on the gelation potential (gelation at 37 °C within 3–5 min). Cu^{2+} and ibuprofen were co-released over time in a sustained manner with a significantly lower burst release compared to MBG particles alone (burst release reduction approx. 85% and 65% for ibuprofen and Cu^{2+}, respectively). Additionally, released Cu^{2+} species triggered polyurethane chemical degradation, thus enabling a possible tuning of gel residence time at the pathological site. The overall results suggest that hybrid injectable thermosensitive gels could be successfully designed for the simultaneous localized co-delivery of multiple therapeutics.

Keywords: polyurethane; injectable hydrogels; ion/drug delivery; mesoporous bioactive glasses; tissue regeneration

1. Introduction

Tissue regeneration is the result of a complex process, which involves the synergistic contributes of cells, biomaterials, and bioactive factors (e.g., drugs, ions, growth factors). In most cases the complete regeneration of tissues fails due to a lack of vascularization; hence insufficient angiogenesis is one of the major current pathological concerns. In addition, the occurrence of drug-resistant infections may lead to several complications and greatly enhance the number of non-healing cases [1]. In order to overcome these issues, much attention has focused on the design of multifunctional systems which are able to simultaneously promote all the processes involved in the tissue/organ healing cascade. The incorporation of drugs, growth factors or inorganic nanomaterials into hydrogels has been extensively investigated [2,3], with the aim of combining in a single formulation several therapeutic abilities (such as pro-angiogenesis, anti-bacterial, anti-inflammatory). Among the approaches already explored, the combination of hydrogels with inorganic nanocarriers, which are able to incorporate and release several agents, represents a promising therapeutic strategy with a high potential for clinical translation. With this perspective, Zhu et al. prepared chitosan/mesoporous silica nanoparticle (MSN) composite hydrogels for localized co-delivery of drugs and macromolecules [4]. Similar hydrogel embedding ibuprofen-loaded MSN was proposed for surface coating of titanium implants [5] and poly(vinyl alcohol) hydrogels carrying oil-loaded MSNs containing curcumin were studied for dermal application [6]. More recently, a similar strategy has been explored to design thermosensitive hydrogels based on a poly(N-isopropylacrylamide) (PNIPAM) copolymer carrying mesoporous silica for dual-responsive (i.e., temperature and light) drug release [7] and for bone tissue regeneration [8].

More recently, in the field of tissue regeneration, considerable attention has been addressed towards the use of mesoporous bioactive glasses (MBGs) enriched with specific metallic ions (i.e., Sr^{2+}, Cu^{2+}, Ag^{+}, Ce^{3+}), as multifunctional therapeutic platforms for advanced medical devices [9,10]. Among the elements which can exert therapeutic properties, copper has been extensively investigated, due to the well documented pro-angiogenic effect alongside its antibacterial potential [11,12]. Several works have evidenced the antibacterial potential of Cu-containing MBGs against both planktonic bacteria and biofilms. In particular, Wu et al. [11] tested the successful antibacterial potential of Cu-containing MBG scaffolds against *Escherichia coli bacteria* and, accordingly with this, the authors recently demonstrated that Cu-containing MBG nanoparticles and their ionic extracts exert promising antibacterial activity against both Gram positive and Gram negative bacteria (*Staphylococcus aureus*, *Staphylococcus epidermidis* and *E. coli*) and are effective in inhibiting and preventing biofilm produced by *S. epidermidis* [13]. In addition to antibacterial abilities, Cu-substituted bioactive glasses were also demonstrated to stimulate angiogenesis and promote revascularization of soft and hard tissues [14,15].

Based on these promising results, we have recently reported the development of hybrid formulations able to combine the injectability of thermosensitive poly(ether urethane)-based (PEU) hydrogels and the capability of MBG particles to release functional copper ions [16]. This system was developed for a prolonged and sustained release of copper ions, with the final aim to promote simultaneously angiogenesis and anti-microbial effect [13,17–19]. However, no effects against persistent inflammation could be achieved with the formulated system. Hence, in order to expand the potentialities of the previously developed hybrid platform, Cu-substituted MBGs, in the form of micro- and nano-particles, were loaded with ibuprofen (Ibu), chosen as an anti-inflammatory agent, and embedded into PEU hydrogels. The resulting hybrid formulations were evaluated in terms of ion/drug co-delivery kinetics, gelation behavior and stability in an aqueous environment.

Since the main goal of this contribution was the definition and the characterization of a novel and adaptable drug/ion release platform, our investigation mainly focused to identify the maximum amount of particles which could be incorporated within the hydrogel and to assess the related effect on the gelation behavior as well as the ion/drug co-release capacity that the final combined formulation could provide. The high potentiality of the proposed strategy lies in its wide versatility, as by tailoring the composition and cargo of MBGs, different ions and bioactive factors can be co-released according to the therapeutic effects required by the final targeted application. Moreover, their release timing and dosages can be finally optimized according to the requirements of each specific application, to make the released payload effective (i.e., releasing kinetics within the therapeutic window), while avoiding cytotoxic concentrations. In addition, the use of a custom-made PEU hydrogel as the vehicle phase of the MBGs provides the final system with a further degree of freedom. The characteristic LEGO-like chemical structure of PEUs allows polymer properties to be finely tuned by changing their constituting building blocks. This versatility opens the way to the design of stimuli-responsive hydrogels with an enhanced control over payload release and/or hydrogel dissolution/degradation, thus improving the release properties compared to those observed for MBGs alone. To the best of our knowledge, this is the first attempt to design a hybrid injectable formulation based on a thermo-sensitive hydrogel embedding MBGs able to co-release therapeutic ions and drugs.

2. Materials and Methods

2.1. Synthesis of Cu-Substituted MBGs

2.1.1. Materials

Cetyltrimethylammonium bromide (CTAB ≥ 98%), ammonium hydroxide solution (NH_4OH), double distilled water (ddH_2O), tetraethyl orthosilicate (TEOS), calcium nitrate tetrahydrate ($Ca(NO_3)_2 \cdot 4H_2O$, 99%), copper chloride ($CuCl_2$, 99%), Pluronic P123 ($EO_{20}PO_{70}EO_{20}$, M_n ~5800 Da), ibuprofen (>98% GC), Trizma® base, primary standard and buffer, ≥99.9% (titration) were purchased from Sigma Aldrich, Milan, Italy and used as received. All solvents were purchased from Sigma Aldrich (Milan, Italy) in analytical grade.

2.1.2. Cu-Substituted MBG Nanoparticles

Cu-substituted MBGs nanoparticles (nominal molar ratio Cu/Ca/Si = 2/13/85, hereafter named as MBG_Cu2%_SG) were prepared using a base-catalyzed template sol-gel synthesis, following an already optimized protocol [16]. In brief, 6.6 g of CTAB, acting as template, and 12 mL of NH_4OH were dissolved in 600 mL of ddH_2O under magnetic stirring (350 rpm) for 30 min. Then, 30 mL of TEOS, 4.888 g of $Ca(NO_3)_2 \cdot 4H_2O$ and 0.428 g of $CuCl_2$ were added, and the obtained solution was vigorously stirred for 3 h at room temperature (RT). The powder was collected by centrifugation (Hermle Labortechnik Z326, Hermle LaborTechnik GmbH, Wehingen, Germany) at 10,000 rpm for 5 min, washed twice with ddH_2O and once with absolute ethanol. The final precipitate was dried at 70 °C for 12 h and calcined at 600 °C in air for 5 h at a heating rate of 1 °C min^{-1} using a Carbolite 1300 CWF 15/5 (Carbolite Ltd., Hope Valley, UK), in order to completely remove CTAB.

2.1.3. Cu-Substituted MBG Microspheres

Cu-substituted MBG microspheres (nominal molar ratio Cu/Ca/Si=2/13/85, hereafter named as MBG_Cu2%_SD) were synthesized following the aerosol-assisted spray-drying method reported by Pontremoli, Boffito et al. [16]. Briefly, 2.030 g of Pluronic P123, acting as template, were dissolved in 85 g of ddH_2O. Simultaneously, in a separate batch, a solution of 10.73 g of TEOS was pre-hydrolyzed under acidic conditions using 5 g of an aqueous HCl solution at pH 2. The solution with TEOS was then added drop-wise into the template solution and kept stirring for half an hour. Then, 0.163 g of $CuCl_2$ and 1.86 g of $Ca(NO_3)_2 \cdot 4H_2O$ were added. The final solution was stirred for 15 min and then sprayed with a Büchi, Mini Spray-Dryer B-290 (Büchi Labortechnik AG, Flawil, Switzerland), using N_2 as atomizing gas (inlet temperature 220 °C, N_2 pressure 60 mm Hg, feed rate 5 mL min^{-1}). The obtained powder was finally calcined at 600 °C in air for 5 h at a heating rate of 1 °C min^{-1} using a Carbolite 1300 CWF 15/5 (Carbolite Ltd., Hope Valley, UK).

2.1.4. Ibuprofen Loading Procedure

Ibuprofen was loaded into MBG_Cu2%_SG and MBG_Cu2%_SD through the incipient wetness method [20]. In brief, 0.1 g of both Cu-substituted MBGs were impregnated several times by dropping consecutive small aliquots of an ibuprofen solution in ethanol (at the final concentration of 30 mg/mL) onto the powders at RT. After each impregnation, ethanol was evaporated at 50 °C for 10 min and the dried powder mixed with a spatula. In order to completely fill the mesopores with ibuprofen the impregnation procedure was carried out with four 100 μL aliquots. Lastly, the obtained powders were dried at 50 °C overnight and named as follows: MBG_Cu2%_SG_Ibu and MBG_Cu2%_SD_Ibu.

2.2. Characterization of Cu-Substituted MBGs Loaded with Ibu

The morphology and particle size of the prepared powders were analyzed by field-emission scanning electron microscopy (FE-SEM) using a ZEISS MERLIN instrument (Oberkochen, Germany). For FE-SEM observations, 10 mg of MBG_Cu2%_SG_Ibu were dispersed in 10 mL of isopropanol using an ultrasonic bath (Digitec DT 103H, Bandelin, Berlin, Germany) for 5 min to obtain a stable suspension. A drop of the resulting suspension was put on a copper grid (3.05 mm Diam. 200 MESH, TAAB), allowed to dry and successively chromium-plated prior to imaging (Cr layer 7 nm). MBG_Cu2%_SD_Ibu sample was dispersed directly onto the double face carbon tape placed on a sample stub and then coated with a Cr layer. Compositional analysis of the powders was performed by energy dispersive spectroscopy (EDS; AZtec EDS, Oxford instruments, Abingdon-on-Thames, UK). EDS spectra were acquired on powder dispersed on carbon tape by analyzing an area of 75 × 50 μm^2. Nitrogen adsorption–desorption isotherms were measured by using an adsorption analyzer ASAP2020 Micromeritics (ASAP 2020 Plus Physisorption, Norcross, GA, USA) at a temperature of −196 °C. Before nitrogen adsorption–desorption measurements, loaded samples were outgassed at 37 °C for 5 h, in order to avoid the degradation of the drug. The Brunauer–Emmett–Teller (BET) equation was used to calculate the specific surface area (SSA_{BET}) from the adsorption data (relative pressures 0.04–0.2). The pore size distribution was calculated through the DFT method (density functional theory) using the NLDFT kernel of equilibrium isotherms (desorption branch). The mesopore structure was investigated by transmission electron microscopy (TEM, Fei Company, Hillsboro, OR, USA) using a FEI Tecnai G2 operated at 200 kV. The samples were prepared by suspending the powders in ethanol and drop-wise placed on carbon coated copper grids. Thermo-gravimetric analysis (TGA) of the samples was performed on a TG 209 F1 Libra instrument from Netzsch (Selb, Germany) over a temperature range of 25–600 °C under air flux at a heating rate of 10 °C min^{-1}. The drug content was determined from the weight loss between 200 and 600 °C, by applying a correction for the weight loss in the same range of temperature due to the surface silanol condensation as recorded on Cu-substituted MBGs before Ibu loading. Fourier transformed infrared (FT-IR) spectra of the drug-loaded samples were collected on a Bruker Equinox 55 spectrometer (Bruker, Billerica, MA, USA) over a range of wavenumbers from 4000 to 400 cm^{-1} (resolution 2 cm^{-1}). In order to assess the amorphous state of incorporated Ibu, X-ray patterns were collected using an X'Pert PRO, PANalytical instrument (X'Pert PRO, PANalytical, Almelo, The Netherlands) (CuKα radiation at 40 kV and 40 mA). Data were obtained from 10° to 80° (diffraction angle 2ϑ) at a step size of 0.0130° and a scan step time of 60 s. Differential scanning calorimetry (DSC) analysis of Ibu-loaded samples was carried out with a DSC 204 F1 Phoenix (Netzsch) instrument ((Selb, Germany). The samples were heated from 37 °C to 200 °C at a heating rate of 10 °C min^{-1} under N_2 flux.

2.3. *Synthesis of Amphiphilic Poly(ether urethane)*

2.3.1. Materials

Amphiphilic water-soluble PEU was synthesized starting from Poloxamer®407 (P407, poly(ethylene oxide)-poly(propylene oxide)-poly(ethylene oxide) PEO-PPO-PEO triblock copolymer, 70% PEO, M_n 12,600 g/mol), 1,6-hexamethylene diisocyanate (HDI) and 1,4-cyclohexanedimethanol (CDM), purchased from Sigma Aldrich, Milan, Italy. Before the synthesis, all the reagents were anhydrified according to the following protocols: P407 was dried for 8 h at 100 °C and then cooled down at 40 °C at a pressure lower than 200 mbar; HDI was distilled under reduced pressure; CDM was vacuum-dried in a dessicator at RT. Dibutyiltin dilaurate (DBTDL) was also purchased from Sigma Aldrich, Milan, Italy and used as received to catalyze the polymerization reaction. Anhydrous 1,2-dichoroethane (DCE, Carlo Erba Reagents, Cornaredo, Italy) was prepared by pouring the solvent over activated (at 120 °C, overnight) molecular sieves (4Å, Sigma Aldrich, Milano, Italy) under N_2 atmosphere for 8 h. All other solvents were purchased from Carlo Erba Reagents, Cornaredo, Italy in the analytical grade and used as received.

2.3.2. Poly(ether urethane) Synthesis

PEU synthesis was carried out according to Boffito et al. [21]. P407, HDI and CDM were reacted at 1:2:1 molar ratio and the synthesis was conducted under N_2 atmosphere following a prepolymerization method with a first reaction between P407 and HDI to form an N=C=O-terminated prepolymer which was then chain extended through the addition of CDM. In detail, the required amount of HDI was added to a 20% *w/v* concentrated P407 solution previously prepared in DCE and equilibrated at 80 °C. DBTDL was also added at 0.1% *w/w* concentration with respect to P407 and the reagents reacted for 150 min. Then, the reaction mixture was cooled down at 60 °C and a 3% *w/v* concentrated CDM solution previously prepared in DCE was added to start the chain extension reaction which lasted 90 min and was finally terminated through the addition of MeOH. The synthesized PEU was collected by precipitation in an excess of petroleum ether (petroleum ether: DCE = 4:1 *v/v*) and further purified by solubilization in DCE (20% *w/v*) and precipitation in a diethyl ether/MeOH mixture (98:2 *v/v*, 5:1 volume ratio with respect to DCE). PEU was collected by centrifugation (MIKRO 220R, Hettich, Tuttlingen, Germany), dried overnight, grinded and stored under vacuum at 4 °C until use. Hereafter, the synthesized PEU will be referred to with the acronym CHP407, where C, H and P407 refer to CDM, HDI and Poloxamer®407, respectively.

2.4. Chemical Characterization of the As-Synthesized PEU

The successful synthesis of CHP407 poly(ether urethane) was assessed by attenuated total reflectance Fourier transformed infrared (ATR-FTIR) spectroscopy and size exclusion chromatography (SEC). CHP407 and P407 ATR-FTIR spectra were registered using a Perkin Elmer (Waltham, MA, USA) Spectrum 100 instrument equipped with an ATR diamond crystal (UATR KRS5). Spectra were obtained as an average of 16 scans registered at RT within the spectral range 4000–600 cm^{-1} (resolution 4 cm^{-1}). CHP407 molecular weight distribution was assessed by SEC using an Agilent Technologies 1200 Series (Agilent Technologies, Inc., Santa Clara, CA, USA) instrument equipped with a refractive index detector (RID) and two columns (Waters Styragel HR1 and HR4, Waters Corporation, Sesto San Giovanni, Italy). Analyses were conducted at 55 °C using N,N-dimethylformammide (DMF, CHROMASOLV Plus, inhibitor-free, for HPLC, Carlo Erba Reagents, Cornaredo, Italy), added with 0.1% *w/v* LiBr (Sigma Aldrich, Milano, Italy) as eluent at a flow rate of 0.5 mL/min. Samples were prepared at 2 mg/mL concentration in the mobile phase and filtered through a 0.45 µm syringe filter (Macherey-Nagel, Düren, Germany poly(tetrafluoro ethylene) membrane) before analysis. The Agilent ChemStation software was finally used to estimate CHP407 Number Average Molecular Weight (M_n) and polydispersity index (Đ) based on a calibration curve previously constructed starting from 10 poly(methyl methacrylate) standards with M_n ranging between 940 and 214,600 g/mol.

2.5. Preparation of CHP407-Based Hydrogels Carrying Cu-MBGs Loaded with Ibu

CHP407 thermosensitive hydrogels were prepared at a final PEU concentration of 15% *w/v* [16,21]. Cu-MBGs loaded with Ibu were encapsulated into CHP407-based hydrogels with final particle concentration of 20 mg/mL. In detail, hybrid hydrogels were prepared by adding an aliquot of particle suspension in ddH_2O (100 mg/mL), previously sonicated for 3 min at 20% amplitude, to a CHP407-based solution prepared in physiological saline solution (0.9% NaCl). The starting concentration of CHP407 solution was determined in order to reach the desired PEU and particle contents in the final system. Both CHP407 solubilization and particle addition were carried out at 4 °C to avoid undesired gelation during sample preparation. Sample mixing was conducted using a Vortex mixer for 30 s to ensure homogeneous particle dispersion within the sol-gel systems. Particles were added to CHP407 aqueous solutions immediately before use to avoid premature payload release before characterization tests. Pure CHP407 sol-gel systems and CHP407 hydrogels loaded with Ibu were also prepared as control samples. Loading of Ibu in CHP407 sol-gel systems was carried out by adding CHP407 aqueous solutions prepared at higher concentration with a predefined volume of an Ibu stock solution (at 40 mg/mL in ethanol) to reach an average content equal to the Ibu amount incorporated into MBG_Cu2%_SG_Ibu and MBG_Cu2%_SD_Ibu, as assessed through TGA analysis. Based on this calculation, CHP407 sol-gel systems were loaded with Ibu at a final concentration of 2.5 mg/mL. Hereafter, the developed sol-gel systems will be referred to with the acronyms reported in Table 1. For all the conducted tests, hydrogels were prepared in Bijou sample containers (inner diameter 17 mm, Carlo Erba Reagents, Cornaredo, Italy) at a final volume of 1 mL, to avoid geometry and volume influence on the performed characterizations.

Table 1. Compositional information and acronyms of the designed sol-gel systems.

Acronym	Composition
CHP407	CHP407 poly(ether urethane) (PEU) at 15% *w/v*
CHP407_Ibu	CHP407 PEU at 15% *w/v* + ibuprofen at a concentration equal to the mean drug amount assessed in MBG_Cu2%_SG_Ibu and MBG_Cu2%_SD_Ibu (i.e., 2.5 mg/mL)
CHP407_MBG_Cu2%_SG_Ibu	CHP407 PEU at 15% *w/v* + MBG_Cu2%_SG_Ibu (20 mg/mL MBG concentration and the corresponding amount of ibuprofen)
CHP407_MBG_Cu2%_SD_Ibu	CHP407 PEU at 15% *w/v* + MBG_Cu2%_SD_Ibu (20 mg/mL MBG concentration and the corresponding amount of ibuprofen)

2.6. Characterization of Hybrid Sol-Gel Systems

The thermosensitive behavior of pure CHP407 (i.e., without particle) and hybrid hydrogels (i.e., containing MBG_Cu2%_SG_Ibu, MBG_Cu2%_SD_Ibu or Ibu) was studied through tube inverting test in temperature ramp mode and isothermal conditions (i.e., at 37 °C) to assess the effects of drug/particle incorporation on hydrogel sol-to-gel transition. In view of their potential clinical application as injectable depots for the prolonged and sustained payload release, the delivery profile of ibuprofen and copper ion was investigated in physiological mimicking conditions. Furthermore, gel swelling potential and residence time in watery environment at physiological temperature were also assessed.

2.6.1. Thermosensitive Behavior of CHP407-Based Sol-Gel Systems

Tube inverting test in temperature ramp conditions was carried out to estimate hydrogel Lower Critical Gelation Temperature (LCGT). In detail, samples were subjected to a step-by-step temperature increase within the range 5–40 °C at a rate of 1 °C/step. In each step, temperature was kept constant for 5 min and the samples were finally inverted for 30 s to visually inspect their sol, biphasic or gel state (in the gel state no flow was observed within the set inversion time). The same test was carried out in isothermal conditions at 37 °C to assess hydrogel gelation time. In this test, instead of temperature, incubation time at 37 °C was varied from 1 to 10 min with a rate of 1 min/step. Each step consisted in sample equilibration in the sol state for 10 min, incubation at 37 °C for the required time, inversion for 30 s and visual inspection.

2.6.2. Gel Swelling and Stability in Aqueous Media

Gel swelling and stability in physiological mimicking conditions, that is, in aqueous media and 37 °C, were investigated on the designed sol-gel systems up to 14 days incubation time [16]. Samples (1 mL) were first incubated at 37 °C (Memmert IF75, Schwabach, Germany) for 15 min to allow a complete gelation and then added with 1 mL of Trizma® (0.1 M, pH 7.4) previously equilibrated at 37 °C. Complete medium refresh was performed every two days. On predefined time points of 6h, 1d, 3d, 7d and 14d three samples were collected, weighted upon removal of the residual buffer, lyophilized (Martin Christ ALPHA 2–4 LSC, Osterode am Harz, Germany) and weighted again. The collected data were finally used to estimate sample percentage of swelling (weight change in wet state %) and polymer weight loss (weight loss in dry state %) [16]. Samples collected on day 14 were also analyzed by SEC to assess changes in polymer molecular weight distribution during incubation in aqueous media at 37 °C.

2.6.3. Payload Release Studies

Payload release studies were carried out on hydrogels incorporating MBG_Cu2%_SG_Ibu, MBG_Cu2%_SD_Ibu and simply dispersed Ibu. Tests were conducted at 37 °C up to 14 days using

Trizma® as release medium. Complete gelation of the samples was ensured through incubation at 37 °C for 15 min; then, 1 mL of release medium (previously equilibrated at 37 °C) was added to each sample and the release test started. Release media were collected and completely refreshed at 1h, 3h, 5h, 1d, 2d, 3d, 4d, 8d, 10d, and 14d incubation time. Released Ibu was quantified through a high performance liquid chromatography (HPLC, Thermo Scientific, Dionex Ultimate 3000, Waltham, MA, USA) instrument equipped with a C18 column (5 μm, 120 Å) according to the protocol described by Alsirawan et al. [22]. A mixture of acetonitrile (ACN, Carlo Erba Reagents, Cornaredo, Italy, HPLC grade) and phosphoric acid solution at 0.03% *w/v* concentration (pH 2.25) at 60/40 *v/v* was used as mobile phase at 1.7 mL/min. Analysis were conducted with an injection volume of 20 μL, at RT and 214 nm for 5 min. In order to prepare the samples, the collected extracts were mixed with ACN at 40/60 volume ratio and filtered through a 0.45 μm syringe filter (Macherey-Nagel, Düren, Germany, poly(tetrafluoro ethylene) membrane). Ibu content was finally quantified with respect to a calibration curve based on ibuprofen standards with concentration ranging between 0 and 2.5 mg/mL. The collected extracts were also characterized by inductively coupled plasma atomic emission spectrometry technique (ICP-AES) (ICP-MS, Thermo Scientific, Waltham, MA, USA) to measure the concentration of released copper ions. In order to express the results in terms of released percentage, the amount of copper initially incorporated into the MBG framework was measured by dissolving MBG_Cu2%_SG_Ibu and MBG_Cu2%_SD_Ibu in a mixture of nitric and hydrofluoric acids (0.5 mL of HNO_3 and 2 mL of HF for 10 mg of powder) and quantifying copper concentration through ICP-AES. To compare the ion/drug release kinetics from MBGs as such and embedded within the hydrogel, release tests from particle alone were also conducted as follows: drug-loaded powders were dispersed in physiological saline solution at 20 mg/mL MBG concentration and Trizma® was added as release medium at 37 °C (1:1 volume ratio), as previously performed with the hydrogel systems. Release tests from free MBG particles were conducted up to 24 h observation time. In order to better characterize copper ion and ibuprofen release mechanism from the hydrogels incorporating MBG_Cu2%_SG_Ibu, MBG_Cu2%_SD_Ibu and Ibu as such, the Korsmeyer-Peppas equation was used in the form recently reported by Boffito et al. [23] to estimate the release exponent n, whose value classifies the type of release. Specifically, an n value of 0.45 or 0.89 is typical of a diffusion- or swelling/relaxation-controlled release, respectively. An n value within the range 0.45–0.89 characterizes an anomalous release, meanwhile when n shows a value higher than 0.89 other processes are ongoing in addition to diffusion and swelling/relaxation.

2.7. Statistical Analysis

Statistical analysis was performed on the data collected from swelling, stability, and payload release tests. All tests were performed in triplicate and results have been reported as mean ± standard deviation. Comparison among results was performed through GraphPad Prism (GraphPad Software, Inc., La Jolla, CA, USA, version 5.03, 2009; http://www.graphpad.com) using a Two-way ANOVA analysis followed by Bofferoni's multiple comparison tests. The degree of statistical difference among the results was defined in accordance to Boffito et al. [21].

3. Results and Discussion

3.1. Cu-MBGs Loaded with Ibu

Morphological and Structural Characterization

FE-SEM images of MBG_Cu2%_SG_Ibu (Figure 1A) and MBG_Cu2%_SD_Ibu (Figure 1C) showed nanoparticles with a monodispersed spherical shape (size range: 150–200 nm) and microspheres in the range of 1–5 μm, respectively. EDS spectra (Figure 1B,D) of both powders confirmed the presence of copper inside the framework, with a Cu/Si molar ratio in good agreement with the nominal ratio for both MBGs.

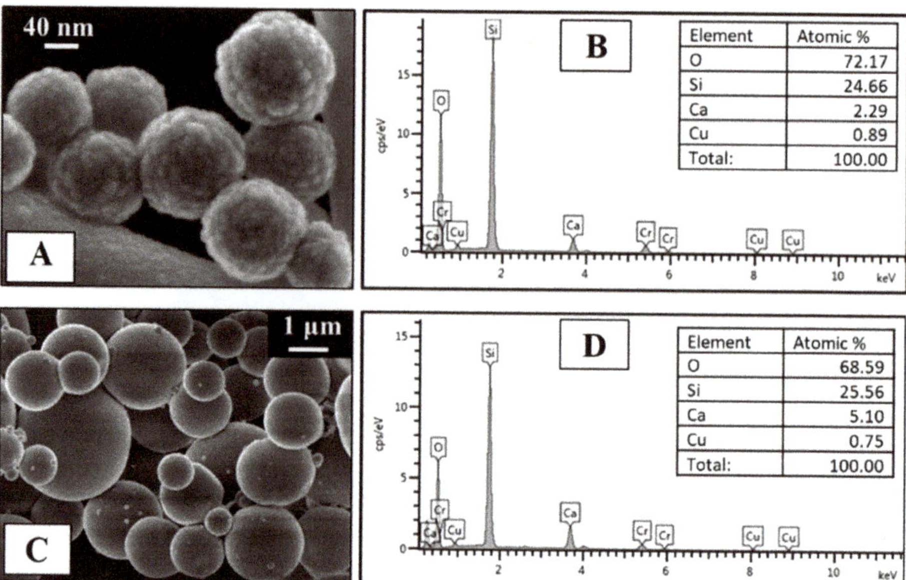

Figure 1. Field-emission scanning electron microscopy (FE-SEM) image of MBG_Cu2%_SG_Ibu (**A**) and MBG_Cu2%_SD_Ibu (**C**). Energy dispersive spectroscopy (EDS) spectrum of MBG_Cu2%_SG_Ibu (**B**) and MBG_Cu2%_SD_Ibu (**D**).

FESEM observations and EDS analysis evidenced that ibuprofen incorporation did not significantly alter the morphological features and the chemical composition of Cu-substituted MBGs, which resulted very similar to those reported for not-loaded samples [16]. In particular, the amount of copper revealed by EDS before and after the drug loading resulted unaffected, evidencing that the loading procedure did not induce any copper release (data not shown).

Figure 2 shows the nitrogen adsorption–desorption isotherms and the pore size distribution of the sample before and after drug loading. As expected, before loading, MBG_Cu2%_SG showed a type IV sorption isotherm, according to the IUPAC classification, with a well-defined step around 0.4 (P/P_0), indicative of uniform mesopores. The specific surface area and pore volume values reported in Table 2 are characteristic of mesoporous materials with uniform pores and remarkable value of specific surface area (SSA_{BET}) [24]. The mesopore size distribution was centered at around 4.2 nm, thus allowing the incorporation of ibuprofen whose molecular size is about 1 nm [25–27]. As expected, drug up-loading induced a significant modification of the adsorption-desorption isotherm (reduction of the adsorbed volume and presence of hysteresis loop) and a drastic reduction of the pore volume, as shown in Figure 2. In particular, the modification of the isotherm curve upon drug incorporation suggested that most of the mesopores were completely filled with Ibu. On the other hand, the remaining population underwent size reduction and shape modification from cylinder to ink-bottle pores, in analogy to the results reported by Hong et al. [28] for similar systems. In addition, the drastic reduction of pore volume as a consequence of drug incorporation was confirmed by the disappearance of the component centered at 4.2 nm (Figure 2B). The isotherm of MBG_Cu2%_SD was a type IV curve (Figure 2C), with H1 hysteresis loop, typical of mesoporous material with pores larger than 4 nm. Although the specific surface area was lower compared to that of MBG_Cu2%_SG, it resulted much higher compared to not-templated sol-gel glasses (few m^2/g), conferring increased surface reactivity to MBGs in the biological environment [29]. The worm-like mesoporous structure was further confirmed by TEM images, reported in Figure S1. The pore size distribution evidenced multisized pores in the range between 8 and 11 nm (Figure 2D), which easily allows the diffusion and incorporation of ibuprofen molecules. A drastic reduction in SSA_{BET} was observed in MBG_Cu2%_SD_Ibu sample, while the pore volume reduction was lower compared to MBG_Cu2%_SG_Ibu sample (Table 2). Hence, in MBG_Cu2%_SD the incorporation of drug molecules occurred without a full occlusion of the available pore volume. The total amount of loaded drug was quantified by TGA analysis on both samples after drug incorporation. As reference, TGA analysis was also conducted on not-loaded MBG samples, proving the complete absence of residual organic species at 600 °C. TGA thermograms of MBG_Cu2%_SG_Ibu and MBG_Cu2%_SD_Ibu exhibited a significant weight decrease between 300 and 400 °C, which can be ascribed to ibuprofen loss [30], due to the rupture of the multiple H-bonding interactions between the drug and the hydroxyl groups of the inner MBG surface, in accordance with Mellaerts and co-workers [26]. The weight percentage of loaded ibuprofen, based on TGA analysis, turned out to be 12% in MBG_Cu2%_SG and 10% in MBG_Cu2%_SD. These results confirmed that the drug loading capacity increases with the increase of MGB surface area and pore volume, according to data reported in the literature for mesoporous silicas [31].

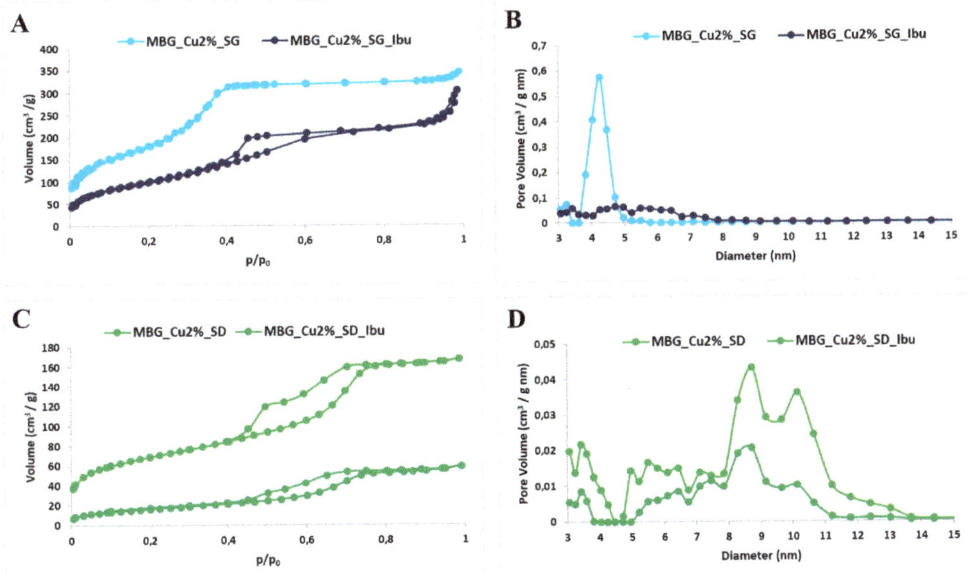

Figure 2. N_2 adsorption–desorption isotherm of MBG_Cu2%_SG and MBG_Cu2%_SG_Ibu (**A**), MBG_Cu2%_SD and MBG_Cu2%_SD_Ibu (**C**). DFT (density functional theory) pore size distribution of MBG_Cu2%_SG and MBG_Cu2%_SG_Ibu (**B**), MBG_Cu2%_SD and MBG_Cu2%_SD_Ibu (**D**).

Table 2. Structural properties (i.e., specific surface area -SSA_{BET}-, pore volume, pore size) of MBG_Cu2%_SG, MBG_Cu2%_SD, MBG_Cu2%_SG_Ibu and MBG_Cu2%_SD_Ibu.

Acronym	SSA_{BET} (m^2 g^{-1})	Pore Volume (cm^3 g^{-1})	Pore Size (nm)
MBG_Cu2%_SG	740	0.65	4.2
MBG_Cu2%_SG_Ibu	330	0.35	4–6
MBG_Cu2%_SD	226	0.24	8–11
MBG_Cu2%_SD_Ibu	54	0.081	8–9

The FTIR spectra of Cu-substituted MBGs before (curves b–d) and after ibuprofen loading (curves c–e) are reported in Figure 3A and compared to the FTIR spectrum of ibuprofen alone (curve a). MBG samples showed the typical adsorption bands of H-bonded hydroxyls (stretching vibration) in the range of 3750–3000 cm^{-1}. For what concerns drug-loaded samples, spectra showed the typical bands of ibuprofen molecule: the absorption bands at 2933 and 2871 cm^{-1} ascribed to C–H stretching modes and the signals at 1475 and 1421 cm^{-1} attributed to C–H bending vibrations. At variance with the spectrum of the drug alone which showed a clear adsorption band at 1706 cm^{-1}, due to the C=O stretching vibration in –COOH groups, for ibuprofen-loaded samples two bands appeared at 1550 and 1407 cm^{-1}, due to the asymmetric (ν_{as}) and symmetric (ν_s) stretching vibration of the carboxylate group COO^-, respectively [32], resulting from proton-transfer reactions from carboxylic moieties to hydroxyl groups at MBG surface [33]. As widely reported in the literature [34,35], drug loading in their amorphous form results in an increased dissolution rates and solubility. Hence, DSC and X-ray powder diffraction (XRD) analyses of ibuprofen-loaded samples were conducted to assess the amorphous state of the drug and exclude the presence of large crystalline aggregates. Figure 3 compares the DSC thermograms of Ibu as such, MBG_Cu2%_SG_Ibu and MBG_Cu2%_SD_Ibu samples. A single endothermic melting peak at 76 °C, ascribed to crystal phase melting, was observed only for ibuprofen as such, confirming the non-crystalline state of ibuprofen loaded into the mesopores. The amorphous state of the drug was also confirmed by XRD analysis (Figure 3D). XRD pattern of ibuprofen powder showed several characteristic X-ray diffraction peaks, which completely disappeared upon drug loading into the pores of Cu-substituted MBGs, which, in accordance with DSC data, strongly suggested that re-crystallization did not occur inside the pores upon solvent evaporation during the incorporation process. This behavior has been previously reported by Bràs et al. [30], who confirmed the amorphous state of ibuprofen confined inside SBA-15 silicas with similar pore size. The proposed attribution is also supported by several authors [36,37] who reported that re-crystallization of the entrapped drug molecules is suppressed below a critical pore diameter, showing that crystallization can occur only when pore size is significantly larger (about 20 times) compared to the drug size [36].

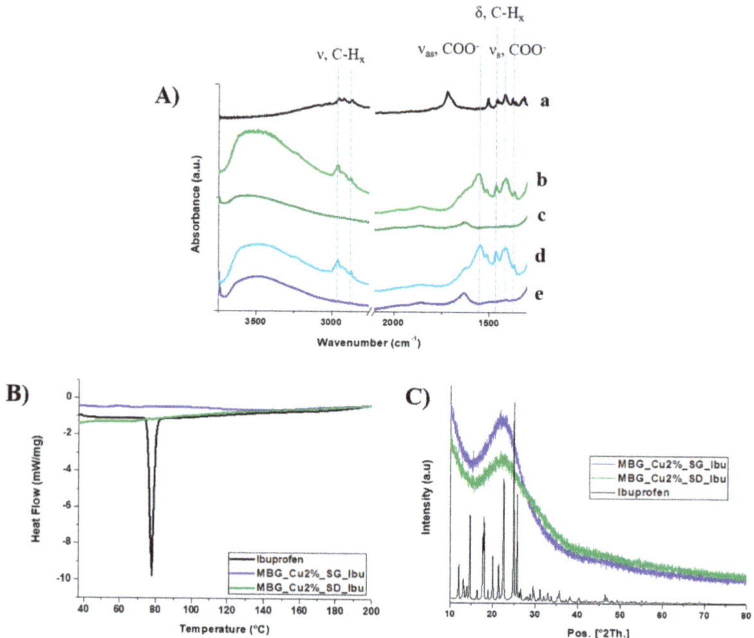

Figure 3. (**A**) Fourier transformed infrared (FTIR) spectra of ibuprofen (**a**), MBG_Cu 2%_SG_Ibu (**b**), MBG_Cu2%_SG (**c**), MBG_Cu 2%_SD_Ibu (**d**) and MBG_Cu2%_SD (**e**). (**B**) Differential scanning calorimetry (DSC) thermograms of ibuprofen, MBG_Cu2%_SG_Ibu and MBG_Cu2%_SD_Ibu samples. (**C**) X-ray powder diffraction (XRD) patterns of ibuprofen, MBG_Cu2%_SG_Ibu and MBG_Cu2%_SD_Ibu samples.

3.2. Poly(ether urethane) Chemical Characterization

The successful synthesis of a PEU carrying P407 blocks in its backbone was confirmed by attenuated total reflectance Fourier transformed infrared (ATR-FTIR) spectroscopy (Figure 4). The appearance of new transmission peaks in CHP407 spectrum, compared to P407 one, clearly proved the formation of urethane bonds among PEU building blocks [16,21]. In detail, the formation of N-H bonds was proved by the appearance of a new peak at 3342 cm^{-1} (stretching vibration), while the signals at 1720 and 1642 cm^{-1} could be ascribed to the stretching vibration of urethane free and bound carbonyl groups (C=O), respectively. N-H bonds also showed a bending vibration at 1540 cm^{-1} concurrent with C-N bond stretching. CHP407 showed M_n of 71,670 Da and a polydispersity index of 1.7, further confirming the success of the polymerization reaction.

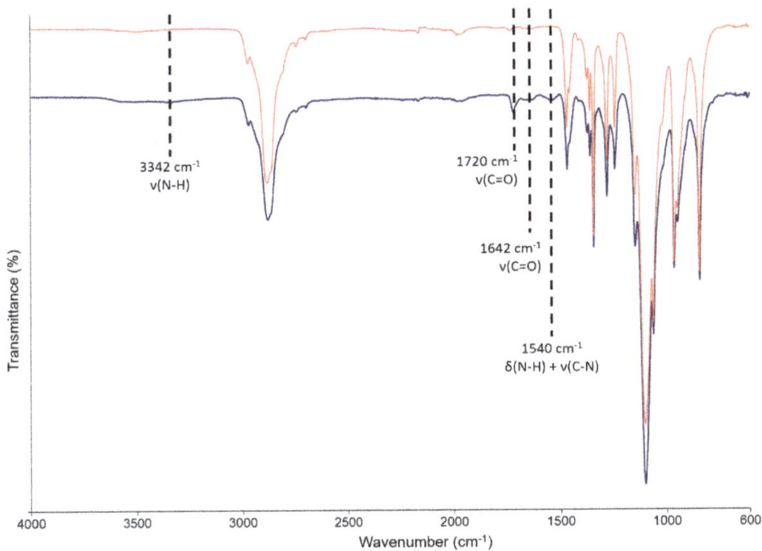

Figure 4. Attenuated total reflectance Fourier transformed infrared (ATR-FTIR) spectra of P407 (red) and CHP407 (blue). Dashed lines identify the characteristic signals of newly formed urethane bonds in CHP407.

3.3. Characterization of Hybrid Sol-Gel Systems

3.3.1. Thermosensitive Behavior of CHP407-Based Hydrogels

The temperature-dependent sol-to-gel transition of the developed systems was characterized through tube inverting tests carried out in temperature ramp mode and in isothermal conditions at 37 °C. The former test allowed the estimation of hydrogel LCGT, while the latter was conducted to evaluate the time required for a complete gelation in physiological conditions. Table 3 reports LCGT values and gelation time in physiological conditions of pure CHP407 hydrogel and CHP407 hybrid hydrogels.

Table 3. Lower critical gelation temperature (LGCT) and gelation time at 37 °C of CHP407, CHP407_Ibu, CHP407_ MBG_Cu2%_SG_Ibu and CHP407_ MBG_Cu2%_SD_Ibu.

Acronym	LCGT (°C) [1]	Gelation Time @ 37 °C (min) [2]
CHP407	28	5
CHP407_Ibu	27	5
CHP407_MBG_Cu2%_SG_Ibu	29	4
CHP407_MBG_Cu2%_SD_Ibu	30	6

[1] Error: ± 0.5 °C. [2] Error: ± 0.5 min.

The incorporation of MBG_Cu2%_SG_Ibu, MBG_Cu2%_SD_Ibu or ibuprofen as such was found to slightly influence the transition kinetics of the designed sol-gel systems, in accordance with our previous work [16]. Particle incorporation marginally increased hydrogel gelation temperature, suggesting that MBG particles act as defects in the gel network, initially hindering and then slowing down the kinetics of the sol-to-gel transition [16]. On the other hand, the slight decrease of gelation time in physiological conditions observed for CHP407 gels containing MBG_Cu2%_SG_Ibu particles could also result from the criterion adopted to define the "sol" and the "gel" states, that is, presence or absence of sample flow within 30 s of vial inversion. Indeed, particle addition to the hydrogels induced an increase in viscosity that, as a consequence, inevitably accounted for the shorter incubation time at 37 °C requested for not observing any flow within the observation time. Similarly, the slightly lower gelation temperature of CHP407_MBG_Cu2%_SG_Ibu compared to CHP407_MBG_Cu2%_SD_Ibu could be correlated to their dimensional differences, which result in different hydrogel viscosity. Indeed, at a fixed MBG concentration of 20 mg/mL, the number of MBG_Cu2%_SG_Ibu contained throughout the gel is expected to be higher compared to MBG_Cu2%_SD_Ibu, due to the lower size of SG particles. Regarding the addition of ibuprofen as such, no effects were observed in gelation time in physiological conditions, while LCGT value slightly decreased. This behavior did not result from the addition of a small volume of EtOH (used to solubilize Ibu) to the sol-gel systems (data not reported), but rather to the intrinsic nature of the drug. In fact, being hydrophobic, ibuprofen is expected to be partly loaded within the core of the forming CHP407 micelles, thus inducing an increase of micelle volume, which then achieves the critical value required for the onset of thermal gelation [8] at a lower temperature. Despite the commented slight changes in LCGT and gelation time at 37 °C, neither the addition of MBG particles of different size nor the incorporation of a hydrophobic drug significantly affected the gelation potential of CHP407-based hydrogels upon temperature increase.

3.3.2. Gel Characterization through Swelling and Stability to Dissolution Tests

CHP407-based gels, both virgin and hybrid formulations, were characterized in terms of aqueous medium absorption (Figure 5A) and dissolution/degradation (Figure 5B) over time in a physiological mimicking environment, that is, at 37 °C in the presence of a physiological-like buffer at pH 7.4. The incorporation of ibuprofen as such within CHP407 gels did not significantly affect gel behavior in aqueous media. Indeed, the percentage of swelling increased overtime up to 7d, followed by a drastic decrease on day 14 for both CHP407 and CHP407_Ibu samples. This change in medium absorption trend can be correlated to two concurrent phenomena occurring in the samples and their balance, namely swelling and dissolution/degradation resulting from the progressive absorption of aqueous media within the gels. The decrease in the swelling percentage observed on day 14 in CHP407 and CHP407_Ibu is thus due to the increasing gel instability in aqueous media over time. This hypothesis was demonstrated by the progressive increase in the percentage of weight loss (statistically significant increase at each time point, with the exception of day 1), that reached a value of 46.9 ± 0.6% and 45.8 ± 1.9% for CHP407 and CHP407_Ibu, respectively, after 14 days incubation in aqueous environment. Differently from ibuprofen, the embedding of MBGs in CHP407 hydrogels significantly affected gel long-term stability in aqueous environment (statistical differences between particle-loaded and not-loaded gels observed from day 3 in terms of swelling and day 1 in terms of weight loss percentage). In fact, both CHP407_MBG_Cu2%_SG_Ibu and CHP407_MBG_Cu2%_SD_Ibu showed negative swelling percentages starting from day 7 of incubation time suggesting that dissolution/degradation had completely prevailed on absorption phenomena. This observation was further proved by weight loss data, with CHP407_MBG_Cu2%_SG_Ibu and CHP407_MBG_Cu2%_SD_Ibu being almost completely dissolved/degraded after 14 days incubation in aqueous medium (weight loss percentage of 83.9 ± 2.2% and 68.9 ± 4.7% for CHP407_MBG_Cu2%_SG_Ibu and CHP407_MBG_Cu2%_SD_Ibu, respectively). The collected data confirmed the behavior previously observed for similar systems [16], with slight differences due to the different sample geometry, that is, gel thickness and extension of the surface in contact with the aqueous medium. More in detail, since day 1, CHP407_MBG_Cu2%_SG_Ibu and CHP407_MBG_Cu2%_SD_Ibu showed statistically different dissolution/degradation, while swelling percentage became significantly different from day 3 on, with hydrogel containing MBG_Cu2%_SG_Ibu exhibiting higher destabilization induced by the progressive absorption of aqueous medium.

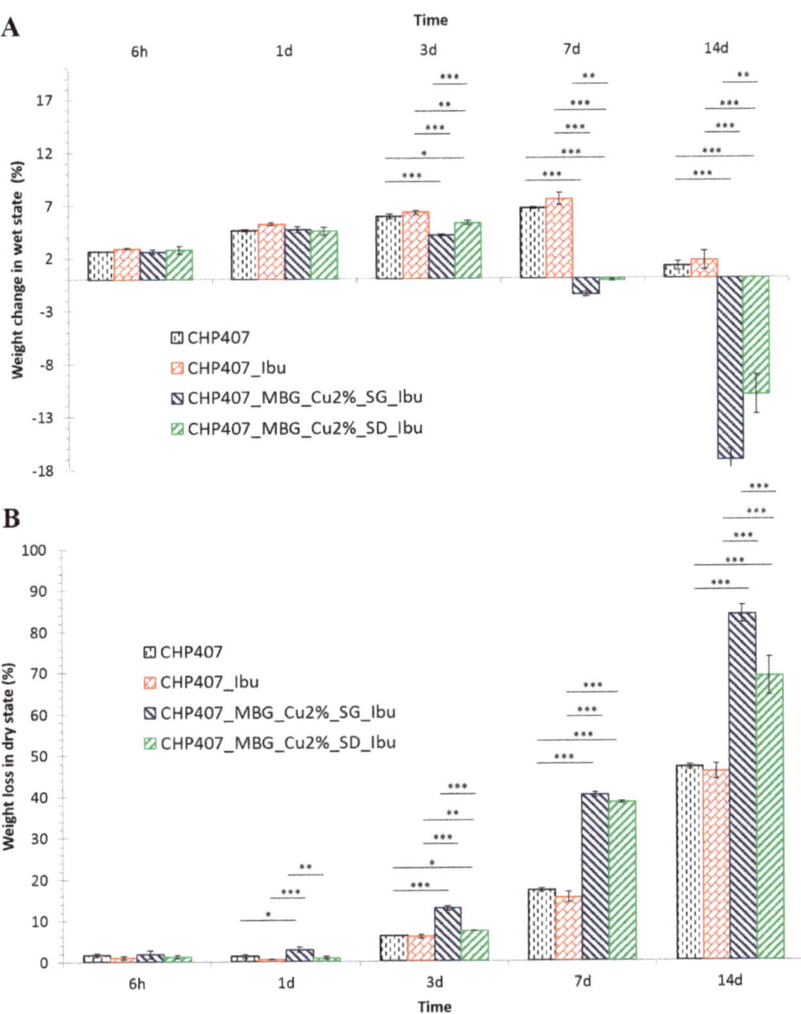

Figure 5. Swelling (**A**) and weight loss (**B**) of CHP407, CHP407_Ibu, CHP407_MBG_Cu2%_SG_Ibu and CHP407_MBG_Cu2%_SD_Ibu gels evaluated at 6 h, 1d, 3d, 7d and 14d.

SEC analysis was also performed on the residual polymer phase collected on day 14 (Figure 6).

A

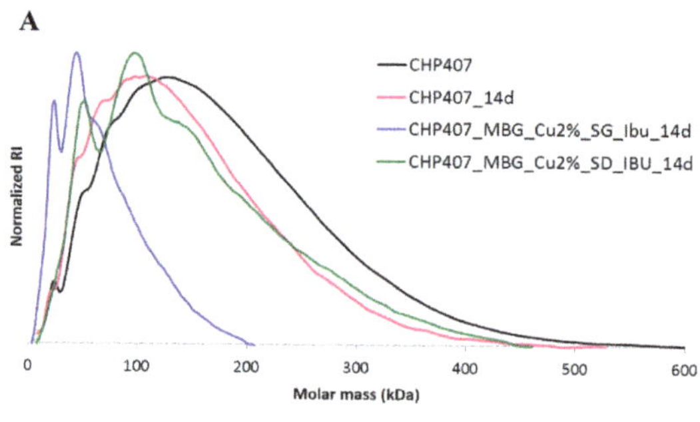

B

	M_n (Da)	D
CHP407	71670	1.7
CHP407_14d	64260	1.6
CHP407_MBG_Cu2%_SG_Ibu_14d	28746	1.6
CHP407_MBG_Cu2%_SD_Ibu_14d	28381	1.8

Figure 6. Molecular weight distribution (normalized refractive index (RI) vs. molar mass) (**A**) and estimated M_n and D values (**B**) of as synthesized CHP407 and residual CHP407 collected from CHP407, CHP407_MBG_Cu2%_SG_Ibu and CHP407_MBG_Cu2%_SD_Ibu gels incubated in aqueous medium for 14 days.

SEC revealed that the high destabilization of MBG containing gels in aqueous medium is ascribable to a progressive polymer chemical degradation, which was almost absent in pure CHP407 and CHP407_Ibu gels. In fact, CHP407 number average molecular weight decreased of about 60% and 10% in MBG-containing and pure hydrogel samples, respectively. In analogy with data concerning the bio-stability for poly(ether urethane)-based biomedical devices, such as pacemaker leads, this behavior can be ascribed to the occurrence of metal ion-mediated oxidation induced by the copper ions progressively released within the gels and the surrounding medium, which trigger the oxidative degradation of the polymer network [38,39].

3.3.3. Ion/Drug Release Test

Ibuprofen release profile from CHP407-based gels was assessed in physiological-like conditions, that is, Trizma® buffer at 37 °C (Figure 7).

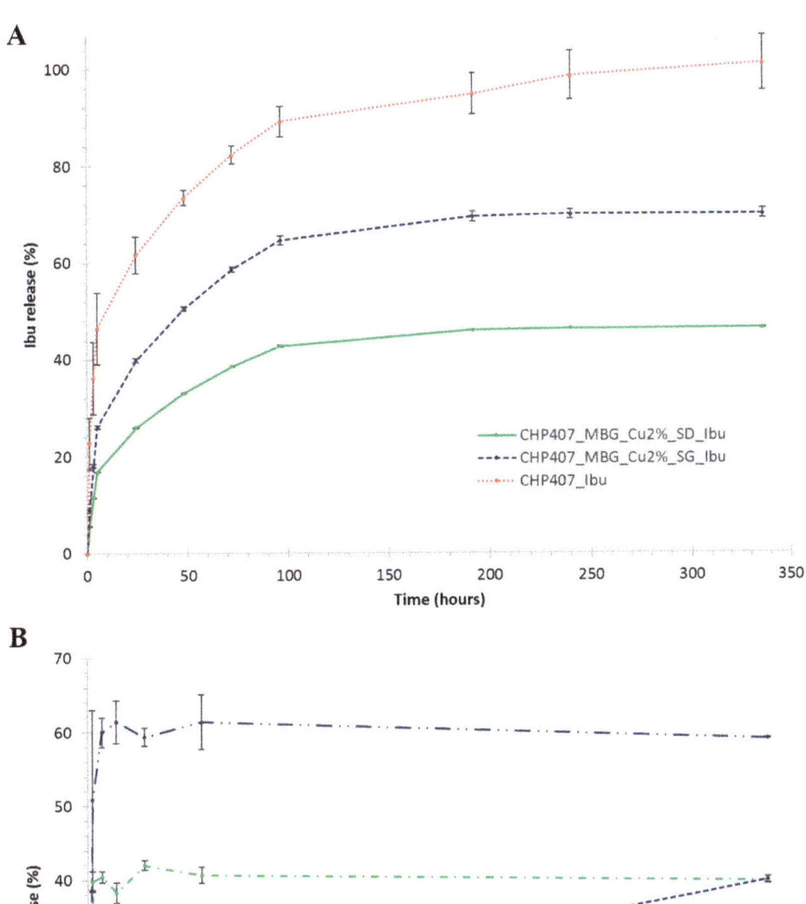

Figure 7. (**A**) Ibuprofen release (%) profile from CHP407_Ibu, CHP407_MBG_Cu2%_SG_Ibu and CHP407_MBG_Cu2%_SD_Ibu hydrogels. (**B**) Comparison among ibuprofen release profiles assessed from CHP407_MBG_Cu2%_SG_Ibu, CHP407_MBG_Cu2%_SD_Ibu, MBG_Cu2%_SD_Ibu and MBG_Cu2%_SD_Ibu.

Figure 7A compares ibuprofen release profiles from CHP407_Ibu, CHP407_MBG_Cu2%_SG_Ibu and CHP407_MBG_Cu2%_SD_Ibu up to 14 days observation time. A complete release of the drug was observed from the gels loaded with ibuprofen as such, with a release of 98.6 ± 5.0% at day

10. On the other hand, MBG-loaded gels showed a sustained and prolonged release of the drug, with a percentage release of 70.0 ± 1.0% and 46.4 ± 0.1% from CHP407_MBG_Cu2%_SG_Ibu and CHP407_MBG_Cu2%_SD_Ibu, respectively, after 14 days incubation. Moreover, starting from day 8 the release profile reached a plateau value for both particle-loaded gels, thus suggesting an incomplete ibuprofen release. However, this phenomenon cannot be avoided, being correlated to the progressive pore occlusion due to the dissolution of silica-based MBG framework and its re-precipitation as silica gel at the pores mouth, in accordance with observations reported by Mortera et al. [40], who, in addition, also reported a significant change of the delivery curves as a function of the release medium volume. To further investigate this aspect, the release profile of ibuprofen from MBG_Cu2%_SG_Ibu and MBG_Cu2%_SD_Ibu alone was studied in the same release conditions (in term of powder/medium volume ratio) adopted for hybrid hydrogel testing. As a matter of fact, both MBG_Cu2%_SG_Ibu and MBG_Cu2%_SD_Ibu showed a plateau ibuprofen release of 61.3 ± 3.7% and 40.6 ± 1.0% after 30 and 10 min incubation time, respectively. The significant difference in the maximum amount released can be attributed to the very high surface reactivity of spray-dried systems when soaked in aqueous media, leading to enhanced dissolution/re-precipitation phenomena and consequent hindered pore accessibility [16]. By comparing ibuprofen release kinetics from free MBGs and hydrogels containing MBG particles within the first 4 h, the role exerted by the polymeric matrix in modulating the release profile of ibuprofen was clearly highlighted (Figure 7B). In fact, a significant reduction ($0.0001 < p < 0.001$) in the initial burst release upon MBG incorporation within CHP407 gels was observed, with approx. 85% burst release reduction for both kinds of particles investigated. The release profile of copper ions from the developed hybrid formulations was also studied, showing a trend similar to that assessed for ibuprofen (Figure 8).

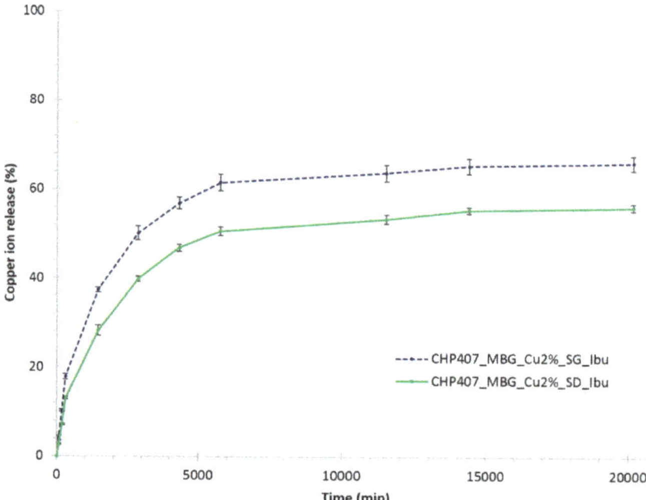

Figure 8. Copper ion release (%) profile from CHP407_MBG_Cu2%_SG_Ibu and CHP407_MBG_Cu2%_SD_Ibu gels.

After 14 days of incubation in aqueous medium, the 66.1 ± 1.6% and 56.1 ± 0.8% (corresponding to 291.5 ± 7.1 ppm and 192.3 ± 2.7 ppm, respectively) of the copper initially present in the MBG framework was released from CHP407_MBG_Cu2%_SG_Ibu and CHP407_MBG_Cu2%_SD_Ibu, respectively. These amounts fully agree with our previous results obtained for Cu-substituted MBGs without ibuprofen [16], highlighting that drug loading within MBG porous structure did not suppress or hinder the capability to release therapeutic ions through ion-exchange reactions. The incorporation of Cu-substituted MBGs within the polymeric phase successfully decreased the undesirable initial burst release of Cu^{2+} species typically observed for particles alone ($0.0001 < p < 0.001$). Indeed, after 1h incubation in similar releasing conditions, CHP407_MBG_Cu2%_SG_Ibu and CHP407_MBG_Cu2%_SD_Ibu released an amount of copper ions approximately 72 % and 61 % lower compared to the corresponding free MBG particles. Both ibuprofen and copper ions were released faster from CHP407_MBG_Cu2%_SG_Ibu compared to CHP407_MBG_Cu2%_SD_Ibu, in accordance with the higher surface area of MBG_Cu2%_SG_Ibu compared to MBG_Cu2%_SD_Ibu, which accounts for a faster molecule diffusion, either ions or drugs, through MBG porous network.

To clarify the ion/drug release mechanism, the equation of Korsmeyer-Peppas was employed to estimate the release exponent n within the timeframe 1–5 h (being Korsmeyer-Peppas equation valid up to 60% release). Ibuprofen delivery from CHP407_Ibu gels turned out to be a purely Fickian diffusion-driven release, with an n exponent of 0.45 ± 0.04. On the other hand, CHP407_MBG_Cu2%_SG_Ibu and CHP407_MBG_Cu2%_SD_Ibu gels released ibuprofen with an anomalous mechanism (n value of 0.66 ± 0.03 and 0.68 ± 0.01 for CHP407_MBG_Cu2%_SG_Ibu and CHP407_MBG_Cu2%_SD_Ibu, respectively), being the drug first diffused from the MBGs within the hydrogel and then through the hydrogel in the surrounding releasing medium. Interestingly, the release exponent clearly highlighted the coexistence of diffusion and ion exchange reactions when copper species were released, being n values higher than 0.89 (n value of 0.94 ± 0.08 and 0.90 ± 0.07 for CHP407_MBG_Cu2%_SG_Ibu and CHP407_MBG_Cu2%_SD_Ibu, respectively).

Taken together the obtained data proved the ability of the developed platform to co-deliver copper ions and ibuprofen with a sustained release profile. The final released concentration can be finely modulated to target the required therapeutic effect with no associated toxicity, by taking advantage of the several degrees of freedom offered by the hybrid system, such as the initial particle concentration within the gel solution (before gelation) and the amount of ion/drug loaded inside the MBG nanocarriers.

4. Conclusions

In this work, a first attempt toward the design of a hybrid sol-gel formulation able to simultaneously co-release copper ions (exerting pro-angiogenic and anti-bacterial effects) and an anti-inflammatory drug (ibuprofen) was successfully achieved. To this aim, Cu-substituted MBG (2% mol) particles were loaded with ibuprofen through an incipient wetness technique, resulting in a high loading capacity, and then embedded within PEU-based thermosensitive gels at the highest possible concentration (20 mg/mL), according to an optimized incorporation protocol. Full characterization of the resulting hybrid systems was performed to highlight the effects of particle encapsulation on gelation kinetics as well as on ions/drug release mechanism. The incorporation of MBG particles within the sol-gel systems did not negatively affect their capability to undergo a temperature-driven sol-to-gel transition within a few minutes. The progressive release of Cu^{2+} species was found to play a significant role on the stability of the gels in an aqueous environment and catalyzed the oxidation of the PEU chains. The co-release of copper ions and ibuprofen from hybrid formulations was sustained and prolonged over time for up to more than one week, with a strongly reduced initial burst effect compared to MBG particles alone (2%–4% vs. 7%–14% Cu^{2+} release and 6%–9% vs. 38%–61% ibuprofen release from hybrid MBG-polyurethane formulations and free MBG particles, respectively). However, the release profile of both copper species and ibuprofen was affected by the progressive occlusion of mesopores resulting from the dissolution and the re-precipitation of silica-based MBG framework in the form of a silica gel at the pore entrance [40–42].

Taken together the obtained data proved the ability of the proposed hybrid thermosensitive formulation to concentrate and maintain the MBG carriers at the pathological site and to guarantee in situ and prolonged co-release of ions and drugs, thus opening the way to the design of multifunctional platforms for advanced treatment of compromised tissue healing. The high versatility of the proposed approach lies in the possibility to modulate the relative amount of the organic and inorganic components, by changing the initial particle concentration within the hydrogel solution (before gelation), and/or the initial drug loading and the chemical composition of MBGs, in order to design systems able to co-release ions and pharmaceutics with concentrations and kinetics adapted to the targeted applications and not producing cytotoxic effects.

The versatility of the herein developed hybrid sol-gel systems could thus pave the way to the treatment of a great variety of pathological conditions of soft (e.g., non-healing wounds) and hard (e.g., delayed bone healing) tissues.

Supplementary Materials: The following are available online at http://www.mdpi.com/1999-4923/11/10/501/s1, Figure S1: TEM image of MBG_Cu2%_SD.

Author Contributions: Conceptualization, M.B., C.P., S.F. and R.L.; methodology, M.B., C.P. and R.L.; validation, M.B., C.P., S.F. and R.L.; formal analysis, M.B., C.P. and R.L.; investigation, M.B., C.P. and R.L.; resources, M.B., C.P. and R.L.; writing—original draft preparation, M.B., C.P., S.F. and R.L.; writing—review and editing, S.F., G.C. and C.V.-B.; visualization, M.B.; supervision, S.F., G.C. and C.V.-B.; funding acquisition, C.V.-B.

Funding: This project has received funding from the European Union's Horizon 2020 research and innovation program under grant agreement No. 685872-MOZART (www.mozartproject.eu).

Conflicts of Interest: The authors declare no competing financial interest.

References

1. Kargozar, S.; Montazerian, M.; Hamzehlou, S.; Kim, H.W.; Baino, F. Mesoporous bioactive glasses: Promising platforms for antibacterial strategies. *Acta Biomater.* **2018**, *81*, 1–19. [CrossRef] [PubMed]
2. Vishnu Priya, M.; Sivshanmugam, A.; Boccaccini, A.R.; Goudouri, O.M.; Sun, W.; Hwang, N.; Deepthi, S.; Nair, S.V.; Jayakumar, R. Injectable osteogenic and angiogenic nanocomposite hydrogels for irregular bone defects. *Biomed. Mater.* **2016**, *11*, 035017. [CrossRef] [PubMed]
3. Lowe, S.B.; Tan, V.T.G.; Soeriyadi, A.H.; Davis, T.P.; Gooding, J.J. Synthesis and High-Throughput Processing of Polymeric Hydrogels for 3D Cell Culture. *Bioconjug. Chem.* **2014**, *25*, 1581–1601. [CrossRef] [PubMed]
4. Zhu, M.; Zhu, Y.; Zhang, L.; Shi, J. Preparation of chitosan / mesoporous silica nanoparticle composite hydrogels for sustained co-delivery of biomacromolecules and small chemical drugs. *Sci. Technol. Adv. Mater.* **2013**, *14*, 045005. [CrossRef] [PubMed]
5. Zhao, P.; Liu, H.; Deng, H.; Xiao, L.; Qin, C. Colloids and Surfaces B: Biointerfaces A study of chitosan hydrogel with embedded mesoporous silica nanoparticles loaded by ibuprofen as a dual stimuli-responsive drug release system for surface coating of titanium implants. *Colloids Surf. B Biointerfaces* **2014**, *123*, 657–663. [CrossRef] [PubMed]
6. Lunter, D.J. Evaluation of mesoporous silica particles as drug carriers in hydrogels. *Pharm. Dev. Technol.* **2017**, *23*, 826–831. [CrossRef] [PubMed]
7. Kim, B.S.; Chen, Y.; Srinoi, P.; Marquez, M.D.; Lee, T.R. Hydrogel-Encapsulated Mesoporous Silica-Coated Gold Nanoshells for Smart Drug Delivery. *Int. J. Mol. Sci.* **2019**, *20*, 3422. [CrossRef] [PubMed]
8. Lei, L.; Liu, Z.; Pingyun, Y.; Yuan, P.; Jin, R.; Wang, X.-d.; Jiang, T.; Chen, X. Injectable colloidal hydrogel with mesoporous silica nanoparticles for sustained co-release of microRNA-222 and aspirin to achieve innervated bone regeneration in rat mandibular defects. *J. Mater. Chem. B* **2019**, *7*, 2722–2735. [CrossRef]
9. Fiorilli, S.; Molino, G.; Pontremoli, C.; Iviglia, G.; Torre, E.; Cassinelli, C.; Morra, M.; Vitale-Brovarone, C. The incorporation of strontium to improve bone-regeneration ability of mesoporous bioactive glasses. *Materials* **2018**, *11*, 678. [CrossRef]
10. Zhou, Y.; Han, S.; Xiao, L.; Han, P.; Wang, S.; He, J.; Chang, J.; Wu, C.; Xiao, Y. Accelerated host angiogenesis and immune responses by ion release from mesoporous bioactive glass. *J. Mater. Chem. B* **2018**, *6*, 3274–3284. [CrossRef]
11. Wu, C.; Zhou, Y.; Xu, M.; Han, P.; Chen, L.; Chang, J.; Xiao, Y. Copper-containing mesoporous bioactive glass scaffolds with multifunctional properties of angiogenesis capacity, osteostimulation and antibacterial activity. *Biomaterials* **2013**, *34*, 422–433. [CrossRef] [PubMed]
12. Wang, X.; Cheng, F.; Liu, J.; Smått, J.H.; Gepperth, D.; Lastusaari, M.; Xu, C.; Hupa, L. Biocomposites of copper-containing mesoporous bioactive glass and nanofibrillated cellulose: Biocompatibility and angiogenic promotion in chronic wound healing application. *Acta Biomater.* **2016**, *46*, 286–298. [CrossRef] [PubMed]

13. Bari, A.; Bloise, N.; Fiorilli, S.; Novajra, G.; Vallet-Regí, M.; Bruni, G.; Torres-Pardo, A.; González-Calbet, J.M.; Visai, L.; Vitale-Brovarone, C. Copper-containing mesoporous bioactive glass nanoparticles as multifunctional agent for bone regeneration. *Acta Biomater.* **2017**, *55*, 493–504. [CrossRef] [PubMed]
14. Li, H.; Li, J.; Jiang, J.; Lv, F.; Chang, J.; Chen, S.; Wu, C. An osteogenesis / angiogenesis-stimulation artificial ligament for anterior cruciate ligament reconstruction. *Acta Biomater.* **2017**, *54*, 399–410. [PubMed]
15. Romero-sánchez, L.B.; Marí-beffa, M.; Carrillo, P.; Ángel, M.; Díaz-cuenca, A. Copper-containing mesoporous bioactive glass promotes angiogenesis in an in vivo zebrafish model. *Acta Biomater.* **2018**, *68*, 272–285. [CrossRef] [PubMed]
16. Pontremoli, C.; Boffito, M.; Fiorilli, S.; Laurano, R.; Torchio, A.; Bari, A.; Tonda-Turo, C.; Ciardelli, G.; Vitale-Brovarone, C. Hybrid injectable platforms for the in situ delivery of therapeutic ions from mesoporous glasses. *Chem. Eng. J.* **2018**, *340*, 103–113. [CrossRef]
17. Vergara-figueroa, J.; Alejandro-mart, S.; Pesenti, H. Obtaining Nanoparticles of Chilean Natural Zeolite and its Ion Exchange with Copper Salt (Cu^{2+}) for antibacterial applications. *Materials (Basel)* **2019**, *12*, 2202. [CrossRef]
18. Rosenberg, M.; Vija, H.; Kahru, A.; Keevil, C.W.; Ivask, A. Rapid in situ assessment of Cu-ion mediated effects and antibacterial efficacy of copper surfaces. *Sci. Rep.* **2018**, *8*, 8172. [CrossRef]
19. Barralet, J.; Gbureck, U.; Habibovic, P.; Vorndran, E.; Gerard, C.; Doillon, C.J. Angiogenesis in Calcium Phosphate Scaffolds by Inorganic Copper Ion Release. *Tissue Eng. Part A* **2009**, *15*, 1601–1609. [CrossRef]
20. Charnay, C.; Bégu, S.; Tourné-Péteilh, C.; Nicole, L.; Lerner, D.A.; Devoisselle, J.M. Inclusion of ibuprofen in mesoporous templated silica: Drug loading and release property. *Eur. J. Pharm. Biopharm.* **2004**, *57*, 533–540.
21. Boffito, M.; Gioffredi, E.; Chiono, V.; Calzone, S.; Ranzato, E.; Martinotti, S.; Ciardelli, G. Novel polyurethane-based thermosensitive hydrogels as drug release and tissue engineering platforms: Design and in vitro characterization. *Polym. Int.* **2016**, *65*, 756–769. [CrossRef]
22. Alsirawan, M.B.; Mohammad, M.A.; Alkasmi, B.; Alhareth, K.; El-Hammadi, M. Development and validation of a simple HPLC method for the determination of ibuprofen sticking onto punch faces. *Int. J. Pharm. Pharm. Sci.* **2013**, *5*, 227–231.
23. Boffito, M.; Brancot, A.G.; Lima, O.; Bronco, S.; Sartori, S.; Ciardelli, G.; Torino, P. Injectable thermosensitive gels for the localized and controlled delivery of biomolecules in tissue engineering/regenerative medicine. *Biomed. Sci. Eng.* **2019**, *3*. [CrossRef]
24. El-Fiqi, A.; Kim, T.-H.; Kim, M.; Eltohamy, M.; Won, J.-E.; Lee, E.-J.; Kim, H.-W. Capacity of mesoporous bioactive glass nanoparticles to deliver therapeutic molecules. *Nanoscale* **2012**, *4*, 7475–7488. [CrossRef] [PubMed]
25. Vallet-Regí, M. Ordered mesoporous materials in the context of drug delivery systems and bone tissue engineering. *Chem. Eur. J.* **2006**, *12*, 5934–5943. [CrossRef] [PubMed]
26. Mellaerts, R.; Jammaer, J.A.; Van Speybroeck, M.; Chen, H.; Van Humbeeck, J.; Augustijns, P.; Van den Mooter, G.; Martens, J.A. Physical state of poorly water soluble therapeutic\rmolecules loaded into SBA-15 ordered mesoporous silica carriers:\rA case study with itraconazole and ibuprofen. *Langmuir* **2008**, *24*, 8651–8659. [CrossRef] [PubMed]
27. Gonzalez, G.; Sagarzazu, A.; Zoltan, T. Infuence of Microstructure in Drug Release Behavior of Silica Nanocapsules. *J. Drug Deliv.* **2013**, *2013*, 803585. [CrossRef] [PubMed]
28. Hong, S.; Shen, S.; Tan, D.C.T.; Ng, W.K.; Liu, X.; Chia, L.S.O.; Irwan, A.W.; Tan, R.; Nowak, S.A.; Marsh, K.; et al. High drug load, stable, manufacturable and bioavailable fenofibrate formulations in mesoporous silica: A comparison of spray drying versus solvent impregnation methods. *Drug Deliv.* **2016**, *23*, 316–327. [CrossRef]
29. López-Noriega, A.; Arcos, D.; Izquierdo-Barba, I.; Sakamoto, Y.; Terasaki, O.; Vallet-Regí, M. Ordered mesoporous bioactive glasses for bone tissue regeneration. *Chem. Mater.* **2006**, *18*, 3137–3144. [CrossRef]
30. Brás, A.R.; Merino, E.G.; Neves, P.D.; Fonseca, I.M.; Dionísio, M.; Schönhals, A.; Correia, N.T. Amorphous ibuprofen confined in nanostructured silica materials: A dynamical approach. *J. Phys. Chem. C* **2011**, *115*, 4616–4623. [CrossRef]
31. Zhang, Y.; Zhi, Z.; Jiang, T.; Zhang, J.; Wang, Z.; Wang, S. Spherical mesoporous silica nanoparticles for loading and release of the poorly water-soluble drug telmisartan. *J. Control. Release* **2010**, *145*, 257–263. [CrossRef]

32. Fiorilli, S.; Onida, B.; Bonelli, B.; Garrone, E. In situ infrared study of SBA-15 functionalized with carboxylic groups incorporated by a Co-condensation route. *J. Phys. Chem. B* **2005**, *109*, 16725–16729. [CrossRef]
33. Cauda, V.; Fiorilli, S.; Onida, B.; Vernè, E.; Vitale Brovarone, C.; Viterbo, D.; Croce, G.; Milanesio, M.; Garrone, E. SBA-15 ordered mesoporous silica inside a bioactive glass-ceramic scaffold for local drug delivery. *J. Mater. Sci. Mater. Med.* **2008**, *19*, 3303–3310. [CrossRef]
34. Wang, F.; Hui, H.; Barnes, T.J.; Barnett, C.; Prestidge, C.A. Oxidized mesoporous silicon microparticles for improved oral delivery of poorly soluble drugs. *Mol. Pharm.* **2010**, *7*, 227–236. [CrossRef]
35. Shen, S.C.; Ng, W.K.; Hu, J.; Letchmanan, K.; Ng, J.; Tan, R.B.H. Solvent-free direct formulation of poorly-soluble drugs to amorphous solid dispersion via melt-absorption. *Adv. Powder Technol.* **2017**, *28*, 1316–1324. [CrossRef]
36. Sliwinska-Bartkowiak, M.; Dudziak, G.; Gras, R.; Sikorski, R.; Radhakrishnan, R.; Gubbins, K.E. Freezing behavior in porous glasses and MCM-41. *Colloids Surf. A Physicochem. Eng. Asp.* **2001**, *187–188*, 523–529. [CrossRef]
37. Radhakrishnan, R.; Gubbins, K.E.; Sliwinska-Bartkowiak, M. Effect of the fluid-wall interaction on freezing of confined fluids: Toward the development of a global phase diagram. *J. Chem. Phys.* **2000**, *112*, 11048–11057. [CrossRef]
38. Thoma, R.J.; Tan, F.R.; Phillips, R.E. Ionic Interactions of Polyurethanes. *J. Biomater. Appl.* **1988**, *3*, 180–206. [CrossRef]
39. Coury, A.J. Chemical and Biochemical Degradation of Polymers Intended to be Biostable. In *Biomaterials Science: An Introduction to Materials in Medicine*, 3rd ed.; Elsevier: Amsterdam, The Netherlands, 2013; ISBN 9780123746269.
40. Mortera, R.; Fiorilli, S.; Garrone, E.; Verné, E.; Onida, B. Pores occlusion in MCM-41 spheres immersed in SBF and the effect on ibuprofen delivery kinetics: A quantitative model. *Chem. Eng. J.* **2010**, *156*, 184–192. [CrossRef]
41. Li, P.; Nakanishi, K.; Kokubo, T.; de Groot, K. Induction and morphology of hydroxyapatite, precipitated from metastable simulated body fluids on sol-gel prepared silica. *Biomaterials* **1993**, *14*, 963–968. [CrossRef]
42. Li, P.; Kangasniemi, I.; de Groot, K.; Kokubo, T.; Yli-Urpo, A.U. Apatite crystallization from metastable calcium phosphate solution on sol-gel-prepared silica. *J. Non-Cryst. Solids* **1994**, *168*, 281–286. [CrossRef]

Article

Silicon Nanofluidic Membrane for Electrostatic Control of Drugs and Analytes Elution

Nicola Di Trani [1,2], Antonia Silvestri [1,3], Yu Wang [1], Danilo Demarchi [3], Xuewu Liu [1] and Alessandro Grattoni [1,4,5,*]

1 Department of Nanomedicine, Houston Methodist Research Institute, Houston, TX 77030, USA; nditrani@houstonmethodist.org (N.D.T.); antonia.silvestri@polito.it (A.S.); ywang2@houstonmethodist.org (Y.W.); xliu@houstonmethodist.org (X.L.)
2 University of Chinese Academy of Science (UCAS), Shijingshan, 19 Yuquan Road, Beijing 100049, China
3 Department of Electronics and Telecommunications, Polytechnic of Turin, 10129 Turin, Italy; danilo.demarchi@polito.it
4 Department of Surgery, Houston Methodist Hospital, Houston, TX 77030, USA
5 Department of Radiation Oncology, Houston Methodist Hospital, Houston, TX 77030, USA
* Correspondence: agrattoni@houstonmethodist.org; Tel.: +1-(713)-441-7324

Received: 8 June 2020; Accepted: 15 July 2020; Published: 19 July 2020

Abstract: Individualized long-term management of chronic pathologies remains an elusive goal despite recent progress in drug formulation and implantable devices. The lack of advanced systems for therapeutic administration that can be controlled and tailored based on patient needs precludes optimal management of pathologies, such as diabetes, hypertension, rheumatoid arthritis. Several triggered systems for drug delivery have been demonstrated. However, they mostly rely on continuous external stimuli, which hinder their application for long-term treatments. In this work, we investigated a silicon nanofluidic technology that incorporates a gate electrode and examined its ability to achieve reproducible control of drug release. Silicon carbide (SiC) was used to coat the membrane surface, including nanochannels, ensuring biocompatibility and chemical inertness for long-term stability for in vivo deployment. With the application of a small voltage (≤ 3 V DC) to the buried polysilicon electrode, we showed in vitro repeatable modulation of membrane permeability of two model analytes—methotrexate and quantum dots. Methotrexate is a first-line therapeutic approach for rheumatoid arthritis; quantum dots represent multi-functional nanoparticles with broad applicability from bio-labeling to targeted drug delivery. Importantly, SiC coating demonstrated optimal properties as a gate dielectric, which rendered our membrane relevant for multiple applications beyond drug delivery, such as lab on a chip and micro total analysis systems (μTAS).

Keywords: electrostatic gating; nanofluidic diffusion; controlled drug release; silicon membrane; smart drug delivery

1. Introduction

Chronic pathologies affect nearly half of the population worldwide [1,2] and represent one of the leading causes of death and disability [3]. Management of chronic conditions is challenged by co-morbidities [4], poor adherence to treatment [5], and a lack of therapeutic technologies suitable to address the complexity of the disease [6]. Long-acting controlled therapeutic administration represents a promising strategy for medical conditions requiring repeated daily dosing [7,8]. In view of this, long-acting platforms for sustained drug release have been developed, leading to significant improvements in the management of conditions, such as hormone deficiency and infectious diseases [9–11]. However, the pathophysiology of most chronic diseases is determined by circadian biological cycles [12], which have a significant impact on the efficacy of treatment and associated adverse effects [13]. This is the case for pathologies, such as diabetes and metabolic disorders [14], hypertension, psychiatric and neurodegenerative conditions [15], rheumatoid arthritis [16], and chronic pain [17], to name a few, where the timing of drug administration is key to elicit the intended therapeutic effect.

Advanced technologies enabling personalized adjustments of therapeutic administration, both in time and dose, represent a desirable but unmet clinical need [18,19]. Ideally, these technologies should incorporate a drug delivery mechanism that can be rapidly and easily tuned to release the required dosage, at the right time, without requiring continuous external stimuli. Further, they should allow for pre-programmed dosing schedules as well as remote control capabilities, to enable healthcare providers to adjust medication through telemedicine approaches [20]. Devices with such capabilities could eradicate treatment compliance issues and dramatically improve the therapeutic index and the quality of life of patients, while substantially reducing healthcare expenditure.

Current approaches developed for controlled drug administration are based on modulation of permeability of membranes via sustained external stimuli. These systems mostly rely on polymeric membrane architectures and achieve changes in pore size and conformation via temperature variation triggered by a magnetic field [21], near-infrared irradiation [22–24], or ultrasound [25]. Other devices use a magnetic field to reversibly or irreversibly obstruct the pores of a membrane using microparticles [26] or low melting temperature polymers [27]. Albeit promising, these strategies are limited by the need for continuous external activation and associated cumbersome external equipment. Electrical actuation offers a solution to these limitations, enabling control via miniaturized circuitry and low energy radio-frequency (RF) communications. In this context, various technologies have been created, either integrating gate electrodes [28] or polypyrrole (PPy) [29,30] on anodic aluminum oxide (AAO) membranes. However, polydispersity in pore size common to AAO membranes represents a limitation to achieve fine control of drug release [28,31].

In this study, we investigated the performance of a silicon nanofluidic membrane that uses electrostatic gating [32] to modulate the transport of charged molecules by modifying nanochannel permeability. Microfabricated using standard semiconductor manufacturing techniques, this membrane features hundreds of thousands of identical slit-nanochannels geometrically distributed across the membrane surface to maximize porosity while maintaining mechanical integrity. A buried polysilicon layer extends over the entire nanochannel surface and acts as a single distributed gate electrode. An outermost layer of biocompatible silicon carbide (SiC) is adopted to bury and insulate the gate electrode and minimize leakage while providing chemical inertness for applications in vivo or in contact with biological fluids. SiC insulation properties were studied in comparison with silicon dioxide (SiO_2), the most common gate dielectric material in metal–oxide–semiconductor field-effect transistor (MOSFET). Further, energy consumption leakage current and gating performance were assessed at different gate potentials. Finally, we adopted two relevant model analytes—methotrexate and quantum dots—to assess the in vitro transport modulation performances. Methotrexate represents an important therapeutic agent commonly used for rheumatoid arthritis [33], whereas quantum dots are adopted for a variety of biomedical imaging applications as well as drug delivery and theranostics [34,35]. In light of the promising results and the ease of integration within implantable devices, our gated membrane might constitute a promising step forward in the development of flexible technologies for the treatment of chronic diseases. Further, our nanofluidic technology could be adopted in other applications for lab on a chip [36] and micro total analysis systems (μTAS) devices for electrokinetic separation processes [37], bio-sample sorting and analysis [38], among others.

2. Materials and Methods

2.1. Nanofluidic Membrane Fabrication

Silicon membranes fabrication was performed using standard semiconductor techniques. The fabrication process is described step-by-step elsewhere [39]. Briefly, a dense array of nanochannels (500 nm width, 6 μm length) was obtained by vertically etching via deep reactive ion etching (DRIE) the device layer (10 μm) of a silicon on insulator (SOI) wafer (total thickness 411 μm). The etching was stopped at the middle oxide layer (1 μm). The handle wafer (400 μm) on the opposite side of the SOI was etched using DRIE up to the oxide layer to create a hexagonal pattern of densely packed circular microchannels. To connect the nanochannels to the microchannels, the buried oxide layer was etched by a buffered oxide etchant solution (BOE). The resulting nanochannels size (770 nm) was reduced by three subsequent processes. First, wet thermal oxidation generated a 175 nm layer of SiO_2, and then a 121 nm layer of polycrystalline silicon (poly-Si) was obtained via low-pressure chemical vapor deposition (LPCVD). Plasma-enhanced chemical vapor deposition (PECVD) was then used for the silicon carbide coating (SiC, 64 nm). Electrode pads were exposed via selective etching of SiC via fluorine-based RIE. Wafers were then diced into individual membranes (ADT 7100 Dicing Saw, Advanced Dicing Technologies, Zhengzhou, China), obtaining individual silicon membranes of 6 mm × 6 mm × 411 μm. Each membrane featured a total of 278,600 identical slit nanochannels 10 μm long and 6 μm wide. Nanochannels were arranged in groups of 1400, where each group led to one circular microchannel on the opposite side of the chip. Finally, microchannels were geometrically organized in a hexagonal pattern to maximize porosity and structural integrity. In this study, membranes with a final layer of SiO_2 were also used and obtained by wet thermal oxidation of poly-Si.

2.2. Assessment of Membrane Structure

Morphological assessment and characterization of the nanochannel multi-layer structure were performed via scanning electron microscopy (SEM). A focused ion beam (FIB) system FEI 235 (Nanofabrication facility of the University of Houston, Houston, TX, USA) was used to simultaneously create nanochannels' cross-sections and acquire images. Gallium ion milling was performed on the micromachine parts of the membrane and to expose the nanochannel cross-section. Imaging was then performed at a 52° angle.

2.3. Electrode Connection

Electrical wires (36 AWG, McMaster Carr, Elmhurst, IL, USA) were epoxied to the membrane pads using a silver-based conductive adhesive (H20E, Epoxy Technology, Inc., Billerica, MA, USA) and cured at 150 °C for 1 h. Electrode insulation was achieved by applying a thin layer of UV epoxy (OG116, Epoxy Technologies, Inc. Billerica, MA, USA) over the conductive pad and UV-curing (UVP UVL-18 EL Series, Analytik Jena US LLC, Upland, CA, USA) for 120 min. The correct electrode connection was tested by measuring the resistance between the two connection pads with a Fluke 177 True RMS Multimeter (Fluke Corporation, Everett, WA, USA).

2.4. Electrochemical Characterization

A custom dual-reservoir polymethyl methacrylate (PMMA) apparatus [39] was employed to perform electrochemical measurements. Membranes were clamped between the two 2 mL reservoirs of the testing apparatus, each containing two Ag/AgCl electrodes (64-1313, Harvard Apparatus, Holliston, MA, USA). All measurements were performed in PBS, except for conductance studies, where KCl solutions at different concentrations (from 10^{-6} to 10^{-1} M) were used. A benchtop electrochemical tester (CH Instruments, Inc. 660E, Austin, TX, USA) was used in either 3 or 4 electrode configurations.

Impedance was measured with a 4-electrodes configuration. A 50 mV perturbation signal was applied through the electrochemical analyzer within a frequency window from 10 mH to 10 kH. The measurements were performed with a superimposed DC voltage in the range −3 V to 3 V in steps of 1 V. Fittings to a Randles cell model were performed with the CHI 660E software (CH Instruments, Inc. 660E, Austin, TX, USA).

Leakage current was measured with a 3-electrodes configuration. Voltages were applied using the CHI 660E between the electrode pad and Ag/AgCl electrodes in solution at a distance of ~1 cm from the membrane. Measurements were performed in the −3 V to +3 V range in 1 V steps, and each step lasted for 120 s, allowing for transient phenomena to resolve and obtain a stable measurement.

For conductance experiments, we employed a 4-electrodes configuration. Measurements were performed for KCl concentrations from 1 μM to 100 mM from the lower to the highest ionic strength. Reservoirs were rinsed with deionized water for 1 min, and the solution was replaced after each measurement. Steps of 400 mV were applied using the CHI 660E from −2 V to 2 V, with 30 s pauses to exhaust possible transient effects. Conductance measurements were performed with a floating gate, and the values calculated for each step and averaged.

Cyclic voltammetry measurements were conducted with a 3-electrodes configuration and a scanning rate of 50 mV/s within the interval −2 V to 2 V. Electrochemical measurements were carried out on membranes with a final dielectric layer of both SiC and SiO_2.

2.5. In Vitro Release Modulation

In vitro release modulation experiments were performed employing a custom dual-reservoir device described in detail elsewhere [40]. Nanochannel membranes were individually clamped between a 250 μL drug reservoir and a UV-Vis transparent macro-cuvette serving as the sink reservoir. Two O-rings were used to prevent fluid leakage between membranes and the reservoir. Fluid evaporation was prevented by sealing a drug reservoir with biocompatible silicone plugs (McMaster Carr, Elmhurst, IL, USA).

Experiments were performed using SiC-coated membranes with ~300 nm nanochannels. To ensure proper channel wetting, membranes were immersed in isopropyl alcohol for 1 h and then rinsed three times in deionized H_2O. Membranes were then placed overnight in 0.01 × PBS or 1 × PBS in preparation for quantum dots and methotrexate release, respectively. Sink reservoirs (4.45 mL) were filled with matching PBS solutions. After fixture assembly, the source reservoir was loaded with either 1 mg/mL 0.01 × PBS solution of quantum dots (CdTe core-type, COOH functionalized, 777978-10MG, Sigma Aldrich, St. Louis, MO, USA) or 2.5 mg/mL PBS solution of methotrexate (13960, Cayman Chemical, Ann Arbor, MI, USA). Both molecules possess a negative charge at pH 7.4, with methotrexate presenting a stable −2q charge (-3.2×10^{-19} C) and quantum dots having a charge that ranges from −5q to −15q depending on pH and ionic strength [41]. Methotrexate has a molar mass of 454 Da and an estimated diameter of 1.6 nm [42], while quantum dots have an estimated molar mass of 200 kDa and an estimated diameter of 4.7 nm [43]. An Ag/AgCl reference electrode (Harvard Apparatus, Holliston, MA, USA) was used and placed in the source drug reservoir.

Absorbance measurements of every sample were performed at 5 min intervals using a custom UV-vis spectrophotometer apparatus consisting of a robotic carousel [44] connected to an Agilent Cary 50 spectrophotometer (Agilent, Technologies, Santa Clara, CA, USA). Sink solution homogeneity was maintained by constant magnetic stirring (600 rpm). Methotrexate absorbance was measured at 373 nm, while quantum dots at 240 nm. An electrical potential (0, −1.5, or −3 V DC) was applied between the Ag/AgCl and the membrane electrodes through a waveform generator (33522A, Keysight Technologies, Santa Clara, CA, USA). Passive (0 V) and active (−1.5 or −3 V) phases were alternated at regular intervals. For methotrexate, phases were alternated every 6 h between passive and active (0 and −3 V DC, respectively). For quantum dots, 12 h passive phases were alternated with 8 h of active applied potential (−1.5 V).

2.6. Statistical Analysis

Statistical analysis was performed using GraphPad Prism 8 (version 8.1.1; GraphPad Software, Inc., San Diego, CA, USA). Mean ± SD values were calculated for all results. Further statistical significance was assessed, adopting the two-tailed paired t-tests (** $p \leq 0.01$; **** $p \leq 0.0001$). Cumulative releases were split into phases, and each fitted by a first-order polynomial (MATLAB® polyfit, MathWorks, Natick, MA, USA). Slopes of cumulative release curves were normalized and displayed as a percentage of the passive release profiles.

3. Results and Discussion

3.1. Nanofluidic Membrane

Prior to investigating the electrical performance of the membranes, we sought to analyze the quality of the membrane fabrication process (Figure 1). Individual silicon membranes were first visually inspected to assess integrity. Figure 1A shows a stereomicroscope picture of a single membrane, highlighting the conductive electrode pads at the top right and bottom left edges. The hexagonal arrangement of microchannels allowed us to maximize packing density without compromising mechanical robustness. By measuring transmembrane nitrogen gas flow and adopting our predictive model for nanofluidic gas transport [45], we obtained an indirect measurement for the size of nanochannels (~300 nm). Sample membranes were further analyzed with SEM imaging. Figure 1B shows the tightly packed nanochannel arranged in arrays of 19 rows and 96 columns with a horizontal pitch of 2 μm and a vertical pitch of 10 μm. No macroscopic defects or pinholes were observed across wafers, which indicated that the fabrication protocol was repeatable.

The analysis of the membrane cross-sections obtained via FIB milling was performed to evaluate the uniformity of layer deposition at different nanofabrication steps. SiO_2 growth via thermal oxidation resulted in a highly uniform layer along the whole length of the vertical nanochannels (Figure 1C). Thermal oxidation is a slow process that enables precise control over layer thickness. Thus, it allowed us to accurately and homogeneously reduce the size of nanochannels. The subsequent deposition of poly-Si (Figure 1D) was used to create a gate electrode that coats the whole nanofluidic structure with the objective of maximizing the electrostatic gating performances. Uniform poly-Si deposition through the chemical vapor deposition-based (CVD) process in high-aspect-ratio hollow structures can be challenging. However, our imaging analysis showed that the deposited layer was uniform (Figure 1D), except for a slight increase in thickness at the nanochannel outlet (bottom right). Finally, a thin layer of SiC (Figure 1E) was used to coat the conductive poly-Si and act as an insulating and chemical inert layer. Despite the high-aspect-ratio of the slit nanochannels, the deposition of SiC was also achieved with good uniformity (Figure 1E). Slight material accumulations at the inlet and outlet of nanochannels were expected. While we did not generate these intentionally, we noted that a local restriction at the nanochannel extremities could improve gating performance.

All materials used for the fabrication of the nanofluidic membrane have previously been demonstrated to be biocompatible using ISO 10993 standards by Kotzar et al. [46]. A subset of these materials has also been investigated in vivo in rodents and has shown biocompatibility and low biofouling [47]. Moreover, in our fabrication protocol, silicon carbide completely encapsulates the membrane and, therefore, is the only material exposed to the environment. Silicon carbide was specifically chosen for this encapsulation purpose as it's considered a versatile material for biomedical applications where extended exposure to physiological fluids is needed [48,49]. Additionally, in vivo biocompatibility of SiC was demonstrated by Cogan et al. [50], who subcutaneously implanted SiC discs in New Zealand White rabbit, the histological evaluation showed no chronic inflammatory response, and a capsule thickness comparable to controls was found.

Furthermore, silicon carbide has previously been shown to offer reduced biofouling when compared to other biocompatible materials, such as silicon or silicon dioxide [51]. Although complete protection against protein adsorption could not be achieved [52], we previously showed that biofouling did not negatively affect the function of our devices. Specifically, nanofluidic membranes, similar to the one presented in this study, have been used in-vivo in rats for up to 6 months [53] and in non-human primates for up to 4 months [54] with no alteration of drug release from biofouling or fibrotic tissue encapsulation.

When compared to membrane architectures previously developed in our lab [55–57], this membrane presented a less cumbersome fabrication process, thanks to the direct alignment of nanochannels and microchannels [58,59]. Further, a substantially higher nanochannels density [55,60] was achieved. In contrast with other gated membrane based on porous alumina (AAO) [29], presenting an irregular pore size distribution [28], our structure achieved a monodispersed nanochannel size that could aid in better control of molecular transport. In its current configuration, featuring 278,600 nanochannels, our membrane configuration was designed to achieve high mass transport rates per unit surface area. This is typically preferable in the context of implantable drug delivery application, where miniaturization is a requirement [39]. However, in light of its modular structure, the same fabrication process could be employed to create alternative configurations with a different number of channels for adoption in electrokinetic-enabled molecular manipulation or sorting applications. For these purposes, the large gate electrode surface area might provide increased electrostatic control of fluid molecules as compared to common Polydimethylsiloxane-glass (PDMS-glass) systems [61,62] which feature localized gate electrodes.

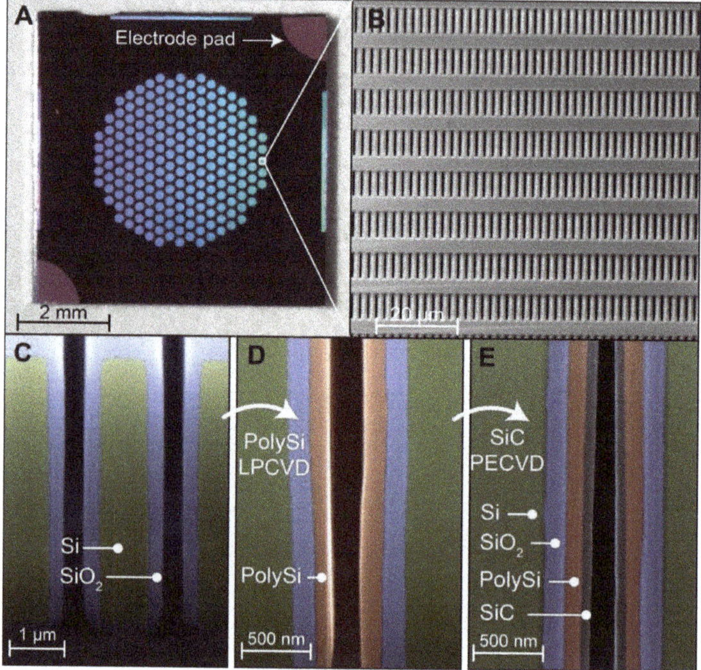

Figure 1. Nanochannel membrane structure. (**A**) Optical image of a silicon nanofluidic membrane, presenting electrode pads with exposed conductive polysilicon. (**B**) SEM micrograph, showing the array of nanochannel inlets. (**C**,**D**,**E**) Vertical cross-section image (SEM) obtained along the length of nanochannel, showing the membrane fabrication at different stages. Micrographs were color-enhanced for clarity of visualization. (**C**) Thermally grown SiO_2 layer (~175 nm, blue); (**D**) Low-pressure chemical vapor deposition (LPCVD)-deposited poly-Si layer (~121 nm, red); (**E**) Plasma-enhanced-CVD deposited SiC coating (~64 nm, gray). Images **C**, **D**, and **E** do not picture the same membrane location.

3.2. *Solid–Liquid Interface, SiO_2 vs SiC*

To evaluate SiC properties as a gate dielectric in contact with ionic solutions, we compared its insulation performance to SiO_2, which is a broadly used gate dielectric in solid electronics [63]. SiO_2 and other metal oxides, such as alumina and hafnium dioxide, owe their success to their high dielectric constants that allow for low leakage currents. Even though these materials excel in solid electronic manufacturing, they either lack biocompatibility or chemical inertness and durability in aqueous environments [50].

Leakage current measurements (Figure 2A) performed with our membranes did not show substantial differences between SiO_2 and SiC, except for 3 V. However, the steep increase in leakage observed for SiC between 2 and 3 V, emphasized by the electrolytic solution environment, suggests that the molecular arrangement in the dielectric layer is not ideal [64]. The literature on gate dielectric leakage in ionic solutions is scarce, and the available models for a solid-state field-effect transistor (FET) are unable to account for the effect of the electrolyte solution environment. In aqueous solutions, currents in the order of μA were measured for electric fields as low as 0.5 MV cm^{-1} (Figure 2A). In contrast, for solid-state FET, currents in the order of magnitude of μA are only expected for electric fields greater than 15 and 2 MV cm^{-1} for SiO_2 and SiC, respectively. High leakage currents are usually attributed to the formation of conductive filaments within the oxide, whereby electrons are trapped and form clusters within defects in the material. When clusters are at tunneling distance, a conductive path can form, leading to high leakage currents [65,66]. The proportional increase in leakage currents at increasing ionic strength of the solution, previously reported by this group [39], provides further support for this phenomenon.

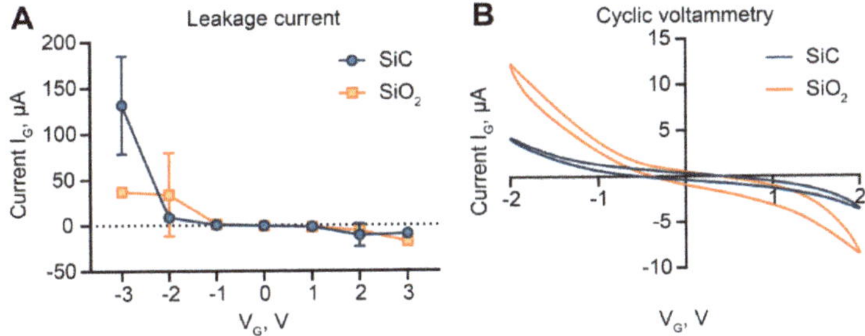

Figure 2. Leakage current and cyclic voltammetry. (**A**) Comparison of gate leakage current for SiO_2 and silicon carbide (SiC) dielectric. (**B**) Cyclic voltammetry comparison between SiO_2 and SiC.

In the voltage range between −2 and 2 V, SiC and SiO_2 exhibited similar values of leakage currents. Thus, to closer investigate differences in performances, we used cyclic voltammetry (CV). As compared to SiO_2-coated membranes, lower currents were measured for SiC at each applied voltage (Figure 2B). Interestingly, we observed a non-linear proportional relationship between voltage and current for both materials. SiC exhibited a steep increase in current for voltages higher than 1 V in absolute value. This suggested that for small applied voltages, no faradaic currents occurred, and the material behaved almost as an ideal capacitor. For voltages above ± 1 V, electrochemical reactions between the surface groups (C, SiO^-) and reactive species in the electrolyte solution (Cl^-, HO^-) led to increased currents.

In contrast, the significant current increase observed for the leakage currents (Figure 2A) for voltages over 2 V was likely related to material deterioration and conductive filament formation. The asymmetry between results obtained with positive and negative voltages provided further support for this theory. Higher currents for negative applied voltages were observed in both measurements. For negative voltages, positive species were attracted to the surface. The percolation model suggests that in the presence of strong electrostatic attraction, protons can diffuse in the insulator, starting a percolating path that can lead to the formation of a conductive filament [65]. Instead, for positive potentials, proton repulsion may cause a reversible interruption of the conductive filament, effectively decreasing leakage [67]. Additionally, the difference in hysteresis between the two CV profiles (Figure 2B) was suggestive of differences in surface charge accumulation between the two materials. A thinner CV profile usually correlates with low charge accumulation. Collectively, the results showed that SiC suffered lower leakage currents in the −2 V to 2 V range, exhibiting better insulation performance than SiO_2.

3.3. Electrochemical Characterization: Conductance

To further investigate the surface properties of SiC, we performed conductance measurements of SiC-coated membranes in the ionic concentration range between 1 μM and 100 mM. We employed a custom fixture [39] that allowed us to limit wetting to the nanochannel part of the membrane. The results are shown in Figure 3A. At high ionic strengths (>10^{-4} M), conductance measurements displayed a linear dependence on the ionic strength. In these conditions, the Debye length was significantly smaller than the size of nanochannels. Accordingly, the results were consistent with the bulk electrolyte conductance (red dashed line in Figure 3A). In contrast, at low ionic strengths (≤10^{-4} M), we observed a plateau in conductance (in the log-log scale). This occurred when the Debye length approached the nanochannel dimension, and the excess of counter-ions balanced the surface charge, reaching channel electroneutrality [68]. Here, as it directly related to the conductance, the surface charge could be calculated by fitting the results to the equation [69]:

$$\frac{I}{V} = 2F\mu\sqrt{\left(\frac{\Sigma}{2}\right)^2 + c_0^2}\frac{wh}{l} \quad (1)$$

In Equation (1), F, μ, and Σ are the Faraday's constant, ionic mobility, and the volume charge density, respectively. Further, c_0 is the solution molarity, and w, h, and l are the nanochannels' width, height, and length, respectively. Using the relation $zF\Sigma = -2\sigma_s/h$, we obtained a surface charge value of $\sigma_s = 1.81$ μC/m^2, which was consistent with the previously reported data for SiC surfaces [70]. Our SiC coating exhibited a surface charge orders of magnitude smaller than SiO_2 (1–100 mC/m^2) [71], which correlated with better performance in electrostatic gating control. In fact, chemically reactive surfaces act as charge buffers. An externally applied electric field is quickly compensated by protonation or deprotonation of reactive groups on the surface, limiting charge rearrangement in the electrical double layer (EDL) [72]. Thus, to minimize surface charge, materials are often artificially treated [28].

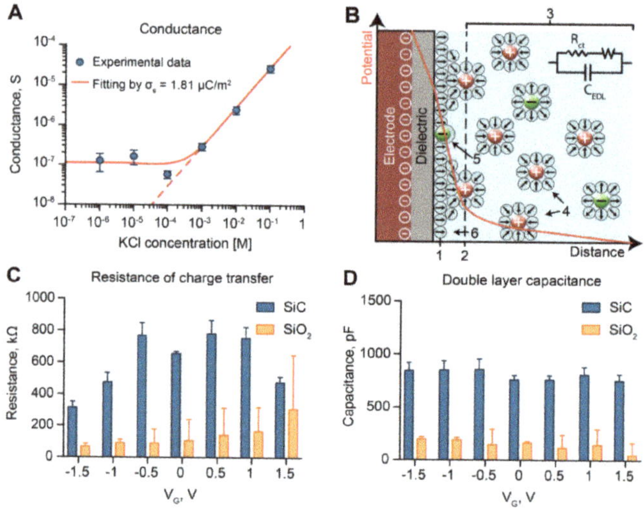

Figure 3. Electrochemical measurements. (**A**) Measured transmembrane ionic conductance. (**B**) Schematic of the electric double layer and relative model. (1) Inner Helmholtz plane; (2) Outer Helmholtz plane; (3) Diffuse layer; (4) Solvated ion; (5) Specifically adsorbed ion; (6) Molecules of the electrolyte solvent. (**C**) Fitted resistance of charge transfer (R_{ct}) of SiO_2-coated membranes versus SiC-coated membranes. (**D**) Fitted double-layer capacitance (C_{dl}) of SiO_2-coated membranes versus SiC-coated membranes.

3.4. Electrochemical Characterization: Electrochemical Impedance Spectroscopy

To investigate dielectric/liquid interface properties with the application of an external voltage, we performed electrochemical impedance spectroscopy (EIS) measurements. Specifically, we compared the resistance to charge transfer and the double layer capacitance at different gate voltages. A comparative assessment was conducted using SiC- and SiO_2-coated chips. Figure 3B shows a schematics of the electrical double layer (EDL), which described the ionic distribution that occurres at the solid–liquid interface of a charged surface to maintain local electroneutrality. The EDL is usually described by that Grahame model, which identifies three main layers consisting of 1) non-hydrated ions adsorbed to the surface, ii) hydrated immobile ions, iii) free moving hydrated ions [73]. The first and second layers of immobile ions are often referred to as the Stern layer. The EDL region is modeled by a series of capacitors, referred to as double-layer capacitance (C_{EDL}), where the Stern layer (~0.2 nm) [74] corresponds to the most significant contribution. As current can flow across the interface upon application of a DC potential, a resistive path is considered in parallel to the capacitance. This is usually referred to as a charge-transfer resistance (R_{ct}). R_{ct} can vary substantially depending on the material ability to exchange electrons with the electrolyte solution. Upon application of an external DC potential, if electrons cannot be easily exchanged, an overpotential builds up at the interface. In non-polarizable materials, such as Ag/AgCl, small R_{ct} permits high currents. In contrast, polarizable materials present high R_{ct}, and the current exchange is limited.

By fitting our EIS measurements to the model described above (Figure 3B), we calculated R_{ct} and C_{EDL} for both a SiO_2- and a SiC-coated membrane at different gate voltages (V_G) applied (Figure 3C,D). Depending on the applied potential, SiC showed an R_{ct} 1.5 to 8 times bigger than SiO_2 (Figure 3C). Interestingly, both materials showed a clear dependence of R_{ct} with the applied voltage. SiO_2 exhibited a monotonic increase of R_{ct} with the applied voltage, where more positive voltages resulted in higher R_{ct}. In contrast, SiC showed a decrease in R_{ct} proportional to the absolute value of the applied voltage. We attributed these phenomena to the difference in surface charge between SiO_2 and SiC. In fact, a higher number of available SiO^- sites on the SiO_2 surface allowed for increased electron exchange.

C_{EDL} did not exhibit a correlation with the applied gate voltage for either material (Figure 3D). Moreover, we unexpectedly found six times higher C_{EDL} for SiC with respect to SiO_2. As C_{EDL} mainly depends on the surface area and EDL thickness, our results could be explained in the context of the material porosity [75]. By presenting a larger surface area, pores displayed increased capacity. Overall, the low surface charge exposed by SiC and the high resistance to charge transfer qualified SiC as a polarizable interface suitable for electrostatic gating.

3.5. Mechanism of Analyte Flow Control through Electrostatic Gating

Nanofluidic systems present high surface to volume ratios. In light of this, charged species diffusing in nanoconfinement exhibit unique behaviors [76,77]. Electrostatic, steric, and hydrodynamic interactions with the nanochannel walls influence local molecular concentration and effective diffusivity. Depending on solution properties, such as ionic strength, pH, and surface charge density, the EDL can extend from a fraction to hundreds of nm in the fluid. Both SiC and SiO_2 surface expose native silanol groups, resulting in a net negative surface charge at pH 7.4 [78]. In proximity to the surface, charged species redistribute to reach electroneutrality [73]. While counter-ions concentration increases, co-ions are depleted following distribution with a characteristic dimension equal to the Debye length.

Once solution properties are defined, the surface charge is the only parameter that has a significant effect on the distribution of charges in the fluid. Thus, nanochannel charge-selectivity can be altered by controlling the channel surface charge. An applied difference in potential between a buried gate electrode and an electrode in solution creates an overpotential at the surface. We employed this strategy to modulate the diffusive transport of analytes through our nanofluidic membrane. With no applied voltage, molecules diffused through the channel unperturbed. By applying a negative gate potential, the transmembrane transport of co-ions was substantially reduced.

3.6. In Vitro Release Modulation of Methotrexate

To investigate the effectiveness of electrostatic gating on controlling trans-membrane transport of a small charged analyte, we performed an in vitro diffusion study using methotrexate. Methotrexate has a molecular weight of 454 Da and is a good representative of small molecules (<900 Da) therapeutics, which accounts for the majority of pharmaceuticals [79]. Clinically, methotrexate is used as a chemotherapeutic agent for the treatment of various cancers, as well as in the management of rheumatoid arthritis [33].

Figure 4A shows the normalized release rates for four consecutive cycles alternating between passive and active phases. During the passive phases, negatively charged molecules (−2q for methotrexate) diffused trough the nanochannels freely, largely unaffected by the low native charge of the SiC surfaces. When a negative voltage was applied (−3 V), an increase in negative surface charge repelled methotrexate molecules, reducing their release. The four alternation cycles between passive and active phases demonstrated that electrostatic gating allowed for repeatability of release modulation.

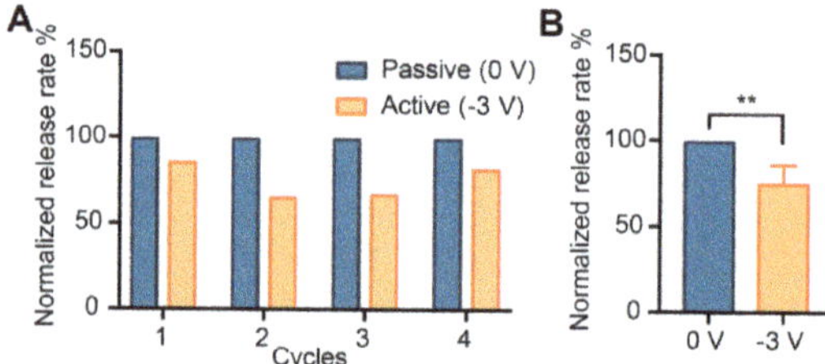

Figure 4. Electrostatically controlled release of methotrexate. (**A**) The normalized release rate of methotrexate for four cycles between free diffusion (Passive) and gated diffusion (Active). (**B**) Release rates grouped by phase typology (** $p \leq 0.01$).

We observed a statistically significant (** $p \leq 0.01$) difference in release rate between active and passive phases, whereby the applied potential −3 V yielded a decrease in the release rate of ~35%. During the passive phase, an average release rate of 10 μg/day was obtained, which was consistent with daily doses used to treat rheumatoid arthritis in pre-clinical testing [80]. Other small molecule therapeutics, including glucocorticoids [81], hormone therapeutics [82], and antivirals [83], present effective daily doses in the order of micrograms. This indicates that the current membrane architecture could, in principle, be adopted for various therapeutic applications. However, further testing with different pharmaceutical agents is warranted.

3.7. In Vitro Controlled Release of Quantum Dots

To assess the ability of our membrane to modulate the release rate of larger molecules, we performed an in vitro release study with quantum dots. Quantum dots possess broad applicability in bioengineering, including imaging [84], theranostics [85], cell labeling for in vivo tracking [86], tissue staining [87]. They have also been investigated as biomarkers for cancer detection and for targeted drug delivery [35]. Figure 5A shows the normalized release rate of each phase, where passive (0 V) and active phases (−1.5 V) were alternated over three cycles.

The application of the negative gate potential drastically reduced the release of quantum dots from the membrane. Subsequent cycles demonstrated consistent and reproducible release rate reduction, suggesting that the membrane and the gating performance were consistent over time. A statistically significant difference (**** $p \leq 0.0001$) in the release between active and passive phases (84%) was observed (Figure 5B). When compared to methotrexate, quantum dots clearly showed a more effective electrostatic modulation, which could be attributed to higher particle charge and lower ionic strength of the solution. Specifically, the high exposed charge is due to the carboxylic functionalization, where several groups result in a negative net charge that ranges from −5 to −15 depending on pH and ionic strength [41]. Moreover, the low ionic strength solution (0.01 × PBS) has a Debye length 10 times greater than the 1 × PBS. These two properties contribute to enhance the electrostatic interactions between the wall and the solute. Thus, the application of the gate potential resulted in increased efficacy of release modulation.

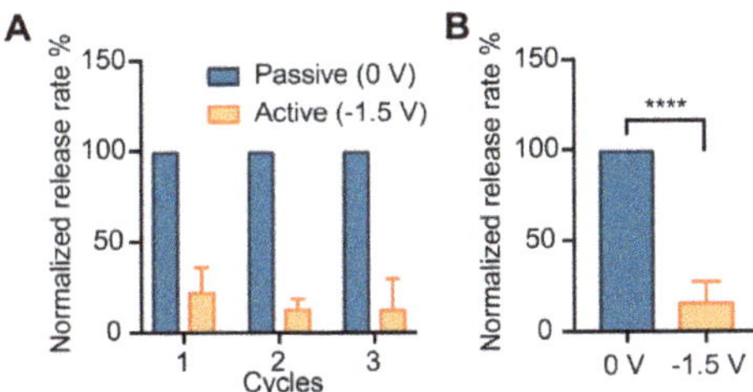

Figure 5. Electrostatically controlled release of quantum dots. (**A**) The normalized release rate of quantum dots for three cycles between free diffusion (Passive) and gated diffusion (Active). (**B**) Release rates grouped by phase typology (**** $p \leq 0.0001$).

3.8. Considerations on Electrostatic Gating Performance

To achieve efficient devices for tunable molecular diffusion via electrostatic gating, various parameters need to be optimized. Of utmost importance is the choice of dielectric material to insulate the buried gate electrode. In this study, we investigated SiC as it conciliates the need for low leakage currents, with a dielectric constant similar to SiO_2 (4.4–4.9) [88], and offers chemical inertness in aqueous solutions [48]. Moreover, SiC offers a low native charge; therefore, it minimizes unwanted non-linearities connected to the buffer capacity of strongly charged surfaces [72].

Further, efficient electrostatic flow modulation is strictly connected to the nanochannel size to the Debye length ratio (h/λ). Our membrane was designed for medical applications, where the ionic strength and pH were bound to physiological values. Our future investigations focus on manufacturing membranes with smaller nanochannels to be able, in principle, to completely stop analyte diffusion. Finally, as flow control trough electrostatic gating is mainly based on coulombic interactions, analytes that expose high surface charges are more suitable for gate modulation. Therefore, drug encapsulation with highly charged polymers can significantly improve administration control of small analytes.

4. Conclusions

In this work, we investigated a SiC-coated nanofluidic membrane capable of the reproducible control of analyte transport via electrostatic gating. The application of a low-intensity electrical potential to the gate electrode allowed us to alter nanochannel surface charge, leading to tunable membrane charge-selectivity, and control over the release of methotrexate and quantum dots. Electrochemical characterization showed that SiC dielectric coating exhibited low leakage current and reduced intrinsic charge as compared to SiO_2. Moreover, SiC offered chemical bioinertness, which rendered it an ideal candidate for use in biomedical devices for therapeutic delivery based on electrostatic-gating. In this context, our membranes could be employed as actuators for remotely controlled drug delivery systems. The low voltage needed to modulate the release rate could be provided via small scale and low-power circuitry. This investigation might pave the way for the development of the next generation of drug delivery systems, enabling pre-programmed or remotely managed pharmaceutic administration. Further, our gated nanofluidic membrane might find applicability in molecular sieving and lab on a chip diagnostic.

5. Patents

Grattoni, A.; Liu, X.; Ferrari, M. Gated Nanofluidic Valve For Active And Passive Electrosteric Control Of Molecular Transport, And Methods Of Fabrication, U.S. Provisional Pat. Ser. No. 62/961,437, filed Jan 15. (2020).

Author Contributions: Conceptualization, N.D.T., A.S., X.L., and A.G.; Data curation, N.D.T; Formal analysis, N.D.T and A.S.; Funding acquisition, X.L. and A.G.; Investigation, N.D.T., A.S., and Y.W.; Methodology, N.D.T., A.S., Y.W., and X.L.; Project administration, A.G.; Resources, X.L., and A.G.; Software, N.D.T.; Supervision, D.D. and A.G.; Validation, D.D., X.L., and A.G.; Visualization, N.D.T.; Writing—original draft, N.D.T.; Writing—review and editing, A.G. All authors have read and agreed to the published version of the manuscript.

Funding: The support was provided by the Houston Methodist Research Institute and NIH-NIGMS R01GM127558 (A.G.). A.G. and research group received additional support through the Frank J. and Jean Raymond Centennial Chair Endowment, and N.D.T. received funding support from the Chinese Academy of Sciences and The World Academy of Sciences through the CAS-TWAS President's fellowship scholarship.

Acknowledgments: We thank Valentina Serafini for her support in experimental studies and Jianhua (James) Gu from the electron microscopy core of the Houston Methodist Research Institute.

Conflicts of Interest: The authors declare no conflict of interest.

References

1. Hajat, C.; Stein, E. The global burden of multiple chronic conditions: A narrative review. *Prev. Med. Rep.* **2018**, *12*, 284–293. [CrossRef] [PubMed]
2. Yach, D.; Leeder, S.R.; Bell, J.; Kistnasamy, B. Global chronic diseases. *AAAS.* **2005**, *307*, 317. [CrossRef] [PubMed]
3. World Health Organization. *Global Status Report on Noncommunicable Diseases 2014*; WHO Press: Geneva, Switzerland, 2014.
4. Divo, M.; Cote, C.; de Torres, J.P.; Casanova, C.; Marin, J.M.; Pinto-Plata, V.; Zulueta, J.; Cabrera, C.; Zagaceta, J.; Hunninghake, G. Comorbidities and risk of mortality in patients with chronic obstructive pulmonary disease. *Am. J. Respi. Crit. Care Med.* **2012**, *186*, 155–161. [CrossRef] [PubMed]
5. Lemstra, M.; Nwankwo, C.; Bird, Y.; Moraros, J. Primary nonadherence to chronic disease medications: A meta-analysis. *Patient Prefer. Adherence* **2018**, *12*, 721. [CrossRef]
6. García-Lizana, F.; Sarría-Santamera, A. New technologies for chronic disease management and control: A systematic review. *J. Telemed. Telecare.* **2007**, *13*, 62–68. [CrossRef]
7. Desai, T.A.; Hansford, D.J.; Ferrari, M. Micromachined interfaces: New approaches in cell immunoisolation and biomolecular separation. *Biomol. Eng.* **2000**, *17*, 23–36. [CrossRef]
8. Peng, L.; Mendelsohn, A.D.; LaTempa, T.J.; Yoriya, S.; Grimes, C.A.; Desai, T.A. Long-term small molecule and protein elution from TiO2 nanotubes. *Nano Lett* **2009**, *9*, 1932–1936. [CrossRef]
9. Pons-Faudoa, F.P.; Ballerini, A.; Sakamoto, J.; Grattoni, A. Advanced implantable drug delivery technologies: Transforming the clinical landscape of therapeutics for chronic diseases. *Biomed. Microdevices* **2019**, *21*, 47. [CrossRef]
10. Chua, C.Y.X.; Jain, P.; Ballerini, A.; Bruno, G.; Hood, R.L.; Gupte, M.; Gao, S.; Di Trani, N.; Susnjar, A.; Shelton, K.; et al. Transcutaneously refillable nanofluidic implant achieves sustained level of tenofovir diphosphate for HIV pre-exposure prophylaxis. *J. Controlled Release* **2018**, *286*, 315–325. [CrossRef]
11. Ballerini, A.; Chua, C.Y.X.; Rhudy, J.; Susnjar, A.; Di Trani, N.; Jain, P.R.; Laue, G.; Lubicka, D.; Shirazi-Fard, Y.; Ferrari, M. Counteracting Muscle Atrophy on Earth and in Space via Nanofluidics Delivery of Formoterol. *Adv. Ther.* **2020**, *3*, 2000014. [CrossRef]
12. Hermida, R.C.; Ayala, D.E.; Smolensky, M.H.; Mojón, A.; Fernández, J.R.; Crespo, J.J.; Moyá, A.; Rios, M.T.; Portaluppi, F. Chronotherapy improves blood pressure control and reduces vascular risk in CKD. *Nat. Rev. Nephrol.* **2013**, *9*, 358. [CrossRef] [PubMed]
13. Lin, S.-Y.; Kawashima, Y. Current status and approaches to developing press-coated chronodelivery drug systems. *J. Controlled Release* **2012**, *157*, 331–353. [CrossRef] [PubMed]
14. Iwasaki, Y.; Sendo, M.; Dezaki, K.; Hira, T.; Sato, T.; Nakata, M.; Goswami, C.; Aoki, R.; Arai, T.; Kumari, P. GLP-1 release and vagal afferent activation mediate the beneficial metabolic and chronotherapeutic effects of D-allulose. *Nat. Commun.* **2018**, *9*, 1–17. [CrossRef] [PubMed]

15. Fifel, K.; Videnovic, A. Chronotherapies for Parkinson's disease. *Prog. Neurobiol.* **2019**, *174*, 16–27. [CrossRef]
16. Kaur, G.; Phillips, C.; Wong, K.; Saini, B. Timing is important in medication administration: A timely review of chronotherapy research. *Int J. Clin. Pharm.* **2013**, *35*, 344–358. [CrossRef]
17. Sprintz, M.; Tasciotti, E.; Allegri, M.; Grattoni, A.; Driver Larry, C.; Ferrari, M. Nanomedicine: Ushering in a new era of pain management. *Eur. J. Pain Suppl.* **2012**, *5*, 317–322. [CrossRef]
18. Celler, B.G.; Lovell, N.H.; Basilakis, J. Using information technology to improve the management of chronic disease. *Med. J. Aust.* **2003**, *179*, 242–246. [CrossRef]
19. Milani, R.V.; Bober, R.M.; Lavie, C.J. The role of technology in chronic disease care. *Prog. Cardiovasc. Dis.* **2016**, *58*, 579–583. [CrossRef]
20. Coye, M.J.; Haselkorn, A.; DeMello, S. Remote patient management: Technology-enabled innovation and evolving business models for chronic disease care. *Health Aff.* **2009**, *28*, 126–135. [CrossRef]
21. Hoare, T.; Timko, B.P.; Santamaria, J.; Goya, G.F.; Irusta, S.; Lau, S.; Stefanescu, C.F.; Lin, D.; Langer, R.; Kohane, D.S. Magnetically triggered nanocomposite membranes: A versatile platform for triggered drug release. *Nano Lett.* **2011**, *11*, 1395–1400. [CrossRef]
22. Timko, B.P.; Arruebo, M.; Shankarappa, S.A.; McAlvin, J.B.; Okonkwo, O.S.; Mizrahi, B.; Stefanescu, C.F.; Gomez, L.; Zhu, J.; Zhu, A.; et al. Near-infrared-actuated devices for remotely controlled drug delivery. *Proc. Natl. Acad. Sci. USA* **2014**, *111*, 1349–1354. [CrossRef] [PubMed]
23. Kim, K.; Jo, M.-C.; Jeong, S.; Palanikumar, L.; Rotello, V.M.; Ryu, J.-H.; Park, M.-H. Externally controlled drug release using a gold nanorod contained composite membrane. *Nanoscale* **2016**, *8*, 11949–11955. [CrossRef] [PubMed]
24. Kumeria, T.; Yu, J.; Alsawat, M.; Kurkuri, M.D.; Santos, A.; Abell, A.D.; Losic, D. Photoswitchable Membranes Based on Peptide-Modified Nanoporous Anodic Alumina: Toward Smart Membranes for On-Demand Molecular Transport. *Adv. Mater.* **2015**, *27*, 3019–3024. [CrossRef]
25. Ferrara, K.W. Driving delivery vehicles with ultrasound. *Adv. Drug Deliver Rev.* **2008**, *60*, 1097–1102. [CrossRef]
26. Lee, S.H.; Piao, H.; Cho, Y.C.; Kim, S.N.; Choi, G.; Kim, C.R.; Ji, H.B.; Park, C.G.; Lee, C.; Shin, C.I.; et al. Implantable multireservoir device with stimulus-responsive membrane for on-demand and pulsatile delivery of growth hormone. *Proc. Natl. Acad. Sci. USA* **2019**, *116*, 11664–11672. [CrossRef] [PubMed]
27. Farina, M. Remote magnetic switch off microgate for nanofluidic drug delivery implants. *Biomed. Microdevices* **2017**, *19*, 42. [CrossRef]
28. Kim, S.; Ozalp, E.I.; Darwish, M.; Weldon, J.A. Electrically gated nanoporous membranes for smart molecular flow control. *Nanoscale* **2018**, *10*, 20740–20747. [CrossRef]
29. Jeon, G.; Yang, S.Y.; Byun, J.; Kim, J.K. Electrically actuatable smart nanoporous membrane for pulsatile drug release. *Nano Lett.* **2011**, *11*, 1284–1288. [CrossRef]
30. Zhang, Q.; Kang, J.; Xie, Z.; Diao, X.; Liu, Z.; Zhai, J. Highly Efficient Gating of Electrically Actuated Nanochannels for Pulsatile Drug Delivery Stemming from a Reversible Wettability Switch. *Adv. Mater.* **2018**, *30*, 1703323. [CrossRef]
31. Kostaras, C.; Dellis, S.; Christoulaki, A.; Anastassopoulos, D.L.; Spiliopoulos, N.; Vradis, A.; Toprakcioglu, C.; Priftis, G.D. Flow through polydisperse pores in an anodic alumina membrane: A new method to measure the mean pore diameter. *J. Appl. Phys.* **2018**, *124*, 204307. [CrossRef]
32. Grattoni, A.; Liu, X.; Ferrari, M. Gated Nanofluidic Valve For Active And Passive Electrosteric Control Of Molecular Transport, And Methods Of Fabrication. U.S. Patent 62/961,437, 15 January 2020.
33. Lopez-Olivo, M.A.; Siddhanamatha, H.R.; Shea, B.; Tugwell, P.; Wells, G.A.; Suarez-Almazor, M.E. Methotrexate for treating rheumatoid arthritis. *Cochrane Database Syst. Rev.* **2014**. [CrossRef] [PubMed]
34. Matea, C.T.; Mocan, T.; Tabaran, F.; Pop, T.; Mosteanu, O.; Puia, C.; Iancu, C.; Mocan, L. Quantum dots in imaging, drug delivery and sensor applications. *Int J. Nanomed.* **2017**, *12*, 5421–5431. [CrossRef]
35. Cheki, M.; Moslehi, M.; Assadi, M. Marvelous applications of quantum dots. *Eur Rev. Med. Pharmacol. Sci.* **2013**, *17*, 1141–1148. [PubMed]
36. Napoli, M.; Eijkel, J.C.; Pennathur, S. Nanofluidic technology for biomolecule applications: A critical review. *Lab Chip* **2010**, *10*, 957–985. [CrossRef] [PubMed]
37. Lu, Y.; Liu, T.; Lamanda, A.C.; Sin, M.L.; Gau, V.; Liao, J.C.; Wong, P.K. AC electrokinetics of physiological fluids for biomedical applications. *J. Lab. Autom.* **2015**, *20*, 611–620. [CrossRef] [PubMed]

38. Gao, J.; Riahi, R.; Sin, M.L.; Zhang, S.; Wong, P.K. Electrokinetic focusing and separation of mammalian cells in conductive biological fluids. *Analyst* **2012**, *137*, 5215–5221. [CrossRef]
39. Di Trani, N.; Silvestri, A.; Sizovs, A.; Wang, Y.; Erm, D.R.; Demarchi, D.; Liu, X.; Grattoni, A. Electrostatically gated nanofluidic membrane for ultra-low power controlled drug delivery. *Lab Chip* **2020**, *20*, 1562–1576. [CrossRef]
40. Grattoni, A.; Gill, J.; Zabre, E.; Fine, D.; Hussain, F.; Ferrari, M. Device for rapid and agile measurement of diffusivity in micro- and nanochannels. *Anal. Chem.* **2011**, *83*, 3096–3103. [CrossRef]
41. Voráčová, I.; Klepárník, K.; Lišková, M.; Foret, F. Determination of ζ-potential, charge, and number of organic ligands on the surface of water soluble quantum dots by capillary electrophoresis. *Electrophoresis* **2015**, *36*, 867–874. [CrossRef]
42. Swain, M. Chemicalize.org. *J. Chem. Inf. Model.* **2012**, *52*, 613–615. [CrossRef]
43. Haro-González, P.; Martínez-Maestro, L.; Martín, I.; García-Solé, J.; Jaque, D. High-Sensitivity Fluorescence Lifetime Thermal Sensing Based on CdTe Quantum Dots. *Small* **2012**, *8*, 2652–2658. [CrossRef] [PubMed]
44. Geninatti, T.; Small, E.; Grattoni, A. Robotic UV-Vis apparatus for long-term characterization of drug release from nanochannels. *Meas. Sci. Technol.* **2014**, *25*. [CrossRef]
45. Scorrano, G.; Bruno, G.; Di Trani, N.; Ferrari, M.; Pimpinelli, A.; Grattoni, A. Gas Flow at the Ultra-nanoscale: Universal Predictive Model and Validation in Nanochannels of Ångstrom-Level Resolution. *ACS Appl. Mater. Interfaces* **2018**, *10*, 32233–32238. [CrossRef] [PubMed]
46. Kotzar, G.; Freas, M.; Abel, P.; Fleischman, A.; Roy, S.; Zorman, C.; Moran, J.M.; Melzak, J. Evaluation of MEMS materials of construction for implantable medical devices. *Biomaterials* **2002**, *23*, 2737–2750. [CrossRef]
47. Voskerician, G.; Shive, M.S.; Shawgo, R.S.; Von Recum, H.; Anderson, J.M.; Cima, M.J.; Langer, R. Biocompatibility and biofouling of MEMS drug delivery devices. *Biomaterials* **2003**, *24*, 1959–1967. [CrossRef]
48. Oliveros, A.; Guiseppi-Elie, A.; Saddow, S.E. Silicon carbide: A versatile material for biosensor applications. *Biomed. Microdevices* **2013**, *15*, 353–368. [CrossRef] [PubMed]
49. Mahmoodi, M.; Ghazanfari, L. *Physics and Technology of Silicon Carbide Devices*; Hijikata, Y, Ed.; InTech: Vienna, Austria, 2012.
50. Cogan, S.F.; Edell, D.J.; Guzelian, A.A.; Ping Liu, Y.; Edell, R. Plasma-enhanced chemical vapor deposited silicon carbide as an implantable dielectric coating. *J. Biomed. Mater. Res.* **2003**, *67*, 856–867. [CrossRef]
51. Zorman, C.A.; Eldridge, A.; Du, J.G.; Johnston, M.; Dubnisheva, A.; Manley, S.; Fissell, W.; Fleischman, A.; Roy, S. *Amorphous Silicon Carbide as a Non-Biofouling Structural Material for Biomedical Microdevices*; Materials Science Forum, Trans Tech Publications Ltd.: Stafa-Zurich, Switzerland, 2012; pp. 537–540. [CrossRef]
52. Takami, Y.; Yamane, S.; Makinouchi, K.; Otsuka, G.; Glueck, J.; Benkowski, R.; Nosé, Y. Protein adsorption onto ceramic surfaces. *J. Biomed. Mater. Res.* **1998**, *40*, 24–30. [CrossRef]
53. Ferrati, S.; Nicolov, E.; Zabre, E.; Geninatti, T.; Shirkey, B.A.; Hudson, L.; Hosali, S.; Crawley, M.; Khera, M.; Palapattu, G.; et al. The Nanochannel Delivery System for Constant Testosterone Replacement Therapy. *J. Sex. Med.* **2015**, *12*, 1375–1380. [CrossRef]
54. Pons-Faudoa, F.P.; Sizovs, A.; Shelton, K.A.; Momin, Z.; Bushman, L.R.; Chua, C.Y.X.; Nichols, J.E.; Hawkins, T.; Rooney, J.F.; Marzinke, M.A. Preventive efficacy of a tenofovir alafenamide fumarate nanofluidic implant in SHIV-challenged nonhuman primates. *BioRxiv.* **2020**. [CrossRef]
55. Bruno, G.; Canavese, G.; Liu, X.; Filgueira, C.S.; Sacco, A.; Demarchi, D.; Ferrari, M.; Grattoni, A. The active modulation of drug release by an ionic field effect transistor for an ultra-low power implantable nanofluidic system. *Nanoscale* **2016**, *8*, 18718–18725. [CrossRef] [PubMed]
56. Ferrati, S.; Fine, D.; You, J.; De Rosa, E.; Hudson, L.; Zabre, E.; Hosali, S.; Zhang, L.; Hickman, C.; Sunder Bansal, S.; et al. Leveraging nanochannels for universal, zero-order drug delivery in vivo. *J. Controlled Release* **2013**, *172*, 1011–1019. [CrossRef] [PubMed]
57. Di Trani, N.; Jain, P.; Chua, C.Y.X.; Ho, J.S.; Bruno, G.; Susnjar, A.; Pons-Faudoa, F.P.; Sizovs, A.; Hood, R.L.; Smith, Z.W.; et al. Nanofluidic microsystem for sustained intraocular delivery of therapeutics. *Nanomed. Nanotechnol. Biol. Med.* **2019**, *16*, 1–9. [CrossRef] [PubMed]
58. Fine, D.; Grattoni, A.; Hosali, S.; Ziemys, A.; De Rosa, E.; Gill, J.; Medema, R.; Hudson, L.; Kojic, M.; Milosevic, M.; et al. A robust nanofluidic membrane with tunable zero-order release for implantable dose specific drug delivery. *Lab Chip* **2010**, *10*, 3074–3083. [CrossRef] [PubMed]

59. Di Trani, N.; Silvestri, A.; Bruno, G.; Geninatti, T.; Chua, C.Y.X.; Gilbert, A.; Rizzo, G.; Filgueira, C.S.; Demarchi, D.; Grattoni, A. Remotely controlled nanofluidic implantable platform for tunable drug delivery. *Lab.Chip* **2019**, *19*, 2192–2204. [CrossRef] [PubMed]
60. Fine, D.; Grattoni, A.; Zabre, E.; Hussein, F.; Ferrari, M.; Liu, X. A low-voltage electrokinetic nanochannel drug delivery system. *Lab Chip* **2011**, *11*, 2526–2534. [CrossRef]
61. Prakash, S.; Conlisk, A.T. Field effect nanofluidics. *Lab Chip* **2016**, *16*, 3855–3865. [CrossRef]
62. Plecis, A.; Tazid, J.; Pallandre, A.; Martinhon, P.; Deslouis, C.; Chen, Y.; Haghiri-Gosnet, A. Flow field effect transistors with polarisable interface for EOF tunable microfluidic separation devices. *Lab Chip* **2010**, *10*, 1245–1253. [CrossRef]
63. Robertson, J. High dielectric constant gate oxides for metal oxide Si transistors. *Rep. Prog. Phys.* **2005**, *69*, 327–396. [CrossRef]
64. Padovani, A.; Gao, D.Z.; Shluger, A.L.; Larcher, L. A microscopic mechanism of dielectric breakdown in SiO2 films: An insight from multi-scale modeling. *J. Appl. Phys.* **2017**, *121*, 155101. [CrossRef]
65. Yao, J.; Zhong, L.; Natelson, D.; Tour, J.M. In situ imaging of the conducting filament in a silicon oxide resistive switch. *Sci. Rep.* **2012**, *2*, 1–5. [CrossRef] [PubMed]
66. Tung, C.H.; Pey, K.L.; Tang, L.J.; Radhakrishnan, M.K.; Lin, W.H.; Palumbo, F.; Lombardo, S. Percolation path and dielectric-breakdown-induced-epitaxy evolution during ultrathin gate dielectric breakdown transient. *Appl. Phys. Lett.* **2003**, *83*, 2223–2225. [CrossRef]
67. Chen, X.; Wang, H.; Sun, G.; Ma, X.; Gao, J.; Wu, W. Resistive switching characteristic of electrolyte-oxide-semiconductor structures. *J. Semicond.* **2017**, *38*, 8. [CrossRef]
68. Daiguji, H.; Yang, P.; Majumdar, A. Ion transport in nanofluidic channels. *Nano Lett.* **2004**, *4*, 137–142. [CrossRef]
69. Yossifon, G.; Mushenheim, P.; Chang, Y.-C.; Chang, H.-C. Nonlinear current-voltage characteristics of nanochannels. *Phys. Rev. E* **2009**. [CrossRef]
70. Grosjean, A.; Rezrazi, M.; Tachez, M. Study of the surface charge of silicon carbide (SIC) particles for electroless composite deposits: Nickel-SiC. *Surf. Coat. Technol.* **1997**, *96*, 300–304. [CrossRef]
71. Karnik, R.; Fan, R.; Yue, M.; Li, D.; Yang, P.; Majumdar, A. Electrostatic control of ions and molecules in nanofluidic transistors. *Nano lett.* **2005**, *5*, 943–948. [CrossRef]
72. Jiang, Z.; Stein, D. Electrofluidic Gating of a Chemically Reactive Surface. *Langmuir* **2010**, *26*, 8161–8173. [CrossRef]
73. Schoch, R.B.; Han, J.; Renaud, P. Transport phenomena in nanofluidics. *Rev. Mod. Phys.* **2008**, *80*, 839–883. [CrossRef]
74. Herbowski, L.; Gurgul, H.; Staron, W. Experimental determination of the Stern layer thickness at the interface of the human arachnoid membrane and the cerebrospinal fluid. *Z. Med. Phys.* **2009**, *19*, 189–192. [CrossRef]
75. Lu, P.; Dai, Q.; Wu, L.; Liu, X. Structure and Capacitance of Electrical Double Layers at the Graphene–Ionic Liquid Interface. *Appl. Sci.* **2017**, *7*, 939. [CrossRef]
76. Bruno, G.; Di Trani, N.; Hood, R.L.; Zabre, E.; Filgueira, C.S.; Canavese, G.; Jain, P.; Smith, Z.; Demarchi, D.; Hosali, S. Unexpected behaviors in molecular transport through size-controlled nanochannels down to the ultra-nanoscale. *Nat. Commun.* **2018**, *9*, 1682. [CrossRef] [PubMed]
77. Di Trani, N.; Pimpinelli, A.; Grattoni, A. Finite-Size Charged Species Diffusion and pH Change in Nanochannels. *ACS Appl. Mater. Interfaces* **2020**, *12*, 12246–12255. [CrossRef] [PubMed]
78. Behrens, S.H.; Grier, D.G. The charge of glass and silica surfaces. *J Chem. Phys.* **2001**, *115*, 6716–6721. [CrossRef]
79. Veber, D.F.; Johnson, S.R.; Cheng, H.-Y.; Smith, B.R.; Ward, K.W.; Kopple, K.D. Molecular Properties That Influence the Oral Bioavailability of Drug Candidates. *J. Med. Chem.* **2002**, *45*, 2615–2623. [CrossRef]
80. Liu, D.Y.; Lon, H.K.; Wang, Y.L.; DuBois, D.C.; Almon, R.R.; Jusko, W.J. Pharmacokinetics, pharmacodynamics and toxicities of methotrexate in healthy and collagen-induced arthritic rats. *Biopharm. Drug Dispos.* **2013**, *34*, 203–214. [CrossRef]
81. Yasin, M.N.; Svirskis, D.; Seyfoddin, A.; Rupenthal, I.D. Implants for drug delivery to the posterior segment of the eye: A focus on stimuli-responsive and tunable release systems. *J. Controlled Release* **2014**, *196*, 208–221. [CrossRef]
82. Langer, R.D. Efficacy, Safety, and Tolerability of Low-Dose Hormone Therapy in Managing Menopausal Symptoms. *J Am. Board Fam. Med.* **2009**, *22*, 563. [CrossRef]

83. Charles, N.C.; Steiner, G.C. Ganciclovir intraocular implant. A clinicopathologic study. *Ophthalmology* **1996**, *103*, 416–421. [CrossRef]
84. Li, J.; Zhu, J.-J. Quantum dots for fluorescent biosensing and bio-imaging applications. *Analyst* **2013**, *138*, 2506–2515. [CrossRef] [PubMed]
85. Ho, Y.-P.; Leong, K.W. Quantum dot-based theranostics. *Nanoscale* **2010**, *2*, 60–68. [CrossRef]
86. Kim, J.; Song, S.H.; Jin, Y.; Park, H.-J.; Yoon, H.; Jeon, S.; Cho, S.-W. Multiphoton luminescent graphene quantum dots for in vivo tracking of human adipose-derived stem cells. *Nanoscale* **2016**, *8*, 8512–8519. [CrossRef] [PubMed]
87. Bajwa, N.; Mehra, N.K.; Jain, K.; Jain, N.K. Pharmaceutical and biomedical applications of quantum dots. *Artif. Cells Nanomed. Biotechnol.* **2016**, *44*, 758–768. [CrossRef]
88. Hsu, J.-M.; Tathireddy, P.; Rieth, L.; Normann, A.R.; Solzbacher, F. Characterization of a-SiCx: H thin films as an encapsulation material for integrated silicon based neural interface devices. *Thin Solid Films* **2007**, *516*, 34–41. [CrossRef] [PubMed]

pharmaceutics

Article

Cell Theranostics on Mesoporous Silicon Substrates

Maria Laura Coluccio [1,†], Valentina Onesto [1,†], Giovanni Marinaro [2,3], Mauro Dell'Apa [4], Stefania De Vitis [1], Alessandra Imbrogno [5], Luca Tirinato [1], Gerardo Perozziello [1], Enzo Di Fabrizio [6], Patrizio Candeloro [1], Natalia Malara [1,*,‡] and Francesco Gentile [4,*,‡]

1 Department of Experimental and Clinical Medicine, University Magna Graecia, 88100 Catanzaro, Italy; coluccio@unicz.it (M.L.C.); valentina.onesto@unicz.it (V.O.); stefania.devitis24@gmail.com (S.D.V.); tirinato@unicz.it (L.T.); gerardo.perozziello@unicz.it (G.P.); patrizio.candeloro@unicz.it (P.C.)
2 Institute of Process Engineering, Technische Universität Dresden, 01069 Dresden, Germany; giovanni.marinaro@kaust.edu.sa
3 Institute of Fluid Dynamics, Helmholtz-Zentrum Dresden-Rossendorf (HZDR), 01328 Dresden, Germany
4 Department of Electrical Engineering and Information Technology, University Federico II, 80125 Naples, Italy; mauro.dellapa@gmail.com
5 Emerging Materials & Devices, Tyndall National Institute, Cork T12 R5CP, Ireland; alessandra.imbrogno@tyndall.ie
6 Department of Applied Science and Technology, Polytechnic University of Turin, 10129 Torino, Italy; enzo.difabrizio@polito.it
* Correspondence: nataliamalara@unicz.it (N.M.); francesco.gentile2@unina.it (F.G.)
† These authors contributed equally to this work.
‡ These two authors share senior authorship.

Received: 27 April 2020; Accepted: 21 May 2020; Published: 25 May 2020

Abstract: The adhesion, proliferation, and migration of cells over nanomaterials is regulated by a cascade of biochemical signals that originate at the interface of a cell with a substrate and propagate through the cytoplasm to the nucleus. The topography of the substrate plays a major role in this process. Cell adhesion molecules (CAMs) have a characteristic size of some nanometers and a range of action of some tens of nanometers. Controlling details of a surface at the nanoscale—the same dimensional over which CAMs operate—offers ways to govern the behavior of cells and create organoids or tissues with heretofore unattainable precision. Here, using electrochemical procedures, we generated mesoporous silicon surfaces with different values of pore size (PS ≈ 11 nm and PS ≈ 21 nm), roughness (Ra ≈ 7 nm and Ra ≈ 13 nm), and fractal dimension (Df ≈ 2.48 and Df ≈ 2.15). Using electroless deposition, we deposited over these substrates thin layers of gold nanoparticles. Resulting devices feature (i) nanoscale details for the stimulation and control of cell assembly, (ii) arrays of pores for drug loading/release, (iii) layers of nanostructured gold for the enhancement of the electromagnetic signal in Raman spectroscopy (SERS). We then used these devices as cell culturing substrates. Upon loading with the anti-tumor drug PtCl (O,O'-acac)(DMSO) we examined the rate of adhesion and growth of breast cancer MCF-7 cells under the coincidental effects of surface geometry and drug release. Using confocal imaging and SERS spectroscopy we determined the relative importance of nano-topography and delivery of therapeutics on cell growth—and how an unbalance between these competing agents can accelerate the development of tumor cells.

Keywords: nanoporous silicon; gold nanoparticles; drug delivery; cancer cells; theranostics

---

1. Introduction

Tissue engineering is a combination of techniques and materials for the fabrication of scaffolds and devices that, interacting with the cells, can lead to the formation of an analogue of tissues and organs that can improve, assist, or replace those already existing in the human body [1–4]. The biomaterials to be used in tissue engineering should exhibit the most convenient combination of mechanical properties, macro-scale architecture, and nanoscale geometry, to influence the collective behavior of cells and induce cells to form efficient structures. Those structures should be biocompatible, energetically efficient, autonomous, computationally efficient, and should be organized in a way to optimize the exchange of biochemical signals, nutrients, and oxygen between the several different parts of the structures and the external environment [1,2,4,5]. Thus, an ideal scaffold should present details over different hierarchical length scales to enable cell colonization, migration, and organization, and should be preferentially porous to enable the transport of bio-molecules.

At the nanoscale, cell behavior is strongly influenced by their interaction with the surrounding microenvironment [6–11]. Nanomaterials can interact without intermediation with the those adhesion molecules (integrins, cadherins, selectins, the immunoglobulin superfamily of cell surface proteins) involved in several different cell functions, including recognition, binding, adhesion, migration, apoptosis, differentiation, survival, and transcription [7–9,11,12]. Due to this unmediated interaction, nanomaterials can be fine-tuned to manipulate cellular function [13]. Materials with a controlled design at the nanoscale have been demonstrated in applications such as stem cells differentiation [14], the activation of the immune synapse [15], the shaping and signaling in neuronal networks [16], cell adhesion [17–19] and growth [20,21], the manipulation and control of neural polarity [22]. While many of the reported works have focused on bi-dimensional geometries, recent advances in additive manufacturing technologies [23], such as light assisted photopolymerization techniques [24], stereolithography [25], digital light projection, and two-photon polymerization [26], allowed a smooth transition from a 2D to a 3D design of the intended nano-structures [27]. The resulting cell culture models have an increased degree of complexity, an increased number of degrees of freedom, and exhibit a more faithful adherence to the 3D complex architectures of cells in living tissue and organs. This in turn enables to reproduce with an increased level of fidelity the natural microenvironment of cells [28].

Nevertheless, despite important advances in the production of scaffolds for tissue engineering applications, the development of characterization techniques has lagged behind the progress of fabrication. Recently, the demand for new devices for the analysis of the behavior of cells at the length scale of cell receptor is increasing. Those devices may reveal the fundamental biological mechanisms behind cell adhesion, migration, and organization at the cell-adhesion-molecule level, disclosing precious information for those interested in designing the scaffolds in the most efficient way.

In this paper, we present a mesoporous silicon device with details over multiple scales for cell culture, growth, and assembly. The device is functionalized with gold nanoparticle clusters that can be exploited to amplify the Raman signal measured at the cell interface. Thanks to the pores in the silicon matrix, the device can release drugs, growth factors, or other biomolecules to the cells over time. Thus, the device combines the ability of a scaffold to support cell growth with the ability of a drug delivery system to vehicle active-molecules to the cells adhering to the scaffold. The gold nanoparticles on the device enable to examine the combined effects of surface nano-topography and the delivery of drugs on cell adhesion and proliferation. In experiments in which we put in culture cancerous MCF-7 cells on the device, we measured the simultaneous effect of the pore size and of the delivery of an anti-tumor drug on the adhesive and proliferation properties of cells. Moreover, using Raman spectroscopy and a multivariate analysis of data, we mapped the spatial distribution of receptors expressed over the cell surface, and correlated that distribution to the nanoscale architecture of the device. We found that cells exhibit an increased ability to grow and to form clusters on substrates with smaller pore size (PS $\approx$ 11 nm) and roughness ($R_a \approx$ 7 nm), compared to substrates with larger pore size (PS $\approx$ 21 nm) and roughness (Ra $\approx$ 13 nm), in line with previous studies [29,30]. Both substrates deliver their payload efficiently, up to 10 days from the initial release, demonstrating high anti-cancer efficacy and killing up to 90% of cancerous cells on the smaller mesoporous substrate after 72 h from cell culture.

The multi-functional device that we developed can be used to evaluate the coincidental effects of (i) a timely administrated drug or nutrient and of the (ii) nanoscale characteristics of a surface on the efficacy of a therapeutic treatment, the functionalities of a scaffold, or a combination of the two. The device can be potentially used in applications that bridge traditional drug delivery, traditional tissue engineering and regenerative medicine, and diagnostics.

2. Methods

2.1. Fabrication of Mesoporous Silicon Surfaces

A detailed scheme of the fabrication of the Au-functionalized substrates is reported in Figure 1. Silicon substrates were electrochemically etched to obtain porous silicon. Porous silicon is a form of silicon with arrays of pores penetrating through its structure [31]. The average pore size (PS) determines the class of the porous silicon material: substrates with PS < 2 nm, $2 <$ PS < 50 nm, or PS > 50 nm are classified as nanoporous, mesoporous, macroporous silicon substrates, respectively [31]. In this work we produced mesoporous silicon substrates with two different non-overlapping values of pore size: MeP_1 silicon substrates with $PS_1 \approx 11$ nm and MeP_2 silicon substrates with $PS_2 \approx 21$ nm. We used P-type, 100 silicon wafers as a substrate. We cut the originating silicon wafers into regular square chips with a side of $\approx$1 cm. We then positioned the chips in an impermeable electrolytic cell where samples were exposed to a solution of hydrofluoric acid (HF) and ethanol or methanol, under the action of an external electric field [30]. As a result, hydrogen ions in solution were accelerated towards the silicon substrate etching the pores. Substrates with a different pore size were obtained by tuning the parameters of the technique: the intensity of etching current, the concentration of HF in solution, the type of neutral component in solution (ethanol or methanol), and the time of the process. MeP_1 silicon with an average pore size of PS $\approx$ 11 nm was obtained using a mixture of HF, D.I. (de-ionized) water and ethanol in a proportion of 1:1:1 in volume. In the process, a value of current density of I = 20 mA/cm^2 was applied for 5 min at 25 °C. MeP_2 silicon with an average pore size of PS $\approx$ 21 nm was obtained using a mixture of HF, D.I. water, and methanol in a proportion of 5:3:2 in volume. In the process, a value of current density of I = 4 mA/cm^2 was applied for 5 min at 25 °C. In all cases, the thickness of the porous layer is of some tens of micrometers. Since porous silicon is intrinsically hydrophobic [32], samples were oxidized in an oven at 200 °C for 2 h before use. The photoluminescence of mesoporous silicon was verified by imaging the light emission of the samples under UV radiation (365 nm).

2.2. Electroless Deposition of Gold Nanoparticles Clusters

Clusters of gold nanoparticles were deposited on the porous sample surface using electroless deposition techniques. In the technique, metal ions in solution are reduced on an autocatalytic surface to form solid deposits of that metal [33]. Following the methods reported in [34], we treated the porous silicon samples in a solution of HF and gold (III) chloride ($AuCl_3$) in a concentration of 0.15 M (HF) and 1 mM ($AuCl_3$) for 3 min at 50 °C. In solution, the ions of gold react with the exposed silicon surface yielding gold nanoparticles with an average particle size d $\approx$ 20 nm. Samples were then rinsed in D.I. water at room temperature for 30 s.

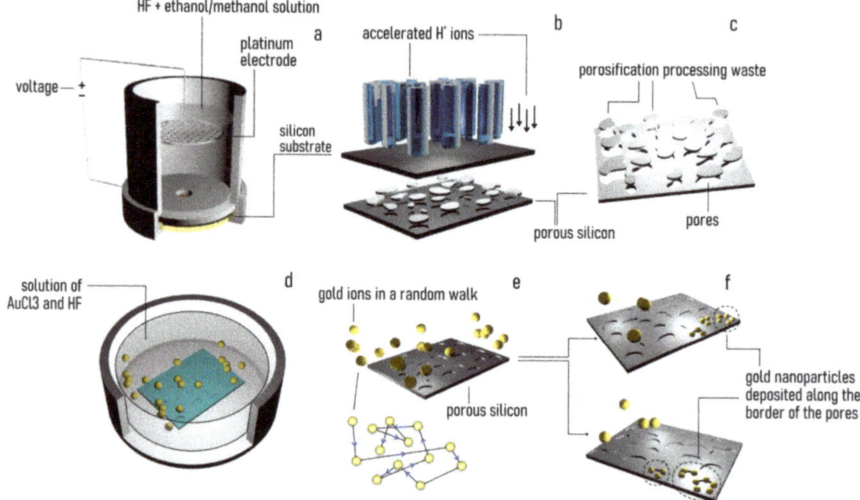

Figure 1. (**a**) An initial silicon chip of approximately 1 × 1 cm is electrochemical etched using a Teflon cell containing a solution of hydrofluoric acid (HF), D.I. (de-ionized) water, and methanol/ethanol in different ratios. (**b**) Upon activation of an external controlled voltage, positive ions in solution are accelerated towards the silicon substrate, creating pores within its structure. (**c**) Depending on the parameters of the process, including etching time, current and voltage intensity, and the concentration of the reagents in solution, one can obtain porous silicon surfaces with a tailored morphology. (**d**) The porous silicon sample is then placed in a baker along with a solution of hydrofluoric acid (HF) and gold(III) chloride ($AuCl_3$). (**e**) The resulting electroless process enables the deposition of gold ions in solution on the autocatalytic porous-silicon surface. (**f**) By varying the parameters of the electroless process, including temperature, concentration, and time, one can produce substrates with controlled gold-nanoparticles shape, size, and density.

2.3. SEM Sample Characterization

SEM (Scanning Electron Microscopy) analysis was conducted with a Zeiss GeminiSEM 500 at Dresden Center for Nanoanalysis (DCN), TU Dresden, Germany. Two types of porous silicon samples were analyzed: mesoporous 1 (MeP_1) and mesoporous 2 (MeP_2). Both samples were provided with and without gold nanoparticles deposited on their surface. Samples were fixed on stubs with a long pin and then mounted on a carousel 9 × 9 mm sample holder. In order to fix the samples, a small amount of silver paint was applied between the edge of the silicon substrate and the stub. A further copper lever was screwed in order to secure the sample on the stub. Several images of the samples were acquired in High Vacuum mode at 3 kV, a magnification factor of 300,000, and a working distance of about 3 mm with an InLens Detector (ZEISS) for secondary electrons. In order to reduce the drift, a frame integration (N = 14) was performed. In this way, every frame was scanned and averaged 14 times.

2.4. AFM Sample Characterization

Sample nanotopography was verified using atomic force microscopy (ICON Atomic Force Microscope, Bruker, Coventry, UK). We measured the surface profile over a sampling area of 1×1 μm^2, in a dynamic tapping mode in air. All measurements were performed at room temperature. During image acquisition, the scan rate was fixed as 0.5 Hz, while images were discretized in 1024 × 1024 points. We used Ultra-sharp Si probes (ACLA-SS, AppNano, Mountain View, CA, USA) with a nominal tip radius less than 5 nm to assure high resolution. Multiple measurements were done in different scan directions to avoid artefacts. At least four images were recorded per sample to reduce uncertainty. After acquisition, images were analyzed using the methods developed in [17] to determine the average surface roughness (Ra) and fractal dimension (Df) for each sample.

2.5. Contact Angle Characterization of Samples

The wettability of the samples was verified using an automatic contact angle meter (KSV CAM 101, KSV Instruments Ltd., Helsinki, Finland). A drop of 5 μL of D.I. water was gently positioned on the sample surface at room temperature. After 5 s from deposition, the contact angle of the drop at the interface with the substrate was measured.

2.6. MCF-7 Cell Culture and Staining

Breast carcinoma MCF-7 cells were grown on the porous silicon surfaces. Cultures were carried out at 37 °C in a humidified 5% CO_2/air atmosphere in a Dulbecco's modified eagle's medium (DMEM, Euroclone) supplied with 10% heat-inactivated fetal bovine serum (Euroclone, Pero (Mi), Italy), streptomycin (0.2 mg/mL) and penicillin (200 IU/mL). When cells on the petri dishes reached 90% confluence, they were dissociated: medium was removed and MCF-7 were treated with a solution of 0.25% Trypsin-0.53mM EDTA (Euroclone) for about 5 min at 37 °C. Trypsin was deactivated by adding medium and completely removed after centrifugation of the cell suspension (1300 rpm, 5 min, 18 °C). Then, trypsin/growth medium solution was removed. Single sterilized porous Si wafer specimens with and without loaded drugs, having a size of around 15 × 15 mm, were individually placed into each well of a 6-well plate (Corning Incorporated) and washed with phosphate-buffered saline solution (PBS, Invitrogen). After that, cells were seeded in complete cell medium and cultured up to 15 days in a humidified incubator at 37 °C with 5% of CO_2. After the incubation period, cell culture medium was removed and the MCF-7 cells were washed twice in PBS, fixed with 4% PFA (paraformaldehyde), and left for 30 min at room temperature (RT). Subsequently, cells were washed twice in PBS and permeabilized with 0.05% triton (Invitrogen, Milano, Italy) for 5 min at RT. Fixed and permeabilized cells were stained with 100 μL DAPI (40, 6-Diamidino-2-phenylindole, Sigma Aldrich, Milano, Italy) solution for 10 min at 4 °C in dark environment. In the end, the DAPI solute ion was removed and each sample was washed with PBS. The total number of cells n_{tot} initially seeded in each well for incubation was approximately $n_{tot} \approx 10^5$. Cells were sub-confluent for the duration of the experiment. MCF-7 cells were chosen as a cell-model because they are characterized by a moderate expression of the integrins. As previously reported [35], the change in the intensity and type of expression of integrin is the basis of the cancer disease progression. Notably, MCF-7 are a secondary cell line. The choice of another cell line, perhaps a primary cell line, while could possibly enhance clustering, may not have—at the same time—the same effect on the expression of integrins in the system.

2.7. Imaging Cells on the Substrates

An inverted Leica TCS-SP2® laser scanning confocal microscopy (Wetzlar, Germany) system was used to image cells adhering on the substrates. All measurements were performed using ArUv laser (Leica, Wetzlar, Germany). The pinhole was set to ≈ 80 μm (1.5 Airy units) and the laser power to 80% of the maximum, these values of the parameters were maintained constant throughout each acquisition. Confocal images of blue (DAPI) fluorescence were acquired using a 405 nm excitation line and a 10× dry objective, so that several cells could be simultaneously imaged in the region of interest, that was of 1174 × 882 μm^2, resulting in a pixel size of ≈ 1.72 μm. For each substrate, a large number of images was taken for statistical analysis. Each image was averaged over four lines and 10 frames to reduce noise. Images were acquired with a resolution of 1024 × 768 pixels, and were exported to a computer for processing and analysis.

2.8. Image Analysis and Topological Characteristics of Cell Networks

Confocal images of cell nuclei stained with DAPI were analyzed with Matlab® to extract the cell positions. Images were preprocessed to enhance contrast and low-pass filtered to remove constant power additive noise. Each image was partitioned into k = k different segments (going gradually from bright, k = 1, to dark, k = k) by k-means segmentation algorithms [30]. The information content of the image was thus associated with a gray level k = t and all the segments brighter than a certain threshold t were considered as background and shifted to 0. The values of the remaining segments, representing the cells, were shifted to 1. After that, the resulting image (g) was downsampled: if f was the average operator, f was shifted over g by steps of size r, where r^2 was the expected area of a cell nucleus in pixels. The pixel intensity (ranging from 0 to 1) of the resulting image indicated the probability that a pixel is a cell. If this probability is greater than a threshold, that pixel is considered being a node of the graph in a bi-dimensional grid. At this point, the links between the nodes can be established using the Waxman model [36], according to which the probability $P(u,v)$ of being a connection between two nodes u and v exponentially decays with their Euclidean distance d. If L is the largest Euclidean distance:

$$P(u,v) = \alpha e^{-d(u,v)/\beta L} \quad (1)$$

where d is the Euclidean distance between nodes u and v, and L is the largest possible Euclidean distance between two nodes of the grid. In the equation, α and β are the Waxman model parameters. According to α and β, which have to be chosen between 0 and 1, the density of links in a graph changes. In particular, low values of these parameters result in a low number of connections. For our study, α = 1 and β = 0.025. The probability P ranges between 0 (if the distance between the pair of nodes is ideally infinite) and 1 (if the distance between the pair of nodes is zero). The connectivity information of a graph is described in the adjacency matrix A. A is a square matrix in which each element a_{ij} indicates whether two nodes i and j are connected ($a_{ij} = 1$) or not ($a_{ij} = 0$). In the analysis, diagonal elements were all zero, since links from a node to itself were not allowed. Moreover, graphs were considered being undirected, so that information could bidirectionally flow from i to j. As a consequence A was symmetric and $a_{ij} = a_{ji}$. As the Euclidean distances d_{ij} in the networks were extracted, we could decide if a pair of nodes is connected by the subsequent formula

$$\alpha e^{-d_{i,j}/\beta L} - R \geq 0 \quad (2)$$

in which R is a constant that we chose as 0.1 so that the probability of being a connection is $P = 0.9$. Once established the connections between the nodes, the network parameters including clustering coefficient, characteristic path length, and small-world-ness can be extracted. The definition and significance of these terms may be found in influential textbooks [37] and papers [38–41]. Once obtained the Cc and Cpl values, we found a precise measure of 'small-world-ness', the 'small-world-ness' coefficient (SW), based on the trade-off between high local clustering and short path length as described in [42]:

$$\text{SW} = \frac{Cc_{graph}}{Cc_{rand}} / \frac{Cpl_{graph}}{Cpl_{rand}} \quad (3)$$

where Cc_{graph} and Cpl_{graph} are the clustering coefficient and the characteristic path length of the graph G under study, and Cc_{rand} and Cpl_{rand} are the equivalent values for a random Erdös-Rényi graph with the same number of nodes and edges of G.

2.9. Raman Analysis of Samples

MCF-7 cells fixed on the Au-mesoporous sample surface were analyzed by a WITec Raman microscope Alpha300 AR equipped with a 50×/0.7 N.A. (Numerical Aperture) objective. The signal was excited by a 633 nm laser, set to a power of 1 mW. For each sample, SERS (Surface Enhanced Raman Spectroscopy) maps of a portion of a cell were acquired in the x-y plane with a 0.5 μm stepsize, with the aim to analyze the SERS spectra coming in particular from the cell membrane, searching the different biochemical composition of each point to evidence the possible presence of adhesion proteins [43]. The Raman spectra were collected in the spectral range from 700 to 3250 1/cm, with an integration time of 1 s.

2.10. Principal Components Analysis of Raman Spectra

Spectra were pre-processed to minimize the effect of the fluorescence of samples by normalization on the total spectrum area. A principal component analysis (PCA) and a clustering analysis were performed on the spectral collection to highlight the chemical differences between the membrane's points [44]. The first five principal components (PCs) accounted for nearly 90% of the total spectral variation and they were then used to implement the clustering analysis by the Kmean method, imposing a number of five classes. All the pre-processing steps, the PCA, and the clustering analysis were carried out using the free software package Raman Tool Set (available on http://ramantoolset.sourceforge.net).

2.11. UV Characterization of Drug Release

To assess the drug-delivery capability of the device we verified the release over time of the anti-tumor drug PtCl(O,O^-acac)(*DMSO*). PtCl(O,O′-acac)(*DMSO*) is a platinum(II) complex containing acetylacetonate (*acac*), characterized by a high toxicity both in immortalized cell lines, as human cervical carcinoma (HeLa) cells or human breast cancer (MCF-7) cells, and in primary cultured human breast epithelial cells [45]. In this work, we incubated the mesoporous silicon samples in 30 μM/50 μM solutions of PtCl(O,O′-acac)(DMSO) in D.I. water for 60 h to load the drug. The release kinetic was tested over time up to 15 days, by immersion of the loaded sample in DI water and monitoring the drug concentration at different time, through the analysis of drug in solution by a spectrophotometer UV/Vis (LAMBDA 25 UV/Vis PerkinElmer), after standard calibration procedures of the samples.

2.12. Statistical Analysis

Data in the article and in the figures are represented as mean ± standard deviation. We used a Student's *t*-test statistics (two-tailed, unpaired) to perform comparison between means of different groups, where we assumed that elements in each group are normally distributed. In performing the test, the null hypothesis is that the means between pairs of samples are equal. Everywhere in the text and the figures the difference between two subsets of data is considered statistically significant if the Student's *t*-test gives a significant level p (p value) less than 0.05.

3. Results

3.1. Producing Gold-Functionalized Mesoporous Surfaces

Using electrochemical etching techniques described in Section 2, we produced porous silicon surfaces. Tuning the parameters of the electrochemical etching, we obtained two different pore morphologies: (i) mesoporous silicon samples with a pore size that oscillates around the central value PS = 11 nm (MeP_1 silicon with a pore size in the lower nanometer range) and (ii) mesoporous silicon samples with an average pore size PS = 21 nm (MeP_2 silicon with a pore size in the higher nanometer range). Scanning electron micrographs (SEM) of MeP_1 and MeP_2 samples taken at different magnifications reveal the morphology of the porous surface at different scales (Figure 2a–f). Pores on the surface of MeP_1 silicon are less uniformly distributed, are less dense, and result in a porosity of the sample of about P ≈ 12% (Figure 2a–c). The porosity or void fraction is a measure of the void spaces in a material, it is a fraction of the volume of voids over the total volume, expressed here as a percentage between 0% and 100%: it was calculated following the image analysis algorithms reported in the Supporting Information 1 of Supplementary Materials. Differently, pores on the surface of MeP_2 silicon are more uniform and are more densely packed compared to the pores found in the MeP_1 morphology: for this category, the porosity of the sample soars to $P \approx 40\%$ (Figure 2d–f).

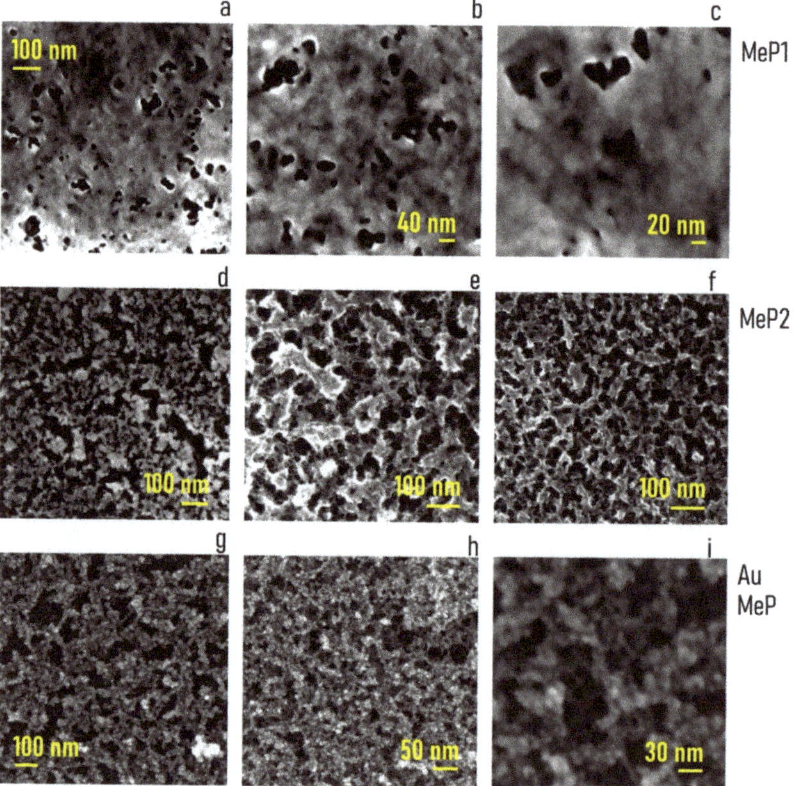

Figure 2. (**a–c**) SEM micrographs of MeP_1 substrates at different levels of magnification, the pores are less uniformly distributed over the sample surface with an average pore size of PS $\approx$ 11 nm. (**d–f**) SEM micrographs of MeP_2 substrates at different levels of magnification, the pores are uniformly distributed over the sample surface with an average pore size of PS $\approx$ 21 nm. (**g–i**) SEM micrographs of clusters of gold nanoparticles deposited over the porous sample surfaces, the particles follow the profile of the samples without occluding the pores, the average particle size is $\approx$8 nm.

Thus, the room potentially available to accommodate drugs or other molecules is significantly larger for MeP_2 silicon than for MeP_1 samples. After porosification, MeP_1 and MeP_2 samples were functionalized with gold nanoparticles using electroless deposition techniques. Electroless deposition is a technique that enables the reduction of metal ions on a solid surface as bulk metal without the application of external electric fields or forces [33]. Samples were treated with a solution of gold chloride and hydrofluoric acid for 3 min at 50 °C (Section 2). The process resulted in the homogeneous deposition of clusters of gold nanoparticles on the surface of the porous samples. SEM images of the sample surface (Figure 2d–f) and a convenient analysis of data (Supporting Information 2 of Supplementary Materials) indicate that the average diameter of the gold nanoparticles is s_{np} = 8 nm, with a small deviation around the mean $\sigma(s_{np}) \approx 1$ nm. In no case do the gold nanoparticles occlude the pores, therefore, preventing the correct release of drugs of molecules from the porous matrix. Porous surfaces functionalized with gold nanoparticle were verified using atomic force microscopy (AFM). AFM imaging enabled to resolve the structure of the samples at the nanoparticle level for both MeP_1 (Figure 3a) and MeP_2 (Figure 3c) silicon. Fast Fourier transform of AFM data enabled to derive the power spectrum associated to MeP_1 (Figure 3b) and MeP_2 (Figure 3d) silicon functionalized with gold. The power spectrum reports the change of information content as a change of size in a bi-logarithmic scale thus indicating how much of the originating complexity is maintained by changing the degree of detail of a surface [46]. Upon analysis of AFM data we found the values of roughness (Ra) and fractal dimension (Df) of the samples as $Ra^{MeP1} = 7 \pm 2$ nm and $Df^{MeP1} = 2.48 \pm 0.4$ for MeP_1, and $Ra^{MeP2} = 13 \pm 3$ nm and $Df^{MeP2} = 2.15 \pm 0.2$ for MeP_2 silicon. Thus MeP_1 samples exhibit values of roughness and fractal dimension larger than the corresponding values found for MeP_2 silicon. For comparison, nominally flat silicon surfaces, used as a control, have significantly smaller values of roughness ($Ra^{Si} = 1 \pm 0.1$ nm) and fractal dimension ($Df^{Si} = 2.1 \pm 0.2$) (Figure 3h). The luminescence properties of porous silicon samples were verified under UV light (Figure 3e). The intense luminescence emission from MeP_1 samples, compared to the low emission of MeP_2 and to the no-emission from simple silicon, indicates that MeP_1 samples may have—in the long tail of their pore size distribution—pores with a size smaller than 2 nm [31]. The wettability of mesoporous silicon samples was verified using contact angle measurements. Before oxidation, porous silicon samples as made exhibit a marked hydrophobicity with values of contact angles (CA) approaching 120 (Figure 3f). After treatment (Section 2), contact angle values measured on the sample surface shift to smaller values (CA ≈ 35°) typical of a hydrophilic surface.

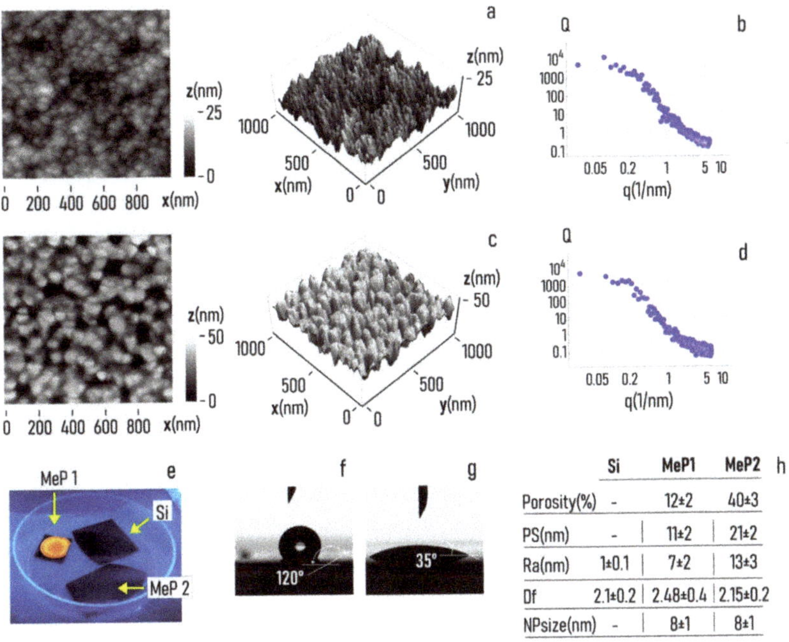

	Si	MeP1	MeP2
Porosity(%)	-	12±2	40±3
PS(nm)	-	11±2	21±2
Ra(nm)	1±0.1	7±2	13±3
Df	2.1±0.2	2.48±0.4	2.15±0.2
NPsize(nm)	-	8±1	8±1

Figure 3. (**a**) Atomic force microscopy (AFM) profile of MeP_1 substrates functionalized with gold nanoparticles imaged over a sampling area of 1 × 1 μm, the height values of the profile fall within the 0–25 nm range. (**b**) Power spectrum density function associated to the topography of the MeP_1 substrate, the slope of the function in the linear regime is indicative of the fractal dimension of the samples. (**c**) AFM profile of MeP_2 substrates functionalized with gold nanoparticles imaged over a sampling area of 1 × 1 μm, the height values of the profile fall within the 0–50 nm range. (**d**) Power spectrum density function associated to the topography of the MeP_2 substrate. (**e**) Luminescence of MeP_1 and MeP_2 samples under UV light, compared to the light emission of silicon. Contact angle of a drop of D.I. water measured on the porous surfaces before (**f**) and after (**g**) sample oxidation. (**h**) Values of porosity, pore size, roughness, fractal dimension, and characteristic size of the gold nanoparticles of the porous surfaces determined through analysis of SEM and AFM images of samples.

3.2. Controlling Cell Organization on Au-Mesoporous Silicon Surfaces

The substrates that we produced exhibit different values of pore size ($PS^{MeP1} \approx 11$ nm, $PS^{MeP2} \approx 21$ nm), gold nanoparticles size ($s_{np} \approx 8$ nm), roughness ($Ra^{MeP1} \sim 7$ nm, $Ra^{MeP2} \approx 13$ nm), and fractal dimension ($Df^{MeP1} \approx 2.48$, $Df^{MeP1} \approx 2.15$). To examine whether the nano-topographical characteristics of the surfaces have the ability to direct cell behavior on the substrate in a controlled way, we put in culture on both Au-MeP_1 and Au-MeP_2 silicon MCF-7 breast cancer cells. We then examined the topological characteristics of the networks that cells formed 24 h from seeding and we correlated them to the topography of the surface. We used nominally flat silicon substrates as a control. Figure 4 shows the spatial layout of cell-nuclei on flat silicon, Au-MeP_1 and Au-MeP_2 silicon imaged 24 h after culture. The initial number of cells deposited in each well for incubation was the same for all the substrates (Section 2). Fluorescence images in Figure 4 show that cells are homogeneously distributed on flat silicon surfaces, showing no preferential points of accumulation. Differently, cells on mesoporous surfaces form complex structures of those cells with a correlation length, cluster size, and topological characteristics that seem to vary from Au-MeP_1 to Au-MeP_2 silicon.

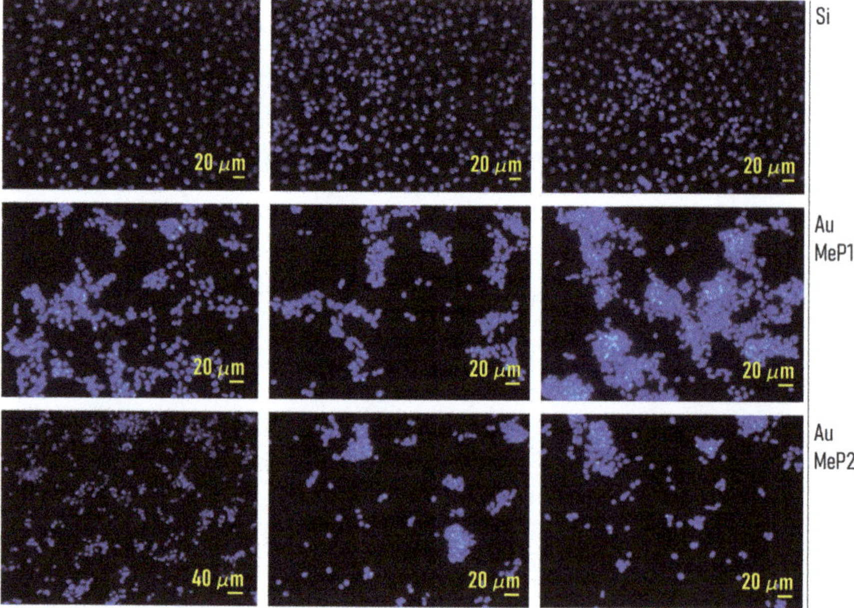

Figure 4. Fluorescence images of MCF-7 cancer cells over MeP_1, MeP_2, and silicon surfaces after 24 h from seeding.

We used image analysis algorithms and the methods of networks analysis, described in [30,40,47,48] and Section 2 of this article, to measure the characteristics of cell-networks quantitatively. Starting from the fluorescence image of a cell configuration, we segmented that image to find the cell-centers. Then, we routed cell-centers using the Waxman algorithm (Figure 5a). The algorithm selected the cell pairs to be connected basing on their distance: cells that were closer than a threshold were connected as described in Section 2. We analyzed more than 30 images per substrate. For each image, we extracted from the resulting network the number of cells in a region of interest (N), the clustering coefficient (Cc), the characteristic path length (Cpl), the small-world-ness (SW). N measures the adhesion strength of cells to a substrate [17]. The clustering coefficient, characteristic path length, and small world coefficient are a quantitative measure of the characteristics of the networks that cells form on a surface.

The clustering coefficient is the ratio of active links to the total combinations of connections that cells, around a reference node, can possibly establish, averaged all over the cells of the network [37]. The characteristic path length is the mean shortest distance between nodes of network [37]. The small-world coefficient, found as a combination of the *Cc* and the *Cpl* [42], is a metric that tests whether the distance between nodes grows with the logarithm of the number of nodes in a graph: $Cpl \propto \log(N)$. Typically, small-world-networks are characterized by a few clusters with a high number of elements for a cluster, the identification of small-world networks is of interest because networks with small-world-characteristics communicate more efficiently than equivalent random or ordered graphs of the same size [48,49].

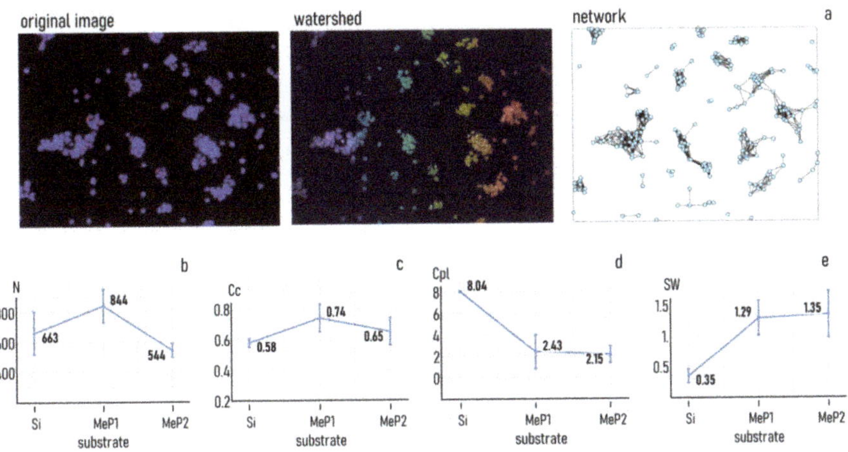

Figure 5. (**a**) Image analysis of fluorescence images: the original images were segmented with a watershed algorithm to identify individual cells, and cell nodes were then connected using the Waxman algorithm to obtain the equivalent graph for each sample. (**b**) Values of adhering cells, (**c**) clustering coefficient, (**d**) characteristic path length, and (**e**) small-world-ness determined for the cultures of MCF-7 cell on MeP_1, MeP_2, and silicon surfaces 24 *h* from seeding.

After network analysis of cell images, we found that the number of adhering cells in a region of interest of 1174 × 882 μm is $N \approx 663$ on flat silicon, $N \approx 884$ for the Au-MeP_1 substrate, $N \approx 544$ for the Au-MeP_2 substrate (Figure 5b). The maximum number of adhering cells is found for the Au-MeP_1 substrate with intermediate values of roughness and higher values of fractal dimension, in line with previous reports [17,29,30,48,49]. Notably, the difference between the number of cells found on Au-MeP_1 and Au-MeP_2 silicon is statistically significant ($p < 0.05$). For the same sets of images, we found that the values of clustering coefficient reach a maximum for the Au-MeP_1 substrate, with $Cc \approx 0.74$, while the clustering coefficient is lower for the Au-MeP_2 substrate ($Cc \approx 0.65$), and significantly lower for simple silicon ($Cc \approx 0.58$) (Figure 5c). At the same time, the values of characteristic path length are nearly identical for the Au-MeP_1 ($Cpl \approx 2.43$) and Au-MeP_2 ($Cpl \approx 2.15$) substrates, and they are statistically different from the values found on flat silicon with $Cpl \approx 8$ (Figure 5d). The resulting small-world-coefficient of cell networks on mesoporous substrates is SW ≈ 1.29 and SW ≈ 1.35 for MeP_1 and MeP_2, respectively, while SW ≈ 0.35 for flat silicon. Thus, cell networks on nanostructured, mesoporous surfaces passed the small-world test, differently from cells on flat silicon that settle on a surface without any appreciable large- or small-scale structure.

3.3. SERS Analysis of Cell Adhesion on Au-Mesoporous Silicon Surfaces

The substrates that we produced induce cell clustering. The increased susceptibility of cells to condensate into compact structures is in turn ascribed to the intermediate values of roughness and large values of fractal dimension of mesoporous silicon compared to flat silicon substrates [17,30]. Since cell clustering and condensation is a side effect of the of the increased adhesive properties of a substrate [48,49], we performed a chemometric analysis of cells cultured on MeP_1 and MeP_2 substrates to examine whether cell adhesion molecules are preferentially expressed from cells on nanostructured surfaces. We mapped the Raman intensity of MCF-7 cells cultivated on MeP_1 and MeP_2 silicon functionalized with gold, compared to the same cells on silicon sputtered with a continuous layer of gold, used as a control (Figure 6a). Raman spectra of cells were acquired following the procedure reported in Section 2 36 h after seeding, that is a sufficiently long time to assure complete adhesion of cells on the substrate. In each point, the Raman maps reported in Figure 6a are proportional to the intensity of the spectra measured at 1569 1/cm, that is typical of integrins as explained in the following of this section. Raman spectra were then subjected to a principal components analysis (PCA), and the principal components resulting from the analysis were in turn clustered into groups using classical k-means algorithms. This allowed to identify in the cell under analysis of five different regions, where points in a region have similar chemometric characteristics (Figure 6b). Moreover, since the principal components are sorted in order of decreasing information content and variance [50], regions in Figure 6b define the portions of the cell that exhibit the more intense and the more vibrantly varying Raman signal.

Integrins are one of four principal cell adhesion molecules families, they play a major role in the process of adhesion of cells to the extracellular matrix (ECM), and especially in tumor cells where they are overexpressed during the process of adhesion [51]. $\alpha5\beta1$ and $\alpha3\beta1$ are integrins specifically expressed by tumour and epithelial cells. In particular, $\alpha3\beta1$ is overexpressed in tumours spreading in ECM with a high content of collagen and laminin, so that an elevated concentration of $\alpha3\beta1$ is a hallmark of cell proliferation and migration [52]. While each of those adhesion molecules have their own distinctive features, nonetheless they exhibit a certain number of peaks that do not vary from spectrum to spectrum representing a fingerprint for those molecules. Those peaks are found at (i) 1126 1/cm related to C-N bond, (ii) 1175 1/cm associated to Tyrosine or Phenylalanine, (iii) 1306 1/cm attributable to amide III, (iv) 1506 1/cm related to Phenylalanine or Hystidine, (v) 1569 1/cm originating from tryptophan, (vi) 1645 1/cm due to the amide I signal [53]. The Raman analysis that we performed on cells on different surfaces was enhanced by the interaction of the electromagnetic (EM) field with gold nanostructures, which amplify the Raman signal by several orders of magnitude in a SERS (surface enhanced Raman spectroscopy) effect [54]. SERS analysis of cells enabled the identification of adhesion markers that are otherwise inaccessible to classical spectroscopy techniques. The Raman intensity profile in Figure 6a for MeP_1 and MeP_2 follows a characteristic and distinguishable spatial distribution. Points in the map with higher values of Raman intensity may be indicative of the expression of integrin cell-adhesion-molecules suggesting that in those spots adhesion is established. The map relative to simple silicon with gold shows less preferential points of adhesion, indicating that smoother unstructured surfaces impair cell adhesion and proliferation. Moreover, SERS maps of cells measured on MeP_1 and MeP_2 substrates show a very high correspondence to the principal components distribution in Figure 6b, and especially to the first two components PC_1 and PC_2. Regions in the maps with greater overlap are at the borders of the cells (Figure 6a,b), with their membrane actively involved in the process of adhesion. The diagram in Figure 6c reports the loading associated to the first principal component measured for the MeP_1, MeP_2, and Si substrates. The loading is a statistical measure of how much different frequencies contribute to a certain principal component. The curves in Figure 6c indicate that the frequency that above all is responsible for the signal is 1569 1/cm for the MeP_1 substrate, while the signal content associated to 1569 1/cm is gradually weaker for the MeP_2 and the simple silicon substrate. Recalling that 1569 1/cm is the distinctive frequency for the integrins, the form of the diagrams of Figure 6c suggests that cell adhesion molecules are preferentially expressed for the foremost on MeP_1 substrates, followed by MeP_2 substrates and by flat silicon. For each substrate we calculated the ratio r between the value of loading intensity measured at 1569 1/cm and the loading averaged over the entire spectral range (Figure 6d). r is a quantitative measure of the relative abundance of integrins at the interface of a cell with a surface. Values of r much larger than one for MeP_1 ($r \approx 4.53$) and MeP_2 ($r \approx 4.34$), compared to the lower value of r measured on silicon ($r \approx 1$), show that the multiscale architecture of mesoporous substrates, with meso-pores functionalized with metal nanoparticles, facilitate cell adhesion compared to flat geometries. The values of r determined for the MeP_1 and MeP_2 substrates are significantly different from that determined for silicon, with $p < 0.05$.

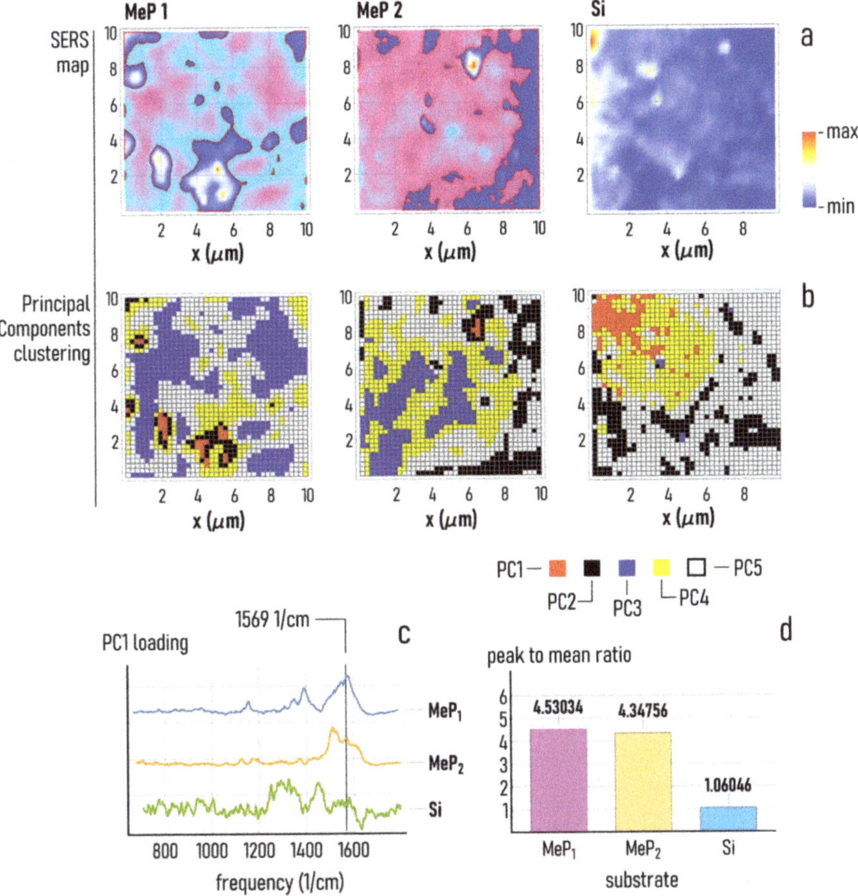

Figure 6. (**a**) Raman maps of MCF-7 cells acquired over a region of interest of 10 × 10 μm for MeP_1, MeP_2, and silicon surfaces, the maps show the Raman intensity measured at 1569 1/cm. (**b**) We show, for each of the considered surfaces (MeP_1, MeP_2, and Si), the first five principal components extracted from the Raman maps, and the spatial distribution of the principal components correlate with the Raman intensity maps previously reported. (**c**) The loading associated to the first principal component measured over MeP_1, MeP_2, and Si surfaces. (**d**) Ratio between the maximum and the mean intensity of the PC1 loading correspondent to MCF-7 cells cultivated over MeP_1, MeP_2, and Si samples.

3.4. *Kinetics of Drug Release from the Mesoporous Silicon Matrices*

The devices that we produced incorporate networks of nano-pores penetrating deep within their structures. We verified the capability of the device to accumulate and consequently release drug molecules over time using UV spectroscopy techniques as described in Section 2. We incubated MeP_1

and MeP_2 silicon substrates with the antitumor drug PtCl(O,O′-acac)(DMSO) for 60 h. We used two different concentrations of the originating drug during the loading process: c_1 = 30 μM and c_2 = 50 μM. We then measured the release of the drug in D.I. water up to 10 days from the activation of the process. The diagram in Figure 7a shows the cumulative dose–response curves for different substrates and different initial values of the loading concentration. The dynamics of release from the MeP_2 silicon system is faster compared to MeP_1 silicon, consistently with the fact the pores in MeP_2 silicon are larger (≈21 nm) than those contained in MeP_1 silicon (≈11 nm). Wanting to approximate the curves of release with a function of time of the form $c(t) = c_o + c_s\left(1 - e^{-t/\tau}\right)$, we found after nonlinear fitting of data the following solutions for c_s and τ: (i) $c_s \approx 4.94$ μM and $\tau \approx 46$ h for MeP_1 silicon loaded with an initial concentration equal to c_1, (ii) $c_s \approx 2.81$ μM and $\tau \approx 50$ h for MeP_1 silicon with an initial c_2 concentration, (iii) $c_s \approx 6.7$ μM and $\tau \approx 38$ h for MeP_2 silicon with an initial c_1 concentration, $c_s \approx 7.2$ μM and $\tau \approx 20$ h for MeP_2 silicon with an initial c_2 concentration. c_s is the steady state value of the concentration increment with respect to a zero reference value. τ is the time constant of the drug delivery system, i.e., the time necessary to the system to reach 66% of its final value of concentration. The values that we found for c_s and τ for the different combinations of substrate morphology and initial loading concentration that we used in our study, indicate that the rate of drug release (1/τ) increases moving from MeP_1 to MeP_2 and, for the same substrate, it is higher for an initial higher concentration of the loaded drug. This behavior can be easily described by the first law of Fick, $J = -D\partial c/\partial x$, where the intensity of the flux (J) is proportional to the gradient of concentration from the substrate to the external environment, and the absolute number of molecules transported through the system per unit time depends on the area of the surface actively releasing the drug. In the equation D is the molecular diffusion coefficient [50]. Moreover, the values of c_s and τ and Figure 7a indicate that the total amount of drug released in a system (c_s) is higher for MeP_2 silicon, that has a higher porosity compared to MeP_1, and is higher for an initial higher concentration of the loaded drug. In the neighbor of $t = 0$, the drug release profile can be expanded in a Taylor series yielding, neglecting higher order terms of the expansion, $c(t) \approx c_s/\tau$: this approximate formula enables calculation of the velocity of release (v) at the early stage of the delivery process. Using data from the model fit, we obtained $v \approx 0.06$ μM/h (MeP_1, c_1), $v \approx 0.11$ μM/h (MeP_1, c_2), $v \sim 0.17$ μM/h (MeP_2, c_1), $v \approx 0.37$ μM/h (MeP_2, c_2). Thus, the kinetics of initial release can be varied in the 0.06-0.37 μM/h interval by changing the parameters of the process. Figure 7b displays the drug released from the systems over time, normalized to the initial concentration of the loaded drug. Values in the figure are a measure of the efficiency of the drug delivery system. Data show that the maximum efficiency of release varies between ≈0.12 for MeP_1 silicon with an initial loading concentration c_2, and ≈0.23 for MeP_2 silicon with an initial loading concentration c_1. While the efficiency of delivery is still larger for MeP_2 silicon with higher values of pore size and porosity, it decreases for increasing values of initial loading concentration, possibly because for larger amounts of initial payload the losses associated to the process are also larger. Thus, one of the points of strength of this bio-chip, is that it can artificially increase the half-life of a drug. The half-life ($t\frac{1}{2}$) is the time required to change the amount of a drug in the body by one-half during elimination. Drug clearance from the body is the result of elimination by renal excretion and by non-renal pathways, the latter most often represent clearance by the liver. Remarkably, the characteristic half-life—the duration of action—of anticancer drugs is, on average, small. Clinical studies and reports [55] indicate that the mean half-life of more than 140 small-molecule drugs approved for oncology indications is ≈15 h, with an even smaller value of median of about ≈5 h. The drug delivery system set-up in this study enables the active release of drugs for more than ≈50 h, depending on the configuration. Thus, the chip that we produced can possibly increase the half-life of most anticancer treatments by 300% on average.

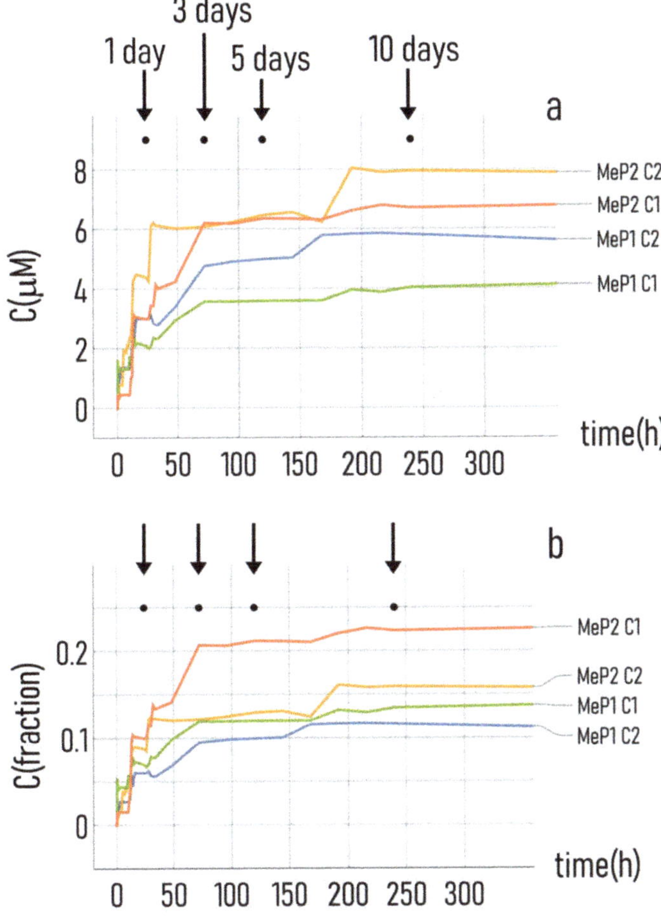

Figure 7. (**a**) Absolute and (**b**) normalized release profile of the anti-tumor drug $PtCl(O, O' - acac)(DMSO)$ measured for MeP_1 and MeP_2 substrates up to 15 days from the beginning of the delivery.

3.5. Timely Delivery of Drugs Impairs Tumor Cells Adhesion

The theranostics device presented in the work has the ability to deliver over time the antitumor drug PtCl (O,O′-acac) (DMSO) for several hours from the initial release. We examined the effects of a controlled release of drugs on the adhesion and growth of tumor MCF-7 cells on the mesoporous substrates. MCF-7 cells were cultured on MeP_1 silicon and MeP_2 silicon functionalized with gold nanoparticles and on flat silicon surfaces used as a control. Half of the mesoporous substrates used in this study were loaded with PtCl (O,O′-acac) (DMSO) in a concentration of 50 μM. We then imaged cell-nuclei on different substrates and at different times from incubation. The different number of cells adhering on the substrates is the effect of a combination of factors: (i) the drug released from the system and (ii) the different nano-topographical characteristics of the substrates. Visual examination of samples reveals that already 6 h from culture the number of adhering cells on flat silicon is lower than that observed on MeP_1 with gold, that is in turn different from the number of cells on MeP_1 silicon loaded with drugs (Figure 8). This difference is exacerbated by time. In Figure 9a we report the bar-chart of the number of cells (N) measured on different substrates 36, 48, and 72 h from incubation. N was estimated from more than 20 images per substrate and five technical repeats per sample.

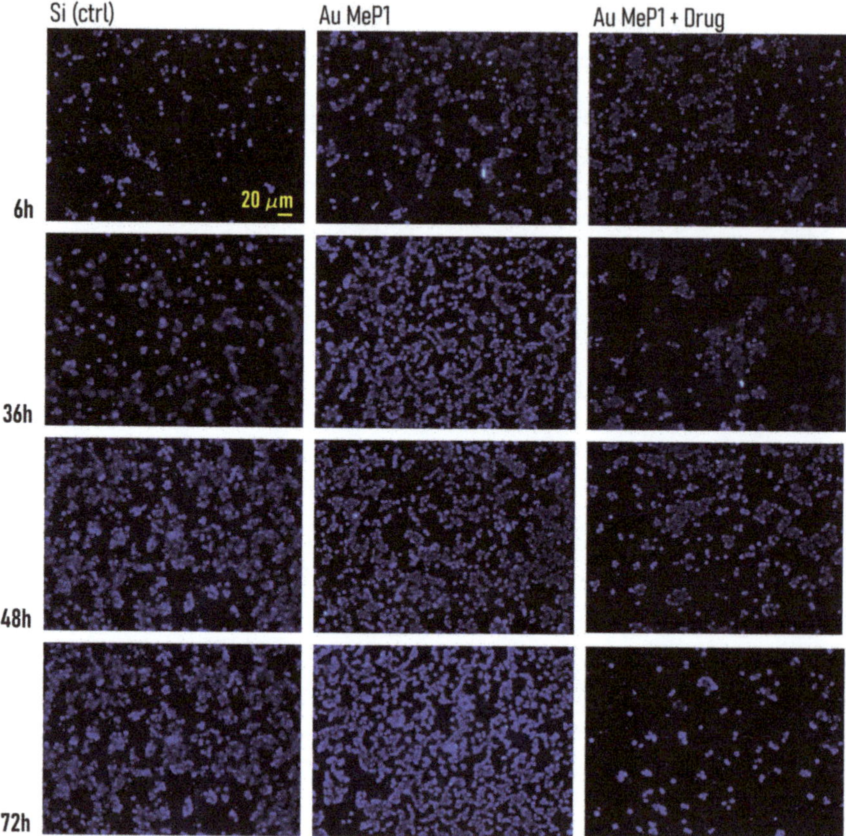

Figure 8. Fluorescence images of MCF-7 cancer cells growing on Au-MeP$_1$ substrates with and without the delivery of the antitumor agent PtCl(O,O′-acac)(DMSO), at different time frames. In the experiments, unfunctionalized silicon was used as a control.

A total of 36 h after culture, the number of cells on MeP$_1$ and MeP$_2$ silicon without drug is not statistically different from that measured on flat silicon, with $N_{MeP1} \approx 1871$, $N_{MeP2} \approx 2100$, and $N_{Si} \approx 1951$. Diversely, N is significantly lower for the mesoporous substrates loaded with drug, being ${N_{MeP1}}^{drug} \approx 675$ and ${N_{MeP2}}^{drug} \approx 424$. A total of 36 h from incubation the effect of topography is negligible compared to the release of drug.

A total of 48 h from incubation, the number of cells on MeP$_1$ silicon without drug increases to $N_{MeP1} \approx 2928$, significantly larger than the number of cells measured at the same time on simple MeP$_2$ and flat silicon, with $N_{MeP2} \approx 1983$ and $N_{Si} \approx 2090$. At this time step, the number of cells measured on the mesoporous substrates loaded with drug falls to ${N_{MeP1}}^{drug} \approx 451$ and ${N_{MeP2}}^{drug} \approx 255$, that are significantly different from the values measured for the unloaded substrates. A total of 48 h after cell culture, the improved adhesive characteristics of MeP$_1$ silicon over simple MeP$_2$ and simple silicon become apparent, while the delivery of drugs from MeP$_2$ silicon achieves maximum effects.

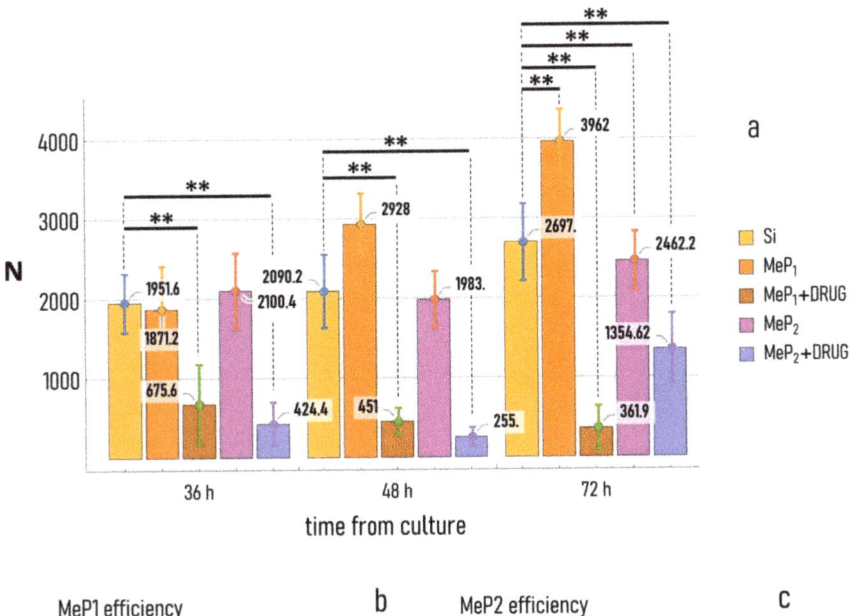

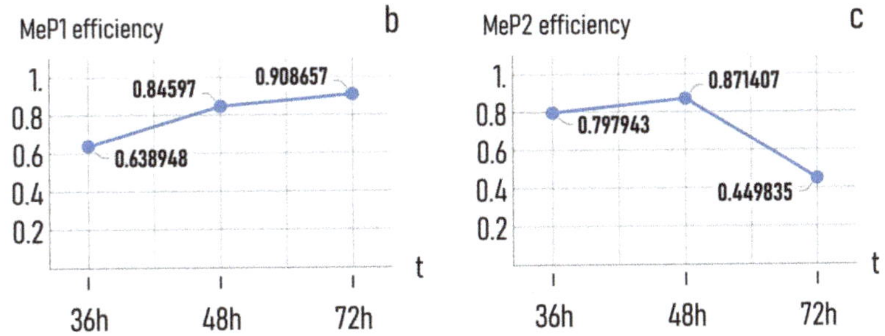

Figure 9. (**a**) Number of adhering cells on Au-MeP$_1$ and Au-MeP$_2$ substrates under and without the influence of the antitumor agent PtCl(O,O′-acac)(DMSO), at 36, 48, and 72 h from cell seeding, compared to plain silicon used as a control. All p values less than 0.05 are summarized with two asterisks. Efficacy of the antitumor drug on the cancer cells at different times from the initial release, for the (**b**) Au-MeP$_1$ (**c**) and Au-MeP$_2$ substrates.

A total of 72 h from incubation, the number of adhering cells on MeP$_1$ silicon with drug hits a minimum, $N_{\text{MeP1}}{}^{\text{drug}} \approx 362$, while N surges to $N_{\text{MeP1}} \approx 3962$ for simple MeP$_1$ silicon without drug. These values are significantly lower and significantly larger than the control: $N_{\text{Si}} \approx 2697$. At this time of the analysis, the MeP$_2$ morphology does not enhance significantly adhesion with respect to the control, with $N_{\text{MeP2}} \approx 2462$, while the release of drug from MeP$_2$ silicon still bears appreciable effects, with $\text{N}_{\text{MeP2}}{}^{\text{drug}} \approx 1355$ significantly lower than the number of cells measured on flat silicon.

Thus, the efficacy of the drug delivery system depends on both of the substrate morphology and the time of the process. For each time, we calculated the efficiency of delivery as the number of adhering cells on the substrate loaded with drug, to the number of cells measured on the same substrate without drug (Figure 9b,c). We observe that the efficiency of the system gradually increases for MeP$_1$ silicon, varying from ≈0.64 at 36 h, to ≈0.84 at 48 h, to ≈0.90 at 72 h from incubation. For this substrate, more than 90% of cells are killed compared to the same substrate in absence of drug delivery. The curve of efficiency for the MeP$_2$ silicon is different. For this system, the efficiency of delivery is sufficiently large already 36 h from the initial release (≈0.80), it attains a maximum value at

48 h (≈0.87), to subsequently drop to ≈0.45 at 72 h from incubation. The different efficiency measured for MeP_1 and MeP_2 silicon reflects the different kinetics of release measured for those devices and reported in previous sections of this work.

4. Discussion

The theranostics substrates that we produced are akin to the cell culture glass coverslips used to plate cells for research and analysis—with the notable exception that they enable the targeted delivery of therapeutics to the cells and cell sensing, simultaneously. Moreover, thanks to the fabrication process capability to attain maximum control over surface morphology at the nanoscale, the substrates can be designed to guide cell adhesion, proliferation, and organization. For these characteristics, this bio-device—and its more sophisticated evolutions that will be developed over time—can be integrated into conventional cell culture dishes or multi-wells to test the adhesion and growth of cells against different external factors, including substrate geometry and a controlled delivery of drugs. Researchers can plate cells over several different replica of the device, each of them with its characteristic topography and drug release profile. Then, the researchers will find the combination of surface topography and device payload that guarantees maximum/minimum cell adhesion and proliferation, depending on whether the aim of the research is optimize a structure for tissue engineering or the effects of a drug for personalized medicine. The search for the optimal values of surface topography and kinetics of release should be possibly conducted within the bounds identified by this and other similar works: where the roughness of the surface and the pore size is varied in the 0–30 nm interval, while drugs are released with a maximum initial rate of ≈0.4 μM/h. The output of the experiment—i.e., cell colonies—can be verified at different time steps from seeding using either confocal microscopy or Raman spectroscopy that is, notably, made possible by the distinctive design of the device. While the first technique provides information about cell adhesion, growth, and clustering, Raman analysis of samples describes the conditions of a cell at the level of its adhesion molecules. Thus, the combination of techniques gives a picture of the evolution of a cell over different scales, bridging the divide between the behavior of cells being observed in isolation (individual behavior of cells) or in-group (collective behavior of cells). Consistency between results may indicate that the substrate operates efficiently towards either improving or impairing cell adhesion and organization. Thus, the device can potentially be the basis for a test campaign aimed to optimize the characteristics of biomaterials for tissue engineering, regenerative medicine, or in-vitro-model applications. After identification of the optimal surface characteristics and drug dosage that assure the wanted effect, these values should be copied to the scaffold intended to support cell-growth, or to the implantable device that will release drugs to a disease, for real-life applications. Nonetheless, this implies a process of engineering of the device, aimed to overtake those complications that can possibly emerge when similar devices are used outside of a research context. A list of possible caveats is identified as follows:

(1) Transition from a 2D to 3D geometry. Results of the work and the way the device operates are restricted to 2D geometries. Cells themselves are plated on a surface and cell clusters are described using bi-dimensional variables. Future research that will be conducted over time shall have to clarify whether, and to which extent, cell behavior changes moving from 2D to 3D scaffolds.

(2) Understanding whether the delivery is active. As demonstrated in the work, the delivery of the drugs from the porous matrix does not last indefinitely. Additionally, there could be cases in which, because of unpredictable leakages or occlusions of the pores, the process of release terminates before the expected time. Thus, the substrate should incorporate a sensor sensitive to the released drug, indicating the rate at which release proceeds, warning of possible malfunctions of the device before cells react to the alterations of delivery.

(3) Switching between on-off delivery states of the device. In this configuration, the release of drugs is always active, being driven by the gradients of concentration in the system—and described by the Fick's laws. In a more sophisticated evolution of the device, one should be able to switch between on-off states: i.e., an active (on) state, in which drug molecules are allowed to flow in the

system, and an inactive (off) state, in which release is temporarily paused. This can be possibly accomplished by varying, in a controlled fashion, the levels of pH and temperature of the system, having previously conjugated the drug with a pH-sensitive cleavable linker as described, as for an example, for injectable nanoparticle generators in [56].

(4) Increasing the efficiency of drug loading, either in terms of pore-capacity and total amount of drug loaded in the pores, and in terms of loading time. The time necessary to load the drug into the device (more than 60 h for the present configuration) is unacceptably long for real life/industrial applications of the device.

(5) Integration. While the device offers complementary ways to measure cell performance, using the bio-chip alternatively in a confocal-microscope or in a Raman setup can take time, and slow down the pace of a test campaign, a trial, or a biology/medicine application of the device. The method can take advantage from the integration of the chip in an automatic multi-well plate reader, where different imaging techniques (confocal imaging and Raman spectroscopy) are combined in the same platform.

(6) Optimization. The device can be especially useful in cancer theranostics. While the results obtained in this work are based on an initial cell density of about 10^5 cells per substrate, this value can be optimized: the total number of cells for which the device has some measurable effects may be significantly lower than the 10^5 limit. Miniaturization has as principal consequence the possibility to use the devices in all those cases in which, because of the early stage of a disease, conventional biopsies are ineffective. For the same reason, the device can be used to perform liquid biopsy. Liquid biopsy is based on the detection and isolation of cancer cells directly from the peripheral blood of the patient [57]. The limit of the method is that, often, cancer cells are too few to be detected. Liquid biopsy could realistically benefit from multi-drug array panels based on mesoporous silicon substrates, designed to evaluate the sensitivity of circulating tumor cells to a test drug.

(7) Generality. Further to the end of liquid biopsy: results of the paper suggest that this theranostics device is also effective towards more aggressive cancer cells, such as triple negative breast cancer cells. The integrin expression that we chose to analyze in this work is believed to be a hallmark of more aggressive forms of breast cancer [58]. In particular, the expression of β1 integrins on the cell surface is a predictive marker of triple negative breast cancer. Since the substrates that we produced induce, for certain configurations, an increased production of integrins on the MCF-7 surface, it is legitimate to hypothesize that this theranostics procedure is effective also in the case of triple negative breast cancer cells. Regarding the impact on normal cells, it has been demonstrated in several studies—some of which have been cited throughout the article—that both the porosity and roughness of nanoscale surfaces affect the adhesion, proliferation, and clustering of cells, depending on their degree of differentiation and replication speed. Thus, the method that we developed is general in scope and can be realistically adapted to several different cell types for different applications.

5. Conclusions

Using electrochemical etching, we produced mesoporous silicon substrates functionalized with gold nanoparticles. Due to their porous structures, the substrates behave as a drug delivery system, where drugs or other agents loaded in the matrix are released over time with a first order kinetics and a time constant that can be varied by tuning the characteristics of the pores. Due to its nanostructure, the device can amplify by several orders of magnitude the signal generated by molecules in the cell membrane during the process of adhesion and migration. These nanodevices combine therapy and diagnostics effects in the same device, with the primary advantage of not being limited to therapy or sensing, as they provide coincident diagnostic information plus delivery of therapeutics. For their capabilities to drive cell growth and high capacities of therapeutic loading, these devices can be possibly used in tissue engineering, regenerative medicine, and nanomedicine. In these fields, the ability to assemble cells together has to be associated with the ability of deliver therapeutics accurately to achieve maximum control over cell fate. The ability of the system to measure adhesion at the nanoscale, is an extra feature that makes this theranostics device the ideal candidate for those who want to design substrates for cell growth and proliferation or, vice versa, induce apoptosis in tumor cells with unprecedented control.

In this work, by varying the parameters of the process we obtained substrates with different values of pore size, porosity, and roughness. MeP_1 substrates exhibit smaller values of pore size (PS $\approx$ 11 nm) and roughness (Ra $\approx$ 7 nm), but have a larger fractal dimension (Df $\approx$ 2.48) compared to MeP_2 substrates, with PS $\approx$ 21 nm, $R_a \approx$ 13 nm, and Df $\approx$ 2.15. Adhesion of MCF-7 cells was accelerated on MeP_1 substrates with larger value of fractal dimension compared to MeP_2 substrates and flat silicon used as a control.

When we considered the effect of the release over time of an anti-tumor drug, we observed that the maximum reduction of cell growth is found for MeP_2 substrates 36 and 48 h after seeding, with a decrease in the number of adhering cells up to 87% with respect to the same substrates without drug. A total of 72 h from cell culture, the therapeutics efficacy of MeP_2 substrates falls to 44%, and is overtaken by MeP_1 silicon, with a cell number contraction of 90%, reflecting the different morphological characteristics of surfaces.

Results suggest that the adhesive properties of mesoporous substrates, the kinetics of drug delivery, and the effects that a combination of the two may have on the adhesion and proliferation of cells, can be conveniently modulated by changing the pore size and roughness in the narrow intervals of PS $\approx$ 11–21 nm and Ra $\approx$ 7–13 nm.

Supplementary Materials: The following are available online at http://www.mdpi.com/1999-4923/12/5/481/s1, Supporting Information 1: Determining the porosity of Mesoporous surfaces. Supporting Information 2: Extracting Au nano-particle diameter.

Author Contributions: Conceptualization, F.G.; methodology, M.L.C., G.M., M.D.A., S.D.V., A.I. and L.T.; software, V.O.; validation, V.O., M.L.C., G.P., E.D.F., P.C., N.M. and F.G.; formal analysis, M.L.C., V.O., E.D.F., N.M. and F.G.; investigation, M.L.C., N.M. and F.G.; resources, N.M. and F.G.; data curation, M.L.C., V.O., N.M. and F.G.; writing—original draft preparation, F.G.; writing—review and editing, M.L.C., V.O., N.M., F.G.; visualization, M.L.C. and F.G.; supervision, G.P., E.D.F., P.C., N.M., F.G.; project administration, N.M. and F.G.; funding acquisition, N.M. All authors have read and agreed to the published version of the manuscript.

Funding: This research received no external funding.

Acknowledgments: We thank the Dresden Center for Nanoanalysis of Technische Universität Dresden for the support provided for the SEM analysis of samples. This publication was supported by the Department of Experimental and Clinical Medicine, University Magna Graecia of Catanzaro.

Conflicts of Interest: The authors declare no conflict of interest.

References

1. Gentile, F. Nanotopographical Control of Cell Assembly into Supracellular Structures. In *Nanomaterials for Advanced Biological Applications*; Rahmandoust, M., Ayatollahi, M., Eds.; Springer: Cham, Switzerland, 2019; Volume 104.
2. Hutmacher, D.W.; Woodfield, T.B.F.; Dalton, P.D. Scaffold Design and Fabrication. In *Tissue Engineering*; Van Blitterswijk, C.A., De Boer, J., Eds.; Academic Press: Boston, MA, USA, 2014; pp. 311–346.
3. Limongi, T.; Schipani, R.; Di Vito, A.; Giugni, A.; Francardi, M.; Torre, B.; Allione, M.; Miele, E.; Malara, N.; Alrasheed, S.; et al. Photolithography and micromolding techniques for the realization of 3D polycaprolactone scaffolds for tissue engineering applications. *Microelectron. Eng.* **2015**, *141*, 135–139. [CrossRef]
4. Lu, T.; Li, Y.; Chen, T. Techniques for fabrication and construction of three-dimensional scaffolds for tissue engineering. *Int. J. Nanomed.* **2013**, *8*, 337–350. [CrossRef]
5. Perozziello, G.; Simone, G.; Candeloro, P.; Gentile, F.; Malara, N.; Larocca, R.; Coluccio, M.L.; Pullano, S.; Tirinato, L.; Geschke, O.; et al. A Fluidic Motherboard for Multiplexed Simultaneous and Modular Detection in Microfluidic Systems for Biological Application. *Micro Nanosyst.* **2010**, *2*, 227–238. [CrossRef]

6. Decuzzi, P.; Ferrari, M. Modulating cellular adhesion through nanotopography. *Biomaterials* **2010**, *31*, 173–179. [CrossRef] [PubMed]
7. Geiger, B.; Bershadsky, A.; Pankov, R.; Yamada, K.M. Transmembrane Extracellular Matrix–Cytoskeleton Crosstalk. *Nat. Rev. Mol. Cell Biol.* **2001**, *2*, 793–805. [CrossRef] [PubMed]
8. Geiger, B.; Spatz, J.P.; Bershadsky, A.D. Environmental sensing through focal adhesions. *Nat. Rev. Mol. Cell Biol.* **2009**, *10*, 21–33. [CrossRef] [PubMed]
9. Kanchanawong, P.; Shtengel, G.; Pasapera, A.M.; Ramko, E.B.; Davidson, M.W.; Hess, H.F.; Waterman, C.M. Nanoscale architecture of integrin-based cell adhesions. *Nature* **2010**, *468*, 580–586. [CrossRef]
10. Perozziello, G.; La Rocca, R.; Cojoc, G.; Liberale, C.; Malara, N.; Simone, G.; Candeloro, P.; Anichini, A.; Tirinato, L.; Gentile, F.; et al. Microfluidic Devices Modulate Tumor Cell Line Susceptibility to NK Cell Recognition. *Small* **2012**, *8*, 2886–2894. [CrossRef]
11. Stevens, M.; George, J. Exploring and engineering the cell surface interface. *Science* **2005**, *310*, 1135–1138. [CrossRef]
12. De Vitis, S.; Coluccio, M.L.; Strumbo, G.; Malara, N.; Fanizzi, F.P.; De Pascali, S.A.; Perozziello, G.; Candeloro, P.; Di Fabrizio, E.M.; Gentile, F. Combined effect of surface nano-topography and delivery of therapeutics on the adhesion of tumor cells on porous silicon substrates. *Microelectron. Eng.* **2016**, *158*, 6–10. [CrossRef]
13. Lane, S.W.; Williams, D.A.; Watt, F.M. Modulating the stem cell niche for tissue regeneration. *Nat. Biotechnol.* **2014**, *32*, 795–803. [CrossRef] [PubMed]
14. Dalby, M.J.; Gadegaard, N.; Oreffo, R.O.C. Harnessing nanotopography and integrin-matrix interactions to influence stem cell fate. *Nat. Mater.* **2014**, *13*, 558–569. [CrossRef]
15. Delcassian, D.; Depoil, D.; Rudnicka, D.; Liu, M.; Davis, D.M.; Dustin, M.L.; Dunlop, I.E. Nanoscale Ligand Spacing Influences Receptor Triggering in T Cells and NK Cells. *Nano Lett.* **2013**, *13*, 5608–5614. [CrossRef]
16. Robinson, J.T.; Jorgolli, M.; Shalek, A.K.; Yoon, M.-H.; Gertner, R.S.; Park, H. Vertical nanowire electrode arrays as a scalable platform for intracellular interfacing to neuronal circuits. *Nat. Nanotechnol.* **2012**, *7*, 180–184. [CrossRef] [PubMed]
17. Gentile, F.; Tirinato, L.; Battista, E.; Causa, F.; Liberale, C.; Di Fabrizio, E.; Decuzzi, P. Cells preferentially grow on rough substrates. *Biomaterials* **2010**, *31*, 7205–7212. [CrossRef] [PubMed]
18. Sorkin, R.; Greenbaum, A.; David-Pur, M.; SaritAnava; Ayali, A.; Ben-Jacob, E.; Hanein, Y. Process entanglement as a neuronal anchorage mechanism to rough surfaces. *Nanotechnology* **2009**, *20*, 015101. [CrossRef] [PubMed]
19. Xie, C.; Hanson, L.; Xie, W.; Lin, Z.; Cui, B.; Cui, Y. Noninvasive Neuron Pinning with Nanopillar Arrays. *Nano Lett.* **2010**, *10*, 4020–4024. [CrossRef]
20. Ankam, S.; Suryana, M.; Chan, L.Y.; Moe, A.A.K.; Teo, B.K.K.; Law, J.B.K.; Sheetz, M.P.; Low, H.Y.; Yim, E.K.F. Substrate topography and size determine the fate of human embryonic stem cells to neuronal or glial lineage. *Acta Biomater.* **2013**, *9*, 4535–4545. [CrossRef]
21. Baranes, K.; Chejanovsky, N.; Alon, N.; Sharoni, A.; Shefi, O. Topographic Cues of Nano-Scale Height Direct Neuronal Growth Pattern. *Biotechnol. Bioeng.* **2012**, *109*, 1791–1797. [CrossRef]
22. Ferrari, A.; Cecchini, M.; Dhawan, A.; Micera, S.; Tonazzini, I.; Stabile, R.; Pisignano, D.; Beltram, F. Nanotopographic Control of Neuronal Polarity. *Nano Lett.* **2011**, *11*, 505–511. [CrossRef]
23. Ligon, S.C.; Liska, R.; Stampfl, J.; Gurr, M.; Mülhaupt, R. Polymers for 3D Printing and Customized Additive Manufacturing. *Chem. Rev.* **2017**, *117*, 10212–10290. [CrossRef] [PubMed]
24. Guvendiren, M.; Molde, J.; Soares, R.M.D.; Kohn, J. Designing Biomaterials for 3D Printing. *ACS Biomater. Sci. Eng.* **2016**, *2*, 1679–1693. [CrossRef]
25. Accardo, A.; Courson, R.; Riesco, R.; Raimbault, V.; Malaquin, L. Direct Laser Fabrication of Meso-Scale 2D and 3D Architectures with Micrometric Feature Resolution. *Addit. Manuf.* **2018**, *22*, 440–446. [CrossRef]
26. Nguyen, A.K.; Narayan, R.J. Two-Photon Polymerization for Biological Applications. *Mater. Today* **2017**, *20*, 314–322. [CrossRef]
27. Fan, D.; Staufer, U.; Accardo, A. Engineered 3D Polymer and Hydrogel Microenvironments for Cell Culture Applications. *Bioengineering* **2019**, *6*, 113. [CrossRef]
28. Accardo, A.; Blatché, M.C.; Courson, R.; Loubinoux, I.; Thibault, C.; Malaquin, L.; Vieu, C. Multiphoton Direct Laser Writing and 3D Imaging of Polymeric Freestanding Architectures for Cell Colonization. *Small* **2017**, *13*, 1700621–1700611. [CrossRef] [PubMed]

29. Gentile, F.; Rocca, R.L.; Marinaro, G.; Nicastri, A.; Toma, A.; Paonessa, F.; Cojoc, G.; Liberale, C.; Benfenati, F.; Di Fabrizio, E.; et al. Differential Cell Adhesion on Mesoporous Silicon Substrates. *ACS Appl. Mater. Interfaces* **2012**, *4*, 2903–2911. [CrossRef]
30. Marinaro, G.; La Rocca, R.; Toma, A.; Barberio, M.; Cancedda, L.; Di Fabrizio, E.; Decuzzi, P.; Gentile, F. Networks of Neuroblastoma Cells on Porous Silicon Substrates Reveal a Small World Topology. *Integr. Biol.* **2015**, *7*, 184–197. [CrossRef]
31. Foll, H.; Christophersen, M.; Carstensen, J.; Hasse, G. Formation and application of porous silicon. *Mater. Sci. Eng.* **2002**, *39*, 93–141. [CrossRef]
32. Gentile, F.; Battista, E.; Accardo, A.; Coluccio, M.; Asande, M.; Perozziello, G.; Das, G.; Liberale, C.; De Angelis, F.; Candeloro, P.; et al. Fractal Structure Can Explain the Increased Hydrophobicity of NanoPorous Silicon Films. *Microelectron. Eng.* **2011**, *88*, 2537–2540. [CrossRef]
33. Coluccio, M.L.; Gentile, F.; Francardi, M.; Perozziello, G.; Malara, N.; Candeloro, P.; Di Fabrizio, E. Electroless Deposition and Nanolithography Can Control the Formation of Materials at the Nano-Scale for Plasmonic Applications. *Sensors* **2014**, *14*, 6056–6083. [CrossRef] [PubMed]
34. Battista, E.; Coluccio, M.-L.; Alabastri, A.; Barberio, M.; Causa, F.; Netti, P.-A.; Di Fabrizio, E.; Gentile, F. Metal enhanced fluorescence on super-hydrophobic clusters of gold nanoparticles. *Microelectron. Eng.* **2017**, *175*, 7–11. [CrossRef]
35. Li, W.; Liu, C.; Zhao, C.; Zhai, L.; Lv, S. Downregulation of β3 integrin by miR-30a-5p modulates cell adhesion and invasion by interrupting Erk/Ets 1 network in triple-negative breast cancer. *Int. J. Oncol.* **2016**, *48*, 1155–1164. [CrossRef] [PubMed]
36. Waxman, B. Routing of multipoint connections. *IEEE J. Sel. Areas Commun.* **1988**, *6*, 1617–1622. [CrossRef]
37. Barabási, A.-L. *Network Science*; Cambridge University Press: Glasgow, UK, 2016; p. 475.
38. Barabási, A.-L.; Albert, R. Emergence of scaling in random networks. *Science* **1999**, *286*, 509–512. [CrossRef]
39. Barabási, A.-L.; Oltvai, Z.N. Network biology: Understanding the cell's functional organization. *Nat. Rev. Genet.* **2004**, *5*, 101–113. [CrossRef]
40. Watts, D.J. *Small Worlds: The Dynamics of Networks between Order and Randomness*; Princeton University Press: Woodstock, UK, 2003.
41. Watts, D.J.; Strogatz, S.H. Collective dynamics of 'small-world' networks. *Nature* **1998**, *393*, 440–442. [CrossRef]
42. Humphries, M.D.; Gurney, K. Network 'Small-World-Ness': A Quantitative Method for Determining Canonical Network Equivalence. *PLoS ONE* **2008**, *3*, e0002051. [CrossRef]
43. Perozziello, G.; Catalano, R.; Francardi, M.; Rondanina, E.; Pardeo, F.; De Angelis, F.; Malara, N.; Candeloro, P.; Morrone, G.; Di Fabrizio, E. A microfluidic device integrating plasmonic nanodevices for Raman spectroscopy analysis on trapped single living cells. *Microelectron. Eng.* **2013**, *111*, 314–319. [CrossRef]
44. Candeloro, P.; Grande, E.; Raimondo, R.; Di Mascolo, D.; Gentile, F.; Coluccio, M.L.; Perozziello, G.; Malara, N.; Francardi, M.; Di Fabrizio, E. Raman database of Amino Acids solutions: A critical study of Extended Multiplicative Signal Correction. *Analyst* **2013**, *138*, 7331–7340. [CrossRef]
45. Muscella, A.; Vetrugno, C.; Migoni, D.; Biagioni, F.; Fanizzi, F.; Fornai, F.; De Pascali, S.; Marsigliante, S. Antitumor activity of [Pt(O,O′-acac)(γ-acac)(DMS)] in mouse xenograft model of breast cancer. *Cell Death Dis.* **2014**, *5*, e1014. [CrossRef] [PubMed]
46. Gentile, F.; Medda, R.; Cheng, L.; Battista, E.; Scopelliti, P.; Milani, P.; Cavalcanti-Adam, E.; Decuzzi, P. Selective modulation of cell response on engineered fractal silicon substrates. *Sci. Rep.* **2013**, *3*, 1461. [CrossRef] [PubMed]
47. Strogatz, S.H. Exploring complex networks. *Nature* **2001**, *410*, 268–276. [CrossRef] [PubMed]
48. Onesto, V.; Cancedda, L.; Coluccio, M.; Nanni, M.; Pesce, M.; Malara, N.; Cesarelli, M.; Fabrizio, E.D.; Amato, F.; Gentile, F. Nano-topography Enhances Communication in Neural Cells Networks. *Sci. Rep.* **2017**, *7*, 1–13. [CrossRef] [PubMed]
49. Onesto, V.; Villani, M.; Narducci, R.; Malara, N.; Imbrogno, A.; Allione, M.; Costa, N.; Coppedè, N.; Zappettini, A.; Cannistraci, C.; et al. Cortical-like mini-columns of neuronal cells on zinc oxide nanowire surfaces. *Sci. Rep.* **2019**, *9*, 4021–4017. [CrossRef]
50. Di Mascolo, D.; Coclite, A.; Gentile, F.; Francardi, M. Quantitative micro-Raman analysis of micro-particles in drug delivery. *Nanoscale Adv.* **2019**, *1*, 1541–1552. [CrossRef]

51. Jin, H.; Varner, J. Integrins: Roles in cancer development and as treatment targets. *Br. J. Cancer* **2004**, *90*, 561–565. [CrossRef]
52. Koistinen, P.; Heino, J. Integrins in Cancer Cell Invasion. In *Madame Curie Bioscience Database [Internet]*; Landes Bioscience: Austin, TX, USA, 2013. Available online: https://www.ncbi.nlm.nih.gov/books/NBK6070/ (accessed on 15 April 2020).
53. Chowdhury, M.H.; Gant, V.A.; Trache, A.; Baldwin, A.M.; Meininger, G.A.; Coté, G.L. Use of surface-enhanced Raman spectroscopy for the detection of human integrins. *J. Biomed. Opt.* **2006**, *11*, 024004. [CrossRef]
54. Gentile, F.; Coluccio, M.L.; Zaccaria, R.P.; Francardi, M.; Cojoc, G.; Perozziello, G.; Raimondo, R.; Candeloroa, P.; Fabrizio, E.D. Selective on site separation and detection of molecules in diluted solutions with superhydrophobic clusters of plasmonic nanoparticles. *Nanoscale* **2014**, *6*, 8208–8225. [CrossRef]
55. Liston, D.R.; Davis, M. Clinically Relevant Concentrations of Anticancer Drugs: A Guide for Nonclinical Studies. *Clin. Cancer Res.* **2017**, *23*, 3489–3498. [CrossRef]
56. Xu, R.; Zhang, G.; Mai, J.; Deng, X.; Segura-Ibarra, V.; Wu, S.; Shen, J.; Liu, H.; Hu, Z.; Chen, L.; et al. An Injectable Nanoparticle Generator Enhances Delivery of Cancer Therapeutics. *Nat. Biotechnol.* **2016**, *34*, 414–418. [CrossRef] [PubMed]
57. Malara, N.; Givigliano, F.; Trunzo, V.; Macrina, L.; Raso, C.; Amodio, N.; Aprigliano, S.; Minniti, A.; Russo, V.; Roveda, L.; et al. In vitro expansion of tumour cells derived from blood and tumour tissue is useful to redefine personalized treatment in non-small cell lung cancer patients. *J. Biol. Regul. Homeost. Agents* **2014**, *28*, 717–731. [PubMed]
58. Yin, H.-L.; Wu, C.-C.; Lin, C.-H.; Chai, C.-Y.; Hou, M.-F.; Chang, S.-J.; Tsai, H.-P.; Hung, W.-C.; Pan, M.-R.; Luo, C.-W. β1 Integrin as a Prognostic and Predictive Marker in Triple-Negative Breast Cancer. *Int. J. Mol. Sci.* **2016**, *17*, 1432. [CrossRef] [PubMed]

pharmaceutics

Review

Drug Delivery Applications of Three-Dimensional Printed (3DP) Mesoporous Scaffolds

Tania Limongi [1,*], Francesca Susa [1], Marco Allione [2] and Enzo di Fabrizio [1]

1 Dipartimento di Scienza Applicata e Tecnologia, Politecnico di Torino, Corso Duca Degli Abruzzi 24, 10129 Torino, Italy; francesca.susa@polito.it (F.S.); enzo.difabrizio@polito.it (E.d.F.)
2 SMILEs Lab, PSE Division, King Abdullah University of Science and Technology, Thuwal 23955-6900, Saudi Arabia; marco.allione@kaust.edu.sa
* Correspondence: tania.limongi@polito.it

Received: 30 June 2020; Accepted: 5 September 2020; Published: 8 September 2020

Abstract: Mesoporous materials are structures characterized by a well-ordered large pore system with uniform porous dimensions ranging between 2 and 50 nm. Typical samples are zeolite, carbon molecular sieves, porous metal oxides, organic and inorganic porous hybrid and pillared materials, silica clathrate and clathrate hydrates compounds. Improvement in biochemistry and materials science led to the design and implementation of different types of porous materials ranging from rigid to soft two-dimensional (2D) and three-dimensional (3D) skeletons. The present review focuses on the use of three-dimensional printed (3DP) mesoporous scaffolds suitable for a wide range of drug delivery applications, due to their intrinsic high surface area and high pore volume. In the first part, the importance of the porosity of materials employed for drug delivery application was discussed focusing on mesoporous materials. At the end of the introduction, hard and soft templating synthesis for the realization of ordered 2D/3D mesostructured porous materials were described. In the second part, 3DP fabrication techniques, including fused deposition modelling, material jetting as inkjet printing, electron beam melting, selective laser sintering, stereolithography and digital light processing, electrospinning, and two-photon polymerization were described. In the last section, through recent bibliographic research, a wide number of 3D printed mesoporous materials, for in vitro and in vivo drug delivery applications, most of which relate to bone cells and tissues, were presented and summarized in a table in which all the technical and bibliographical details were reported. This review highlights, to a very cross-sectional audience, how the interdisciplinarity of certain branches of knowledge, as those of materials science and nano-microfabrication are, represent a growing valuable aid in the advanced forum for the science and technology of pharmaceutics and biopharmaceutics.

Keywords: drug delivery; three-dimensional porous scaffolds; electron beam melting; selective laser sintering; stereolithography; electrospinning; two-photon polymerization; osteogenesis; antibiotics; anti-inflammatory

1. Introduction

Recently, one of the main thrusts of the micro and nano technologies application in the biomedical and clinical field has certainly been observed in the pharmaceutical drug delivery technologies optimization. Whether it is based on active or passive drug delivery, the way in which drugs are delivered substantially impact their efficacy and toxicity affecting their biocompatibility, pharmacokinetics, and pharmacodynamics. Drugs and active molecules can be introduced into the body via a number of administration routes such as buccal/sublingual, nasal, ocular, oral, pulmonary, anal/vaginal, transdermal and parenteral drug delivery [1–3]. Since a high percentage of the active pharmaceutical ingredients settled by the pharmaceutical production are precluded for a classical administration route, due to their low bioavailability [4], novel technologies assist modern drug delivery. As a result, an increase is observed in the effectiveness and reduction of side effects of the formulations in relation to patient compliance and costs reduction. In the past years, many drug delivery systems as organic and inorganic micro- and nanoparticulated systems as nanoparticles, micelles, liposomes, extracellular vesicles, nanotubes, metal–organic frameworks (MOF) and hydrogels have been used to deliver drugs at their therapeutic concentration to specific cell types and tissues [5–8]. Both material and design should be taken into account when optimizing a drug delivery carrier able to guarantee tuneable release (sustained, controlled, or pulsed), to act as a temporary reservoir, to increase the solubility of hydrophobic formulations, to float in the gastrointestinal tract and to protect the biological cargo from degradation [9].

Porous carriers have been successfully used as drug delivery matrices for their surface properties, high surface area and tuneable pore dimensions [10,11]. According to their pore sizes, porous materials are classified into three different categories, namely microporous, mesoporous, and macroporous [9,12]. Microporous materials such as MOFs and zeolites, are characterized by a well-interconnected network of pores less than 2 nm in size and high thermal stability and catalytic activity [13]. In macroporous materials, pores dimension ranges between 50 and 1000 nm [14] while in mesoporous materials pore size is between 2 to 50 nm [15]. In more details, mesoporous materials with a narrow pore dimension distribution and high surface area can be considered valuable candidates in drug delivery applications [16,17]. In the wide category of mesoporous, many materials are included such as mesoporous silica, hydroxyapatite and carbon, hydrogel and nanogel, metal and metal-doped nanoparticles. These materials have great versatility since their actions can be regulated by tuning the chemical environment optimizing the loading and consequent release of the chosen drug [18–20]. The drug incorporation into a mesoporous material is usually carried out by embedding the matrix in a concentrated solution of the drug and by a successive drying step. The size of the absorbable molecule (from small active molecules to proteins) is related to the dimension of the pore, and generally, a pore/drug size ratio >1 allows the adsorption of active molecules inside the pores. By using polymeric structure-directing agents, varying the chain length of surfactant or solubilizing supplementary substances into micelles, mesopores sizes can be adjusted from some nanometres to several tens of nanometers [21].

Recent advancement in micro/nano-fabrication techniques, materials science, chemistry and pharmacology has allowed the development of a number of mesoporous materials for drug delivery application characterized by evident structural advancement such as tuneable pore sizes, different grade of skeleton rigidity and two/three dimensional (2D–3D) architectures arrangement [22–26].

Hard (nanocasting) or soft templating approaches are applied to produce ordered mesostructured porous materials. The templated synthesis usually requires three successive steps: template preparation, template-directed synthesis and template removal. Hard templating leads to very robust structures containing several constituents as carbon, and metals (oxides, nitrides and sulphides) [27,28]. It is a synthetic method based on the deposition of the targeted materials into the narrowed spaces of the template, resulting in a reversed copy of the mold. The pores of these templates are soaked with a precursor of the looked-for product (e.g., a metal salt for metal oxides) which is in situ thermally transformed to the final product. When the template is removed, mesoporous material remains as the negative replica of the hard template [29].

Soft-templating techniques allow direct synthesis of porous materials through block copolymers including blocks of ionic and non-ionic oligomers, amphiphilic surfactants employed as structure-directing agents (SDAs) and through the addition of precursors as metal salts for metal oxide nanomaterials and organosilanes or triethoxysilane for SiO_2-based nanomaterials. Soft-templating techniques are those in which small sub-units self-assemble to define the final structure, which is an aggregate of these starting units, which are not embedded in other matrices or removed as in the techniques described above. Upon self-assembly in a solvent, a micellar structure is realized by the fact that the hydrophobic sides of the molecules of the amphiphilic surfactants point inward and the hydrophilic ones outward in case the solvent is polar, while the opposite occurs if the solvent is non-polar. After this step, micelles are functionalized on their external corona structure using functional groups, frequently polymeric oligomers. Finally, it is the cross-linking of these external terminations which assemble the micelles in a mesoporous superstructure [30].

Producing porous hierarchical materials by integrating macropores in mesoporous tools manifestly increases their practical drug delivery applicability since macropores increase mass transport decreasing diffusion restraints characterizing purely mesoporous materials, while the mesopores empower great surface area [31,32].

Many methodologies have been optimized to engineer the hierarchically structured mesoporous solutions. The dual-templating synthesis method, applying colloidal crystal (opal) hard-templating and soft-templating techniques, is employed for realizing, as schematized in Figure 1, 3D macro/mesoporous materials for a wide range of applications, including the drug delivery ones [33,34].

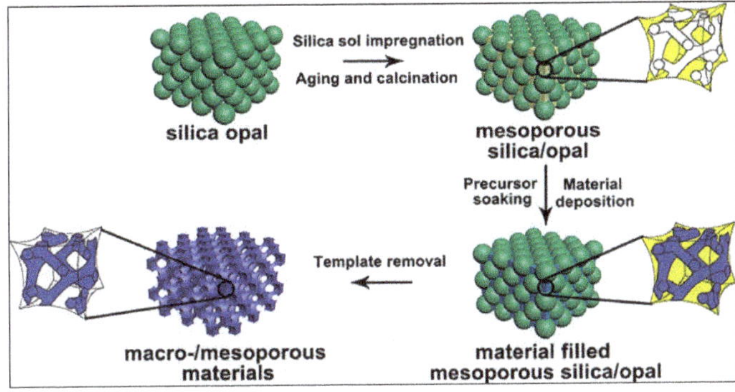

Figure 1. Schematic representation of 3D macro/mesoporous materials preparation reproduced with permission from [34], Chemistry of Materials, 2018.

2. 3D Printed (3DP) Mesoporous Scaffolds Fabrication Technique

The idea of realizing a macroscopic object via a bottom-up approach has been attractive for a long time but recently, the advancement of both the materials to be used and the techniques to be exploited have made possible the fabrication of 3D printers able to produce any shape in many different natural [35], synthetic, plastic and metallic materials, at variable size scales and with potentially very high accuracy in positioning [36–38]. This has pushed some researchers towards the idea to explore the possibility to use these techniques to realize solutions with different designs, characterized by being made of different types of mesoporous materials [21]. 3D porous substrates, used with or without further functionalization or engineering, are used more and more frequently in in vitro and in vivo drug delivery studies to assist cell growth or tissue regeneration ensuring the right degree of asepticity and differentiation [39–41].

2D and 3D printing tools are appealing for drug delivery applications since state-of-the-art equipment allows the deposition of liquid, gel, and solid constituents enclosing a wide range of pharmaceutics according to predefined schemes. The layer-by-layer assembling mode to print scaffold allows exact control of the design and of the geometry of the internal pores system, which consequently leads to tune the strength of the final products [42,43].

3DP technology can successfully assist engineers, pharmacologists and clinicians in the design and realization of 3D mesoporous scaffolds to be used for different medical applications such as tissue engineering and regenerative medicine implants characterized by the adjustable loading and unloading activity of pharmacologically active substances such as, antibiotics, growth and differentiation factors (Figure 2) [44].

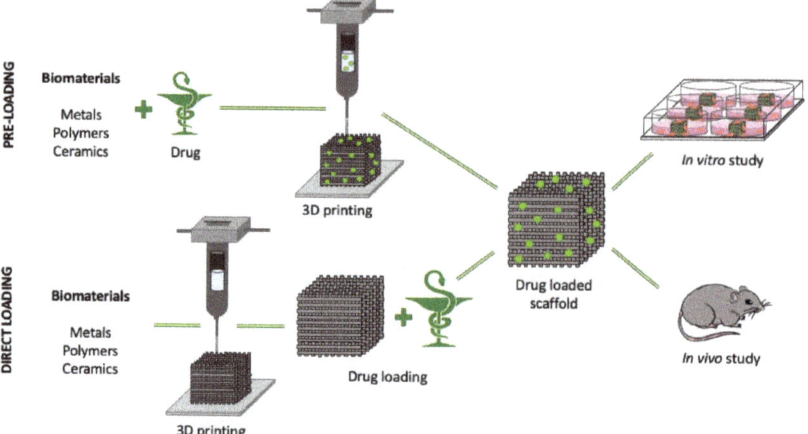

Figure 2. Schematic layout summarizing pre-loading and direct loading 3DP porous substrate fabrication for in vitro and in vivo drug delivery applications.

These active substances can be incorporated inside the mesoporous 3D structures in two different main steps: during the manufacturing process (pre-loading, PL) by mixing the substances with the printable material and then proceeding with the 3DP technique in mild conditions (i.e., electrospinning or inkjet printing), or at the end of the printing step (direct loading, DL), by soaking the 3D-printed scaffold in a solution of the molecule to be loaded as reported for bone morphogenetic protein-2 (BMP-2) mesoporous calcium silicate (MesoCS) 3D-printed scaffold [45]. PL methods are usually applied for the production of scaffolds able to locally deliver antibiotiotics [46], but unfortunately, antibiotics such as those of the cephalosporin family have significantly reduced efficiency when exposed to heat and, consequently, the DL method is definitely applied to sensitive molecules when the 3DP process is carried out at high temperatures or pressures [47].

There are many 3DP strategies available to the scientific community that allow the realization of mesoporous scaffolds under computer aids combining different processes and materials like carbon nanotubes, nanoparticles, nanofibers, polymers with active biomolecules with or without live cells. These 3DP fabrication techniques, as summarized in Figure 3, include fused deposition modeling (FDM), material jetting as inkjet printing (IP), electron beam melting (EBM), selective laser sintering (SLS), stereolithography (SLA) and digital light processing, electrospinning, and two-photon polymerization (TPP).

2.1. Fused Deposition Modeling

FDM is one of the most inexpensive nozzle-based deposition systems that allows direct printing of 3D CAD designed layer by layer objects. Thermoplastic degradable (polylactic acid, PLA, poly(ε-caprolactone), PCL, polyvinyl alcohol, PVA) and non-degradable (acrylo-nitrile butadiene styrene, ABS, ethylene vinyl acetate, EVA, poly methyl methacrylate, PMMA) polymer filament are pushed into the heater block to melt before extruding from a high-temperature nozzle solidifying onto the previous layer on the build plate [48].

The easiest method of loading target drugs into the thermoplastic polymer filament is the impregnation obtained leaving the just printed device in a concentrated drug solution (mostly ethanol or methanol) followed by a drying step [49,50].

2.2. Inkjet Printing

The inkjet-based non-contact printing technology reproduces digital patterns with tiny ink drops through thermal, piezoelectric and magnetic approaches. The thermal stimulation, reaching until

100–300 °C, nucleates a bubble and directly leads to droplet expulsion from the printhead. The size of droplets is related to the temperature gradient and ink viscosity employed. Likewise, the ink drop generation can be produced by the pulse strain and acoustic waves generated from a piezoelectric actuator and larger size ink droplets can be produced by means of electromagnetic filed [51–53].

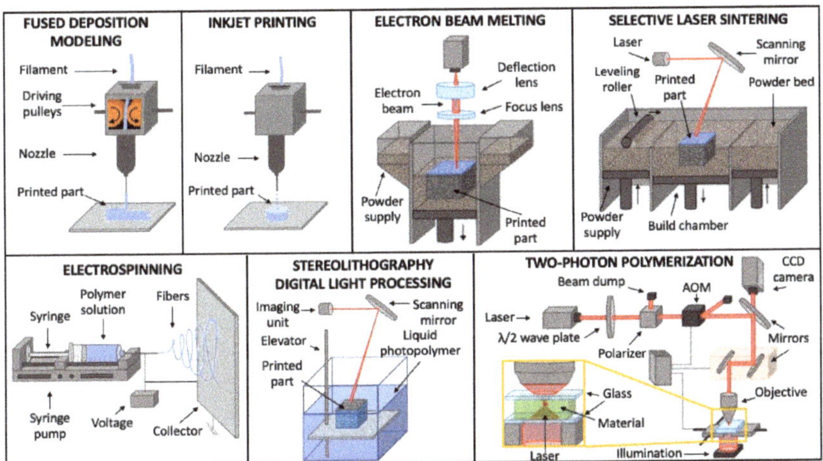

Figure 3. Schematic illustrations of the most diffused 3D printing fabrication techniques for porous scaffolds manufacturing: fused deposition modeling, inkjet printing, electron beam melting, selective laser sintering, stereolithography, electrospinning, two-photon polymerization.

2.3. Electron Beam Melting

EBM is a modern fast solution to manufacture metal parts on a layer-by-layer basis through an electron beam that, bombarding the metal powders, melts them, constructing 3D geometries. This technique compared with ones using a laser, are characterized by high energy utilization and material absorption rate, improved stability, and reduced maintenance fees [54]. Although this technique is successfully used for the realization of porous orthopedic and dental implants made of metallic biomedical alloys as Ti6Al4V [55], to date, there are no applications of EBM for the production of 3D printed mesoporous devices for drug delivery application. This is due to the fact that these kinds of scaffolds are characterized by large surface roughness since EBM microfabrication accuracy ranges from 0.3–0.4 mm.

2.4. Selective Laser Sintering

SLS operates without a mold through a computer-controlled laser beam, powder bed, a piston assuring a vertical movement, and a roller to spread continuously powder layers [55]. This technique allows the realization of polymeric, metallic, and ceramic parts. SLS implies solid and semisolid consolidation procedures at a sintering temperature usually lower than the melting point. In the semisolid process suitable for treating low melting point polymer, as PCL, polyglycolide, PLA and poly(L-lactic) acid (PLLA), partially melted powder particles produce a certain volume of the liquid phase, which glues other solid elements. Microsphere-based hydroxyapatite (HA)/PCL scaffolds realized by SLS, shows a highly ordered porous structure [56]. Polyamide/HA composite platforms with porosities ranging from 40% to 70% and with a maximum tensile strength of 21.4 MPa were obtained by SLS [55,57]. Although the low near-infrared laser absorptivity of oxide ceramics, the direct SLS of ceramics throughout powder coating adds to the low melting point or composites ceramics has been done [58]. Many sacrificial binders as waxes, thermoplastics, long-chain fatty acids or sometimes a combination of binders as thermoset/semi-crystalline PA-11 or wax/PMMA are used for the realization of porous 3D structured materials as graphite and composite ceramic Al_2O_3-ZrO_2-TiC [59–61]. A high-energy laser beam increasing the temperature on the surface promotes the particle interaction to each other before sintering together, while the material on the grain borderline continues to diffuse into the pores, stimulating densification activities. Since SLS is characterized by a high heating rate and short holding time, it results as an excellent alternative in producing scaffolds supported by low-dimensional nanomaterials such as graphene and carbon nanotubes [62].

2.5. Electrospinning

Nanodimensional high specific surface area devices can be fabricated with bioactive loaded polymer through the electrospinning technique [63]. In a conventional electrospinning system, generally comprising of a high-voltage power supply, syringe, pump and a collector, polymeric nanofibers, inorganic nanofibers, and composite nanofibers are ejected into a sequence of droplets forming steady fiber [64,65].

2.6. Stereolithography and Digital Light Processing

Stereolithographic (SLA) and Digital Light Processing 3DP allow the layer by layer realization of 3D mesoporous stuff by cross-linking photo-sensitive materials using laser light or digital light projection technique, respectively [66]. The curing stereolithographic step, both in single-photon and two-photon polymerization, is actuated by tuning the incidence, the intensity, and the duration of near-infrared, visible or UV light.

2.7. Two-Photon Polymerization

While single-photon polymerization requires one-photon absorption, in TPP, the molecule simultaneously absorbs two photons. By employing a focused femtosecond near-IR, TPP stereolithography, processing biocompatible synthetic or natural hydrogels or polymers, grants the ultra-fast production of 3D structures with submicron resolution [67].

3. Applications of 3DP Mesoporous Material for Drug Delivery

Doing a search on the Web of Science and on Pubmed at the beginning of June 2020, resulted that in the last decades a fair number of publications are strictly related to the specific topic covered in this review and, more in details, related to mesoporous 3D printed materials for drug delivery applications. As highlighted in Table 1, most of the results focalized on the application of these 3D porous materials for tissue engineering bone substitute realization. Their porosity, by mimicking the bone structure, allows nutrient transport, waste removal, cell migration, angiogenesis and differentiation phenomena, assisting bone regeneration in bone defects related to traumatic events or pathologies.

3.1. Bone Regeneration

3.1.1. Growth Factors and Peptides

Several FDA approved growth factors as BMP-2 have been used in clinics for bone and cartilage regeneration. In a mesoporous bioactive glass (MBG) covered silicate 1393 bioactive glass scaffolds candidate for bone repairing application, BMP-2 release was higher than that of DNA and dexamethasone. MBG successfully physically absorbed and released the active molecule without upsetting its pharmacological activity [68]. Fish hydrogel-based mesoporous strontium-doped calcium silicate scaffolds were proved to be efficient BMP-2 carriers for in vitro human Wharton jelly mesenchymal stem cell differentiation [69]. Customized 3D-printed osteoinductive implants were realized integrating porous silicon BMP-2 carriers within a 3D-printed PCL patient-specific implant [70]. FDM 3D MesoCS scaffolds combined with PCL were presented as odontoinductive biomaterial with efficient BMP-2 delivery capability [71]. In vitro tested BMP-2 pre-loaded mesoporous calcium silicate/PCL scaffolds, even if not suitable for clinical applications, exhibited high biocompatibility and sustained drug delivery pattern compared to the ones directly immersed with BMP-2 after the FDM fabrication [45].

As some authors reported some relevant side-effect related to the use of BMP-2 [72], 3D dipyridamole-coated hydroxyapatite (HA)/beta-tri-calcium phosphate (β-TCP) scaffolds were successfully used to promote bone regeneration in critical bone defects as well as BMP-2 [73].

Since vascularization is a key step of the osteogenesis process, 3DP dimethyloxallyl glycine loaded MBGs and poly(3-hydroxybutyrate-*co*-3-hydroxyhexanoate) polymers scaffolds and results showed that dimethyloxallyl glycine was effectively released improving angiogenesis and osteregeneration in the bone faults [74]. Vascular endothelial growth factor (VEGF), was well encapsulated in chitosan/dextran sulfate microparticles and mixed into a calcium phosphate paste for the 3D plotting of growth factor loaded calcium-phosphate-based scaffolds applicable for bone tissue engineering [75].

Materials for tissue regeneration can be functionalized with engineered peptides able to regulate bone healing and regeneration. In vitro tests with naringin and calcitonin gene-related peptide-loaded 3DP MBG/sodium alginate/gelatin scaffolds showed that their high porosity assure efficient sustained drug delivery [76]. Peptide osteostatin and Zn^{2+} ions loaded meso-macroporous 3D scaffolds based on MBGs, exhibited a synergistic effect improving human mesenchymal stem cell growth, promoting their osteogenic differentiation [77]. SLS 3DP poly(3-hydroxybutyrate) scaffolds, when post-printing loaded with osteogenic growth peptide, exhibited the ability to support cell growth and tissue restoration [78].

3.1.2. Anti-Inflammatory and Antibiotics

Since any bone loss such as that following trauma, bone diseases and surgery, potentially provides suitable conditions for the onset of chronic infections or biofilm, it is highly desired the realization of anti-inflammatory and antibiotic-eluting scaffolds for sustained release without side effect in osteointegration, osteogenesis and osteoconduction processes. Dexamethasone loaded mesoporous $CaSiO_3$/calcium sulfate hemihydrate (MCS/CSH) cement scaffolds have been realized by 3D printing. Compared to the tissue culture plates control, MCS/CSH scaffolds exhibited a good in vitro OCT-1 cells response, an extra balanced degradation rate and capacities to slowly release the uploaded drug in targeted sites [79].

3DP high porosity dual-drug delivery layered MBG/sodium-alginate (SA)–SA scaffolds were successfully fabricated enriching the printing step with SA cross-linking. They resulted able to stimulate proliferation and osteogenic differentiation of human bone marrow-derived mesenchymal stem cells, furthermore, bovine serum albumin (BSA) and ibuprofen were successfully loaded in SA layer and the MBG of MBG/SA layer, respectively, resulting in a quite fast BSA release due to the macroporous network of SA, and in a constant release of ibuprofen due to the retention effect of the mesoporous channels of MBG [80].

It is well-known that inflammation phenomena thwart bone regeneration in transplanted loci and the local effect of short-term corticosteroid administration increase the effectiveness of bone tissue engineering [81]. Dexamethasone-loaded polydopamine-functionalized MBG was incorporated into polyglycolic acid/poly-1-lactic acid (PGPL) to fabricate a 3D mesoporous scaffold via laser additive manufacturing able to stimulate cell differentiation, biomineralization [82]. Loaded dexamethasone electrospun fibrous scaffolds of PCL-gelatin, incorporating MBG nanoparticles (MBGn), were presented excellent valid 3D platforms for bone tissue engineering [83].

In the case of dexamethasone-loaded 3DP strontium-containing MBG scaffolds the mesoporous matrix with enhanced mechanical strength to ensure great bone-growing bioactivity together with marked drug delivery capability [84]. 3DP scaffolds realized by using MBG and concentrated alginate pastes efficiently delivered dexamethasone in an in vitro test with human bone marrow-derived mesenchymal stem cells thanks to their matrix characterized by a well-ordered network of nano-channels and micro and macro-pores [85]. Poly(1,8-octanediol-*co*-citrate) and β-tricalcium phosphate (β-$Ca_3(PO_4)_2$), together with ibuprofen-loaded SiO_2 were made-up by micro-droplet jetting

3DP technique. Their hierarchically macro/mesoporous extremely interconnected pore matrix made them a valid antimicrobial bioengineered solution for bone regeneration [86,87].

The antimicrobials local application usually provides higher drug delivery than those attained with the intravenous application [88,89] and many 3DP macro/meso-porous composite scaffolds, are at the moment used to support a reproducible safe a better and well-regulated in situ antibiotics delivery. Some doxorubicin-loaded 3DP magnetic Fe_3O_4 nanoparticles containing mesoporous bioactive glass/polycaprolactone composite scaffolds enhanced osteogenic activity also assured sustained local anticancer delivery coupled with magnetic hyperthermia treatment [90].

Multidrug-loaded scaffold undoubtedly improves the applicability of 3D rapid prototype implants to ward off biofilm growth and drug resistance. Antibiotics are usually locally delivered via PMMA bone cement spacers [91,92] compatible with a restricted number of antibiotics and characterized by having low release profiles. Mesoporous bioactive glass/metal-organic framework and macro/meso-porous composite bioactive ceramics bound with poly (3-hydroxybutyrate-*co*-3-hydroxyhexanoate) scaffolds loaded with high dosages of isoniazid and/or rifampin, anti- tuberculosis drugs, had good biocompatibility and bioactivity when tested for long-term therapy after osteoarticular tuberculosis debridement surgery. Hierarchical 3DP multidrug scaffolds built with nanocomposite bioceramic and PVA were coated of gelatin-glutaraldehyde (Gel-Glu). Levofloxacin was loaded into the mesopores of the bioceramic part, vancomycin was packed into the biopolymer portion while rifampin in the external layer of Gel-Glu. The early delivery of rifampin followed by a sustained release of vancomycin and levofloxacin, represented an excellent and encouraging alternative for bone infection management [93]. 3DP rifampin- and vancomycin-loaded calcium phosphate scaffolds, used in a mouse model implant-associated staphylococcus aureus bone infection, proved that the concomitant local delivery of rifampin and vancomycin significantly improves the outcomes of the implant compared to PMMA spacers which cannot carry rifampin [94]. Gelatine and Si-doped hydroxyapatite porous 3D scaffolds were successfully loaded with vancomycin since they were rapidly prototyped fabricated at room temperature and apart from by increasing in vitro pre-osteoblastic MC3T3-E1 cell differentiation they also inhibit bacterial growth [95].

3.1.3. Metallic Ions and Trace Elements

Recently, many metallic ions such as zinc, copper, silver, cerium, strontium and cobalt, were combined with bioactive glasses to improve osteogenesis and angiogenesis [96–98]. Silver, among all, is the one that stands out for its strong antibacterial qualities. Silver/graphene oxide homogeneous nanocomposites were modified on 3DP β-tricalcium phosphate bioceramic scaffolds leading to a bifunctional scaffold with, just test in vitro, antibacterial and osteogenic activity were realized and in vitro tested [99].

In addition to the direct effect that a drug-loaded on a scaffold can have at the implantation site, several authors highlighted that also the integration of trace elements such as strontium, zinc, magnesium, calcium, copper, boron and cerium in 3DP mesoporous bioactive glass scaffolds enhance in vitro and in vivo osteogenic and differentiation activity [100–102].

3.2. *Other Applications*

Apart from the numerous applications in the bone regeneration field, mesoporous 3DP scaffolds including also mesoporous elastomer characterized by ordered and aligned nanofibrillar architecture that can be rapidly managed into multifaceted objects are starting to be more and more widespread even in other branches of biomedical research and medical clinic [103].

Coaxial electrospinned silk fibroin-based scaffolds are successfully tested as a potential brain-derived neurotrophic factor and VEGF delivery carrier in nerve repair and reconstruction applications [104].

Anti-HIV-1 drugs, including emtricitabine, tenofovir disoproxil fumarate and efavirenz were successfully loaded in a 24-layered rectangular prism-shaped 3DP controlled release fixed-dose combination tablets able to control the intestinal release of the active molecules [105].

Table 1. Applications of 3DP porous materials for tissue engineering and bone substitute realization.

Material	3D Printing Method	Drug	Drug Loading Method	Application	Reference
Mesoporous strontium substitution calcium silicate/recycled fish gelatin 3D cell-laden scaffold.	IP	BMP-2	DL	Bone tissue engineering. In vitro	[69]
PCL 3DP patient-specific implant, with degradable porous silicon-based carriers.	SLA		DL	Bone graft for critical size bone defects. In vitro	[70]
Mesoporous calcium silicate 3DP scaffold.	FDM		DL	Bone regeneration. In vitro	[45]
Hierarchical 3D multidrug scaffolds based on nanocomposite bioceramic and PVA with an external coating of gelatin-glutaraldehyde.	IP	Dipyridamole and BMP-2	DL	Bone tissue engineering. In vivo (mice)	[73]
Scaffold is composed of MBG and poly(3-hydroxybutyrate-*co*-3-hydroxyhexanoate) polymers.	IP	Dimethyloxallyl glycine	PL	Angiogenesis and osteogenesis for bone tissue engineering. In vivo (rats)	[74]
MBG with sodium alginate and gelatin.	IP	Naringin and calcitonin gene-related peptide	DL	Bone repair. In vitro	[76]
MBG.	IP	Peptide osteostatin and Zn^{2+} ions	DL	Bone grafts with enhanced osteogenic capacity. In vitro	[77]
Poly(3-hydroxybutyrate) scaffold.	SLS	Osteogenic growth peptide and its C-terminal sequence (10–14)	DL	Bone tissue engineering. In vitro	[78]
Calcium phosphate cement scaffolds by 3D plotting with growth factors encapsulating chitosan/dextran sulfate microparticles mixed into the paste.	IP	BSA and VEGF	PL	Encapsulate growth factors in a cement. In vitro	[75]
Layered MBG/SA.	IP	BSA and ibuprofen	PL	Stimulate human bone mesenchymal stem cells (hBMSCs) adhesion, proliferation and osteogenic differentiation. In vitro	[80]
Integrate MBG with 3D printing basic 1393 bioactive glass scaffolds.	IP	Dexamethasone and BMP-2	DL	Bone repair and relative bone disease treatment. In vivo (rats)	[68]

Table 1. *Cont.*

Material	3D Printing Method	Drug	Drug Loading Method	Application	Reference
MBG is functionalized with polydopamine and PGPL.	SLS	Dexamethasone	PL	Osteogenic differentiation and biomineralization. In vitro	[82]
Calcium sulfate hemihydrate cement is incorporated in mesoporous calcium silicate.	IP		PL	Bone tissue engineering. In vitro	[79]
Electrospun fibrous scaffolds of PCL-gelatin incorporating mesoporous bioactive glass nanoparticles.	ES		PL	Bone regeneration. In vivo (rats)	[83]
MBG with strontium.	IP		PL	Bone regeneration. In vitro	[84]
Hierarchical scaffolds of MBG and concentrated alginate pastes.	IP		PL	Bone tissue engineering. In vitro	[85]
3D magnetic Fe_3O_4 nanoparticles containing MBG/PCL composite scaffolds.	IP	Doxorubicin	PL	Osteogenic activity, local anticancer drug delivery and magnetic hyperthermia. In vitro	[90]
Hollow mesoporous structure of silica (SiO_2) microspheres loaded in a Poly(1,8-octanediol-*co*-citrate) and β-tricalcium phosphate scaffold.	IP	Ibuprofen	PL	Bone regeneration of infected bone defects. In vitro	[86]
Poly(1,8-octanediol-*co*-citrate) and β-tricalcium phosphate scaffold.	IP		PL	Bone defect repair. In vitro	[87]
MBG with MOFs and PCL.	IP	Isoniazid	PL	Osteoarticular tuberculosis treatment. In vitro	[106]
Carboxylic MBG and methyl-functionalized mesoporous silica nanoparticles.	IP	Isoniazid and rifampin	PL	Filler after surgical treatment of osteoarticular tuberculosis. In vivo (rabbits)	[107]
Hierarchical 3D multidrug scaffolds based on nanocomposite bioceramic and PVA with an external coating of Gel-Glu.	IP	Rifampin, levofloxacin and vancomycin	PL	Destroy Gram-positive and Gram-negative bacteria biofilms for local bone infection therapy. In vitro	[93]
3D printed calcium phosphate scaffolds.	IP	Rifampin and vancomycin	PL	Treat an implant-associated Staphylococcus aureus bone infection. In vivo (mice)	[94]
Porous 3-D scaffolds consisting of gelatine and Si-doped hydroxyapatite.	IP	Vancomycin	PL	Pre-osteoblastic MC3T3-E1 cell differentiation. In vitro	[95]

Table 1. *Cont.*

Material	3D Printing Method	Drug	Drug Loading Method	Application	Reference
β-tricalcium phosphate bioceramic scaffolds with a homogeneous nanocomposite made of silver nanoparticles and graphene oxide.	IP	Silver nanoparticles and graphene oxide	PL	Bone grafts with good antibacterial performance. In vitro	[99]
Borosilicate MBG.	IP	Boron and silicon ions	PL	Repair bone defects. In vivo (rats)	[100]
3D porous composite scaffolds made of cerium oxide, mesoporous calcium silicate and PCL.	IP	Cerium ions	PL	Bone regeneration. In vitro	[101]
MBG with strontium.	IP	Strontium ions	PL	Bone regeneration. In vivo (rats)	[102]
MBG modified β-tricalcium phosphate.	IP	Calcium, phosphorus and silicon ions	PL	Angiogenesis and osteogenesis for bone tissue engineering. In vivo (rabbits)	[108]
Mesoporous calcium silicate 3D-printed scaffold.	IP	BMP-2	PL	Odontoinductive biomaterial in regenerative endodontics. In vitro	[71]
Silk fibroin porous scaffold.	ES	Brain-derived neurotrophic factor and VEGF	PL	Cavernous nerve regeneration. In vivo (rats)	[104]
Nanostructured ordered mesoporous elastomers composed of molecular double networks (poly(ethylene oxide)–poly(propylene oxide)–poly(ethylene oxide) pluronic copolymers, and PMMA.	IP	Ibuprofen and vancomycin	DL	Biomedical and engineering applications as the need for high mechanical performance coexisting with precise nano-microstructural features. In vitro	[103]
Humic acid-polyquaternium 10 tablet.	IP	Efavirenz, tenofovir disoproxil fumarate and emtricitabine	PL	Anti-HIV-1 controlled drug delivery. In vivo (pigs)	[105]
Mesoporous iron oxide nanoraspberry inside microneedles.	DLP	Minoxdil	DL	Treatment of androgenetic alopecia. In vivo (mice)	[109]
Porous poly(ethylene glycol) dimethacrylate devices.	TPP	Rhodamine B as model drug	PL	Different biomedical applications. In vitro	[110]

4. Conclusions

For many years, 3D devices have been assisting research in very different areas, ranging from simple cell cultures to tissue engineering and drug delivery applications. 2D cell culture represents a chief tool in molecular and cellular biology due to its fast, ease, reproducibility and cheap distinctive characteristic. However, it is now universally accepted that 2D cell culture methods understate the live cells in vivo setting unlike reported for last-generation 3D biomaterials which, on the contrary, are able to mimic in a much more realistic way the environment required for a whole range of biomedical and clinical applications. The development of three-dimensional supports has even greater resonance in tissue engineering and regenerative medicine applications since, in those cases, the function of tissues or organs must be restored ensuring the spatial and functional interconnection between different cell types, in order to guarantee the exchange of gas, nutrients or drugs and the elimination of waste products. In this review, we wanted to highlight how these characteristics can be optimized by merging together the need to provide solid supports capable of assisting cell growth at the level of tissue and organ and, at the same time, the right degree of porosity of the materials that in the specific case of 3D biomaterials offers a whole series of drug delivery capabilities worthy of study and implementation. The way an active molecule is carried to a specific region or cellular type can impact on its interaction efficacy. Each drug has a therapeutic window in which health benefits must be maximized and side effects minimized. This need has materialized in the ever-stricter demand of a multidisciplinary approach for the implementation of new materials and methods for an effective in vitro and in vivo drug delivery. Materials science, chemistry and micro/nanofabrication offer both original and effective solutions applicable in research and clinical areas. The rapid and often inexpensive fabrication of 3DP structures enhances the performance of devices no longer used only as structural supports for tissue regeneration and differentiation thanks to the optimization of their intrinsic and tuneable porosity. 3DP mesoporous devices allow an effective drug delivery of personalized therapy, customizable both from the geometric point of view and from the point of view of pharmacological requests for each individual patient. With the topics covered in this review, we want to highlight how 3D printing techniques allow the production of CAD designing structures that fully correspond to the request of each patient in response to needs following trauma or pathologies. The future implementation of new biodegradable biopolymers and of multi-step etching processes for post-printing functionalization/modification, will also allow more efficient drug delivery application of scaffolds as the 3D EBM produced ones, by conferring them the not-yet optimized degree of mesoporosity.

Author Contributions: T.L., F.S., M.A. and E.d.F. conceived the review, analyzed the data of literature and wrote the paper. All authors have read and agreed to the published version of the manuscript.

Funding: This research received no external funding.

Conflicts of Interest: The authors declare no conflict of interest.

References

1. Anselmo, A.C.; Gokarn, Y.; Mitragotri, S. Non-invasive delivery strategies for biologics. *Nat. Rev. Drug Discov.* **2019**, *18*, 19–40. [CrossRef]
2. Thabet, Y.; Klingmann, V.; Breitkreutz, J. Drug Formulations: Standards and Novel Strategies for Drug Administration in Pediatrics. *J. Clin. Pharmacol.* **2018**, *58*, S26–S35. [CrossRef]
3. Moon, C.; Smyth, H.D.C.; Watts, A.B.; Williams, R.O. Delivery Technologies for Orally Inhaled Products: An Update. *AAPS PharmSciTech* **2019**, *20*, 117. [CrossRef] [PubMed]
4. Santos, A.; Veiga, F.; Figueiras, A. Dendrimers as Pharmaceutical Excipients: Synthesis, Properties, Toxicity and Biomedical Applications. *Materials* **2019**, *13*, 65. [CrossRef] [PubMed]
5. Palazzolo, S.; Bayda, S.; Hadla, M.; Caligiuri, I.; Corona, G.; Toffoli, G.; Rizzolio, F. The Clinical Translation of Organic Nanomaterials for Cancer Therapy: A Focus on Polymeric Nanoparticles, Micelles, Liposomes and Exosomes. *Curr. Med. Chem.* **2018**, *25*, 4224–4268. [CrossRef] [PubMed]

6. Limongi, T.; Rocchi, A.; Cesca, F.; Tan, H.; Miele, E.; Giugni, A.; Orlando, M.; Perrone Donnorso, M.; Perozziello, G.; Benfenati, F.; et al. Delivery of Brain-Derived Neurotrophic Factor by 3D Biocompatible Polymeric Scaffolds for Neural Tissue Engineering and Neuronal Regeneration. *Mol. Neurobiol.* **2018**, *55*, 8788–8798. [CrossRef] [PubMed]
7. Abu, B.; Mohd Salman, K.; Muhammad Zafar, I.; Mohd Sajid, K. Tumor-Targeted Drug Delivery by Nanocomposites. *Curr. Drug Metab.* **2020**, *21*, 1–15. [CrossRef]
8. Dianzani, C.; Foglietta, F.; Ferrara, B.; Rosa, A.C.; Muntoni, E.; Gasco, P.; Della Pepa, C.; Canaparo, R.; Serpe, L. Solid lipid nanoparticles delivering anti-inflammatory drugs to treat inflammatory bowel disease: Effects in an in vivo model. *World J. Gastroenterol.* **2017**, *23*, 4200–4210. [CrossRef] [PubMed]
9. Arruebo, M. Drug delivery from structured porous inorganic materials. *WIREs Nanomed. Nanobiotechnol.* **2012**, *4*, 16–30. [CrossRef]
10. Choi, Y.; Kim, J.; Yu, S.; Hong, S. pH- and temperature-responsive radially porous silica nanoparticles with high-capacity drug loading for controlled drug delivery. *Nanotechnology* **2020**, *31*, 335103. [CrossRef]
11. Hongfei, L.; Jie, Z.; Pengyue, B.; Yueping, D.; Jiapeng, W.; Yi, D.; Yang, Q.; Ying, X. Establishment and In Vitro Evaluation of Porous Ion-responsive Targeted Drug Delivery System. *Protein Pept. Lett.* **2020**, *27*, 1–12. [CrossRef]
12. Ahuja, G.; Pathak, K. Porous carriers for controlled/modulated drug delivery. *Indian J. Pharm. Sci.* **2009**, *71*, 599–607. [CrossRef] [PubMed]
13. Chowdhury, A.H.; Salam, N.; Debnath, R.; Islam, S.M.; Saha, T. Chapter 8—Design and Fabrication of Porous Nanostructures and Their Applications. In *Nanomaterials Synthesis*; Beeran Pottathara, Y., Thomas, S., Kalarikkal, N., Grohens, Y., Kokol, V., Eds.; Elsevier: Amsterdam, The Netherlands, 2019; pp. 265–294. [CrossRef]
14. Solano, V.; Vega-Baudrit, J. Micro, Meso and Macro Porous Materials on Medicine. *J. Biomater. Nanobiotechnol.* **2015**, *6*, 247–256. [CrossRef]
15. Seaton, N.A. Determination of the connectivity of porous solids from nitrogen sorption measurements. *Chem. Eng. Sci.* **1991**, *46*, 1895–1909. [CrossRef]
16. Aquib, M.; Farooq, M.A.; Banerjee, P.; Akhtar, F.; Filli, M.S.; Boakye-Yiadom, K.O.; Kesse, S.; Raza, F.; Maviah, M.B.J.; Mavlyanova, R.; et al. Targeted and stimuli–responsive mesoporous silica nanoparticles for drug delivery and theranostic use. *J. Biomed. Mater. Res. Part A* **2019**, *107*, 2643–2666. [CrossRef] [PubMed]
17. Guimarães, R.S.; Rodrigues, C.F.; Moreira, A.F.; Correia, I.J. Overview of stimuli-responsive mesoporous organosilica nanocarriers for drug delivery. *Pharmacol. Res.* **2020**, *155*, 104742. [CrossRef]
18. Yu, Q.; Deng, T.; Lin, F.-C.; Zhang, B.; Zink, J.I. Supramolecular Assemblies of Heterogeneous Mesoporous Silica Nanoparticles to Co-deliver Antimicrobial Peptides and Antibiotics for Synergistic Eradication of Pathogenic Biofilms. *ACS Nano* **2020**, *14*, 5926–5937. [CrossRef]
19. Quinlan, E.; López-Noriega, A.; Thompson, E.M.; Hibbitts, A.; Cryan, S.A.; O'Brien, F.J. Controlled release of vascular endothelial growth factor from spray-dried alginate microparticles in collagen–hydroxyapatite scaffolds for promoting vascularization and bone repair. *J. Tissue Eng. Regen. Med.* **2017**, *11*, 1097–1109. [CrossRef]
20. Laurenti, M.; Lamberti, A.; Genchi, G.G.; Roppolo, I.; Canavese, G.; Vitale-Brovarone, C.; Ciofani, G.; Cauda, V. Graphene Oxide Finely Tunes the Bioactivity and Drug Delivery of Mesoporous ZnO Scaffolds. *ACS Appl. Mater. Interfaces* **2019**, *11*, 449–456. [CrossRef]
21. Vallet-Regí, M.; Balas, F.; Arcos, D. Mesoporous Materials for Drug Delivery. *Angew. Chem. Int. Ed.* **2007**, *46*, 7548–7558. [CrossRef]
22. Sengottuvelan, A.; Mederer, M.; Boccaccini, A.R. Preparation and characterization of mesoporous calcium-doped silica-coated TiO2 scaffolds and their drug releasing behavior. *Int. J. Appl. Ceram. Technol.* **2018**, *15*, 892–902. [CrossRef]
23. Sun, Y.; Han, X.; Wang, X.; Zhu, B.; Li, B.; Chen, Z.; Ma, G.; Wan, M. Sustained Release of IGF-1 by 3D Mesoporous Scaffolds Promoting Cardiac Stem Cell Migration and Proliferation. *Cell. Physiol. Biochem.* **2018**, *49*, 2358–2370. [CrossRef] [PubMed]
24. Boccardi, E.; Philippart, A.; Juhasz-Bortuzzo, J.A.; Beltrán, A.M.; Novajra, G.; Vitale-Brovarone, C.; Spiecker, E.; Boccaccini, A.R. Uniform Surface Modification of 3D Bioglass(®)-Based Scaffolds with Mesoporous Silica Particles (MCM-41) for Enhancing Drug Delivery Capability. *Front. Bioeng. Biotechnol.* **2015**, *3*, 177. [CrossRef] [PubMed]

25. Davis, M.E. Ordered porous materials for emerging applications. *Nature* **2002**, *417*, 813–821. [CrossRef]
26. Marcos-Hernández, M.; Villagrán, D. 11-Mesoporous Composite Nanomaterials for Dye Removal and Other Applications. In *Composite Nanoadsorbents*; Kyzas, G.Z., Mitropoulos, A.C., Eds.; Elsevier: Amsterdam, The Netherlands, 2019; pp. 265–293. [CrossRef]
27. Pellicer, E.; Cabo, M.; Solsona, P.; Suriñach, S.; Baró, M.; Sort, J. Nanocasting of Mesoporous In-TM (TM = Co, Fe, Mn) Oxides: Towards 3D Diluted-Oxide Magnetic Semiconductor Architectures. *Adv. Funct. Mater.* **2013**, *23*, 900–911. [CrossRef]
28. Yonemoto, B.T.; Hutchings, G.S.; Jiao, F. A General Synthetic Approach for Ordered Mesoporous Metal Sulfides. *J. Am. Chem. Soc.* **2014**, *136*, 8895–8898. [CrossRef]
29. Savic, S.; Vojisavljević, K.; Počuča-Nešić, M.; Živojević, K.; Mladenovic, M.; Knezevic, N. Hard Template Synthesis of Nanomaterials Based on Mesoporous Silica. *Metall. Mater. Eng.* **2018**, *24*. [CrossRef]
30. Zhao, T.; Elzatahry, A.; Li, X.; Zhao, D. Single-micelle-directed synthesis of mesoporous materials. *Nat. Rev. Mater.* **2019**, *4*, 775–791. [CrossRef]
31. Lokupitiya, H.N.; Jones, A.; Reid, B.; Guldin, S.; Stefik, M. Ordered Mesoporous to Macroporous Oxides with Tunable Isomorphic Architectures: Solution Criteria for Persistent Micelle Templates. *Chem. Mater.* **2016**, *28*, 1653–1667. [CrossRef]
32. Zuo, X.; Xia, Y.; Ji, Q.; Gao, X.; Yin, S.; Wang, M.; Wang, X.; Qiu, B.; Wei, A.; Sun, Z.; et al. Self-Templating Construction of 3D Hierarchical Macro-/Mesoporous Silicon from 0D Silica Nanoparticles. *ACS Nano* **2017**, *11*, 889–899. [CrossRef]
33. Manzano, M.; Colilla, M.; Vallet-Regí, M. Drug delivery from ordered mesoporous matrices. *Expert Opin. Drug Deliv.* **2009**, *6*, 1383–1400. [CrossRef] [PubMed]
34. Sun, T.; Shan, N.; Xu, L.; Wang, J.; Chen, J.; Zakhidov, A.A.; Baughman, R.H. General Synthesis of 3D Ordered Macro-/Mesoporous Materials by Templating Mesoporous Silica Confined in Opals. *Chem. Mater.* **2018**, *30*, 1617–1624. [CrossRef]
35. Kang, Y.G.; Wei, J.; Shin, J.W.; Wu, Y.R.; Su, J.; Park, Y.S.; Shin, J.-W. Enhanced biocompatibility and osteogenic potential of mesoporous magnesium silicate/polycaprolactone/wheat protein composite scaffolds. *Int. J. Nanomed.* **2018**, *13*, 1107–1117. [CrossRef] [PubMed]
36. Wijk, A.; van Wijk, I. *3D Printing with Biomaterials: Towards a Sustainable and Circular Economy*; IOS Press: Amsterdam, The Netherlands, 2015; pp. 1–85. [CrossRef]
37. Erokhin, K.S.; Gordeev, E.G.; Ananikov, V.P. Revealing interactions of layered polymeric materials at solid-liquid interface for building solvent compatibility charts for 3D printing applications. *Sci. Rep.* **2019**, *9*, 20177. [CrossRef]
38. Salea, A.; Prathumwan, R.; Junpha, J.; Subannajui, K. Metal oxide semiconductor 3D printing: Preparation of copper(ii) oxide by fused deposition modelling for multi-functional semiconducting applications. *J. Mater. Chem. C* **2017**, *5*, 4614–4620. [CrossRef]
39. Zhang, S.; Xing, M.; Li, B. Recent advances in musculoskeletal local drug delivery. *Acta Biomater.* **2019**, *93*, 135–151. [CrossRef]
40. Trofimov, A.D.; Ivanova, A.A.; Zyuzin, M.V.; Timin, A.S. Porous Inorganic Carriers Based on Silica, Calcium Carbonate and Calcium Phosphate for Controlled/Modulated Drug Delivery: Fresh Outlook and Future Perspectives. *Pharmaceutics* **2018**, *10*, 167. [CrossRef]
41. Liang, F.; Qin, L.; Xu, J.; Li, S.; Luo, C.; Huang, H.; Ma, D.; Li, Z.; Xu, J. A hydroxyl-functionalized 3D porous gadolinium-organic framework platform for drug delivery, imaging and gas separation. *J. Solid State Chem.* **2020**, *289*, 121544. [CrossRef]
42. Rawtani, D.; Agrawal, Y.K. Emerging Strategies and Applications of Layer-by-Layer Self-Assembly. *Nanobiomedicine* **2014**, *1*, 8. [CrossRef]
43. Vanderburgh, J.; Sterling, J.A.; Guelcher, S.A. 3D Printing of Tissue Engineered Constructs for In Vitro Modeling of Disease Progression and Drug Screening. *Ann. Biomed. Eng.* **2017**, *45*, 164–179. [CrossRef]
44. Charbe, N.B.; McCarron, P.A.; Lane, M.E.; Tambuwala, M.M. Application of three-dimensional printing for colon targeted drug delivery systems. *Int. J. Pharm. Investig.* **2017**, *7*, 47–59. [CrossRef] [PubMed]
45. Huang, K.-H.; Lin, Y.-H.; Shie, M.-Y.; Lin, C.-P. Effects of bone morphogenic protein-2 loaded on the 3D-printed MesoCS scaffolds. *J. Formos. Med. Assoc.* **2018**, *117*, 879–887. [CrossRef] [PubMed]

46. Shim, J.-H.; Kim, M.-J.; Park, J.Y.; Pati, R.G.; Yun, Y.-P.; Kim, S.E.; Song, H.-R.; Cho, D.-W. Three-dimensional printing of antibiotics-loaded poly-ε-caprolactone/poly(lactic-co-glycolic acid) scaffolds for treatment of chronic osteomyelitis. *Tissue Eng. Regen. Med.* **2015**, *12*, 283–293. [CrossRef]
47. Visscher, L.E.; Dang, H.P.; Knackstedt, M.A.; Hutmacher, D.W.; Tran, P.A. 3D printed Polycaprolactone scaffolds with dual macro-microporosity for applications in local delivery of antibiotics. *Mater. Sci. Eng. C Mater. Biol. Appl.* **2018**, *87*, 78–89. [CrossRef] [PubMed]
48. Long, J.; Gholizadeh, H.; Lu, J.; Bunt, C.; Seyfoddin, A. Review: Application of Fused Deposition Modelling (FDM) Method of 3D Printing in Drug Delivery. *Curr. Pharm. Des.* **2016**, *22*, 433–439. [CrossRef]
49. Goole, J.; Amighi, K. 3D printing in pharmaceutics: A new tool for designing customized drug delivery systems. *Int. J. Pharm.* **2016**, *499*, 376–394. [CrossRef]
50. Goyanes, A.; Buanz, A.B.M.; Hatton, G.B.; Gaisford, S.; Basit, A.W. 3D printing of modified-release aminosalicylate (4-ASA and 5-ASA) tablets. *Eur. J. Pharm. Biopharm.* **2015**, *89*, 157–162. [CrossRef]
51. Cui, X.; Boland, T.; D'Lima, D.D.; Lotz, M.K. Thermal Inkjet Printing in Tissue Engineering and Regenerative Medicine. *Recent Pat. Drug Deliv. Formul.* **2012**, *6*, 149–155. [CrossRef]
52. Boehm, R.D.; Miller, P.R.; Daniels, J.; Stafslien, S.; Narayan, R.J. Inkjet printing for pharmaceutical applications. *Mater. Today* **2014**, *17*, 247–252. [CrossRef]
53. Mathew, E.; Pitzanti, G.; Larrañeta, E.; Lamprou, D.A. 3D Printing of Pharmaceuticals and Drug Delivery Devices. *Pharmaceutics* **2020**, *12*, 266. [CrossRef]
54. Ni, J.; Ling, H.; Zhang, S.; Wang, Z.; Peng, Z.; Benyshek, C.; Zan, R.; Miri, A.K.; Li, Z.; Zhang, X.; et al. Three-dimensional printing of metals for biomedical applications. *Mater. Today Bio* **2019**, *3*, 100024. [CrossRef] [PubMed]
55. Yang, Y.; Wang, G.; Liang, H.; Gao, C.; Peng, S.; Shen, L.; Shuai, C. Additive manufacturing of bone scaffolds. *Int. J. Bioprint.* **2019**, *2019*, 5. [CrossRef] [PubMed]
56. Du, Y.; Liu, H.; Yang, Q.; Wang, S.; Wang, J.; Ma, J.; Noh, I.; Mikos, A.G.; Zhang, S. Selective laser sintering scaffold with hierarchical architecture and gradient composition for osteochondral repair in rabbits. *Biomaterials* **2017**, *137*, 37–48. [CrossRef] [PubMed]
57. Kumaresan, T.; Gandhinathan, R.; Murugan, R.; Muthusamy, A.; Kamarajan, B. Design, analysis and fabrication of polyamide/ hydroxyapatite porous structured scaffold using selective laser sintering method for bio-medical applications. *J. Mech. Sci. Technol.* **2016**, *30*, 5305–5312. [CrossRef]
58. Gan, M.X.; Wong, C.H. Properties of selective laser melted spodumene glass-ceramic. *J. Eur. Ceram. Soc.* **2017**, *37*, 4147–4154. [CrossRef]
59. Bai, P.K.; Cheng, J.; Liu, B. Selective laser sintering of polymer-coated Al2O3/ZrO2/TiC ceramic powder. *Trans. Nonferrous Metals Soc. China* **2005**, *15*, 261–265.
60. Deckers Jan, P.; Shahzad, K.; Cardon, L.; Rombouts, M.; Vleugels, J.; Kruth, J.-P. Shaping ceramics through indirect selective laser sintering. *Rapid Prototyp. J.* **2016**, *22*, 544–558. [CrossRef]
61. Chang, S.; Li, L.; Lu, L.; Fuh, J.Y.H. Selective Laser Sintering of Porous Silica Enabled by Carbon Additive. *Materials* **2017**, *10*, 1313. [CrossRef]
62. Gao, C.; Feng, P.; Peng, S.; Shuai, C. Carbon nanotube, graphene and boron nitride nanotube reinforced bioactive ceramics for bone repair. *Acta Biomater.* **2017**, *61*, 1–20. [CrossRef]
63. Moulton, S.E.; Wallace, G.G. 3-dimensional (3D) fabricated polymer based drug delivery systems. *J. Controll. Release* **2014**, *193*, 27–34. [CrossRef]
64. Partheniadis, I.; Nikolakakis, I.; Laidmäe, I.; Heinämäki, J. A Mini-Review: Needleless Electrospinning of Nanofibers for Pharmaceutical and Biomedical Applications. *Processes* **2020**, *8*, 673. [CrossRef]
65. Contreras-Cáceres, R.; Cabeza, L.; Perazzoli, G.; Díaz, A.; López-Romero, J.M.; Melguizo, C.; Prados, J. Electrospun Nanofibers: Recent Applications in Drug Delivery and Cancer Therapy. *Nanomaterials* **2019**, *9*, 656. [CrossRef] [PubMed]
66. Melchels, F.P.W.; Feijen, J.; Grijpma, D.W. A review on stereolithography and its applications in biomedical engineering. *Biomaterials* **2010**, *31*, 6121–6130. [CrossRef] [PubMed]
67. Pereira, R.F.; Bártolo, P.J. 3D Photo-Fabrication for Tissue Engineering and Drug Delivery. *Engineering* **2015**, *1*, 90–112. [CrossRef]
68. Wang, H.; Deng, Z.; Chen, J.; Qi, X.; Pang, L.; Lin, B.; Adib, Y.T.Y.; Miao, N.; Wang, D.; Zhang, Y.; et al. A novel vehicle-like drug delivery 3D printing scaffold and its applications for a rat femoral bone repairing in vitro and in vivo. *Int. J. Biol. Sci.* **2020**, *16*, 1821–1832. [CrossRef]

69. Yu, C.-T.; Wang, F.-M.; Liu, Y.-T.; Ng, H.Y.; Jhong, Y.-R.; Hung, C.-H.; Chen, Y.-W. Effect of Bone Morphogenic Protein-2-Loaded Mesoporous Strontium Substitution Calcium Silicate/Recycled Fish Gelatin 3D Cell-Laden Scaffold for Bone Tissue Engineering. *Processes* **2020**, *8*, 493. [CrossRef]
70. Rosenberg, M.; Shilo, D.; Galperin, L.; Capucha, T.; Tarabieh, K.; Rachmiel, A.; Segal, E. Bone Morphogenic Protein 2-Loaded Porous Silicon Carriers for Osteoinductive Implants. *Pharmaceutics* **2019**, *11*, 602. [CrossRef]
71. Huang, K.-H.; Chen, Y.-W.; Wang, C.-Y.; Lin, Y.-H.; Wu, Y.-H.A.; Shie, M.-Y.; Lin, C.-P. Enhanced Capability of Bone Morphogenetic Protein 2–loaded Mesoporous Calcium Silicate Scaffolds to Induce Odontogenic Differentiation of Human Dental Pulp Cells. *J. Endod.* **2018**, *44*, 1677–1685. [CrossRef]
72. James, A.W.; LaChaud, G.; Shen, J.; Asatrian, G.; Nguyen, V.; Zhang, X.; Ting, K.; Soo, C. A Review of the Clinical Side Effects of Bone Morphogenetic Protein-2. *Tissue Eng. Part B Rev.* **2016**, *22*, 284–297. [CrossRef]
73. Ishack, S.; Mediero, A.; Wilder, T.; Ricci, J.L.; Cronstein, B.N. Bone regeneration in critical bone defects using three-dimensionally printed β-tricalcium phosphate/hydroxyapatite scaffolds is enhanced by coating scaffolds with either dipyridamole or BMP-2. *J. Biomed. Mater. Res. B Appl. Biomater.* **2017**, *105*, 366–375. [CrossRef]
74. Min, Z.; Shichang, Z.; Chen, X.; Yufang, Z.; Changqing, Z. 3D-printed dimethyloxallyl glycine delivery scaffolds to improve angiogenesis and osteogenesis. *Biomater. Sci.* **2015**, *3*, 1236–1244. [CrossRef] [PubMed]
75. Akkineni, A.R.; Luo, Y.; Schumacher, M.; Nies, B.; Lode, A.; Gelinsky, M. 3D plotting of growth factor loaded calcium phosphate cement scaffolds. *Acta Biomater.* **2015**, *27*, 264–274. [CrossRef] [PubMed]
76. Wu, J.; Miao, G.; Zheng, Z.; Li, Z.; Ren, W.; Wu, C.; Li, Y.; Huang, Z.; Yang, L.; Guo, L. 3D printing mesoporous bioactive glass/sodium alginate/gelatin sustained release scaffolds for bone repair. *J. Biomater. Appl.* **2018**, *33*, 755–765. [CrossRef] [PubMed]
77. Heras, C.; Sanchez-Salcedo, S.; Lozano, D.; Peña, J.; Esbrit, P.; Vallet-Regi, M.; Salinas, A.J. Osteostatin potentiates the bioactivity of mesoporous glass scaffolds containing Zn^{2+} ions in human mesenchymal stem cells. *Acta Biomater.* **2019**, *89*, 359–371. [CrossRef] [PubMed]
78. Saska, S.; Pires, L.C.; Cominotte, M.A.; Mendes, L.S.; de Oliveira, M.F.; Maia, I.A.; da Silva, J.V.L.; Ribeiro, S.J.L.; Cirelli, J.A. Three-dimensional printing and in vitro evaluation of poly(3-hydroxybutyrate) scaffolds functionalized with osteogenic growth peptide for tissue engineering. *Mater. Sci. Eng. C* **2018**, *89*, 265–273. [CrossRef] [PubMed]
79. Pei, P.; Wei, D.; Zhu, M.; Du, X.; Zhu, Y. The effect of calcium sulfate incorporation on physiochemical and biological properties of 3D-printed mesoporous calcium silicate cement scaffolds. *Microporous Mesoporous Mater.* **2017**, *241*, 11–20. [CrossRef]
80. Fu, S.; Du, X.; Zhu, M.; Tian, Z.; Wei, D.; Zhu, Y. 3D printing of layered mesoporous bioactive glass/sodium alginate-sodium alginate scaffolds with controllable dual-drug release behaviors. *Biomed. Mater.* **2019**, *14*, 065011. [CrossRef]
81. Chihara, T.; Zhang, Y.; Li, X.; Shinohara, A.; Kagami, H. Effect of short-term betamethasone administration on the regeneration process of tissue-engineered bone. *Histol. Histopathol.* **2020**, *35*, 709–717. [CrossRef]
82. Xu, Y.; Hu, Y.; Feng, P.; Yang, W.; Shuai, C. Drug loading/release and bioactivity research of a mesoporous bioactive glass/polymer scaffold. *Ceram. Int.* **2019**, *45*, 18003–18013. [CrossRef]
83. El-Fiqi, A.; Kim, J.-H.; Kim, H.-W. Osteoinductive Fibrous Scaffolds of Biopolymer/Mesoporous Bioactive Glass Nanocarriers with Excellent Bioactivity and Long-Term Delivery of Osteogenic Drug. *ACS Appl. Mater. Interfaces* **2015**, *7*, 1140–1152. [CrossRef]
84. Zhang, J.; Zhao, S.; Zhu, Y.; Huang, Y.; Zhu, M.; Tao, C.; Zhang, C. Three-dimensional printing of strontium-containing mesoporous bioactive glass scaffolds for bone regeneration. *Acta Biomater.* **2014**, *10*, 2269–2281. [CrossRef]
85. Luo, Y.; Wu, C.; Lode, A.; Gelinsky, M. Hierarchical mesoporous bioactive glass/alginate composite scaffolds fabricated by three-dimensional plotting for bone tissue engineering. *Biofabrication* **2012**, *5*, 015005. [CrossRef] [PubMed]
86. Chen, F.; Song, Z.; Gao, L.; Hong, H.; Liu, C. Hierarchically macroporous/mesoporous POC composite scaffolds with IBU-loaded hollow SiO2 microspheres for repairing infected bone defects. *J. Mater. Chem. B* **2016**, *4*, 4198–4205. [CrossRef] [PubMed]
87. Gao, L.; Li, C.; Chen, F.; Liu, C. Fabrication and characterization of toughness-enhanced scaffolds comprising β -TCP/POC using the freeform fabrication system with micro-droplet jetting. *Biomed. Mater.* **2015**, *10*, 035009. [CrossRef]

88. Kluin, O.S.; van der Mei, H.C.; Busscher, H.J.; Neut, D. Biodegradable vs non-biodegradable antibiotic delivery devices in the treatment of osteomyelitis. *Expert Opin. Drug Deliv.* **2013**, *10*, 341–351. [CrossRef]
89. Kamboj, N.; Rodriguez, M.; Rahmani Ahranjani, R.; Prashanth, K.G.; Hussainova, I. Bioceramic scaffolds by additive manufacturing for controlled delivery of the antibiotic vancomycin. *Proc. Est. Acad. Sci.* **2019**, *68*, 185–190. [CrossRef]
90. Zhang, J.; Zhao, S.; Zhu, M.; Zhu, Y.; Zhang, Y.; Liu, Z.; Zhang, C. 3D-printed magnetic Fe3O4/MBG/PCL composite scaffolds with multifunctionality of bone regeneration, local anticancer drug delivery and hyperthermia. *J. Mater. Chem. B* **2014**, *2*, 7583–7595. [CrossRef] [PubMed]
91. Jaeblon, T. Polymethylmethacrylate: Properties and Contemporary Uses in Orthopaedics. *J. Am. Acad. Orthop. Surg.* **2010**, *18*, 297–305. [CrossRef] [PubMed]
92. Darouiche, R.O. Treatment of infections associated with surgical implants. *N. Engl. J. Med.* **2004**, *350*, 1422–1429. [CrossRef]
93. García-Alvarez, R.; Izquierdo-Barba, I.; Vallet-Regí, M. 3D scaffold with effective multidrug sequential release against bacteria biofilm. *Acta Biomater.* **2017**, *49*, 113–126. [CrossRef]
94. Inzana, J.A.; Trombetta, R.P.; Schwarz, E.M.; Kates, S.L.; Awad, H.A. 3D printed bioceramics for dual antibiotic delivery to treat implant-associated bone infection. *Eur. Cell Mater.* **2015**, *30*, 232–247. [CrossRef]
95. Martínez-Vázquez, F.J.; Cabañas, M.V.; Paris, J.L.; Lozano, D.; Vallet-Regí, M. Fabrication of novel Si-doped hydroxyapatite/gelatine scaffolds by rapid prototyping for drug delivery and bone regeneration. *Acta Biomater.* **2015**, *15*, 200–209. [CrossRef] [PubMed]
96. Philippart, A.; Gómez-Cerezo, N.; Arcos, D.; Salinas, A.J.; Boccardi, E.; Vallet-Regi, M.; Boccaccini, A.R. Novel ion-doped mesoporous glasses for bone tissue engineering: Study of their structural characteristics influenced by the presence of phosphorous oxide. *Non-Cryst. Solids* **2017**, *455*, 90–97. [CrossRef]
97. Hoppe, A.; Güldal, N.S.; Boccaccini, A.R. A review of the biological response to ionic dissolution products from bioactive glasses and glass-ceramics. *Biomaterials* **2011**, *32*, 2757–2774. [CrossRef]
98. Kaya, S.; Cresswell, M.; Boccaccini, A.R. Mesoporous silica-based bioactive glasses for antibiotic-free antibacterial applications. *Mater. Sci. Eng. C* **2018**, *83*, 99–107. [CrossRef]
99. Zhang, Y.; Zhai, D.; Xu, M.; Yao, Q.; Zhu, H.; Chang, J.; Wu, C. 3D-printed bioceramic scaffolds with antibacterial and osteogenic activity. *Biofabrication* **2017**, *9*, 025037. [CrossRef]
100. Qi, X.; Wang, H.; Zhang, Y.; Pang, L.; Xiao, W.; Jia, W.; Zhao, S.; Wang, D.; Huang, W.; Wang, Q. Mesoporous bioactive glass-coated 3D printed borosilicate bioactive glass scaffolds for improving repair of bone defects. *Int. J. Biol. Sci.* **2018**, *14*, 471–484. [CrossRef]
101. Zhu, M.; Zhang, J.; Zhao, S.; Zhu, Y. Three-dimensional printing of cerium-incorporated mesoporous calcium-silicate scaffolds for bone repair. *J. Mater. Sci.* **2016**, *51*, 836–844. [CrossRef]
102. Zhao, S.; Zhang, J.; Zhu, M.; Zhang, Y.; Liu, Z.; Tao, C.; Zhu, Y.; Zhang, C. Three-dimensional printed strontium-containing mesoporous bioactive glass scaffolds for repairing rat critical-sized calvarial defects. *Acta Biomater.* **2015**, *12*, 270–280. [CrossRef]
103. Rajasekharan, A.K.; Gyllensten, C.; Blomstrand, E.; Liebi, M.; Andersson, M. Tough Ordered Mesoporous Elastomeric Biomaterials Formed at Ambient Conditions. *ACS Nano* **2020**, *14*, 241–254. [CrossRef]
104. Zhang, Y.; Huang, J.; Huang, L.; Liu, Q.; Shao, H.; Hu, X.; Song, L. Silk Fibroin-Based Scaffolds with Controlled Delivery Order of VEGF and BDNF for Cavernous Nerve Regeneration. *ACS Biomater. Sci. Eng.* **2016**, *2*, 2018–2025. [CrossRef]
105. Siyawamwaya, M.; du Toit, L.C.; Kumar, P.; Choonara, Y.E.; Kondiah, P.P.P.D.; Pillay, V. 3D printed, controlled release, tritherapeutic tablet matrix for advanced anti-HIV-1 drug delivery. *Eur. J. Pharm. Biopharm.* **2019**, *138*, 99–110. [CrossRef]
106. Pei, P.; Tian, Z.; Zhu, Y. 3D printed mesoporous bioactive glass/metal-organic framework scaffolds with antitubercular drug delivery. *Microporous Mesoporous Mater.* **2018**, *272*, 24–30. [CrossRef]
107. Zhu, M.; Li, K.; Zhu, Y.; Zhang, J.; Ye, X. 3D-printed hierarchical scaffold for localized isoniazid/rifampin drug delivery and osteoarticular tuberculosis therapy. *Acta Biomater.* **2015**, *16*, 145–155. [CrossRef] [PubMed]
108. Zhang, Y.; Xia, L.; Zhai, D.; Shi, M.; Luo, Y.; Feng, C.; Fang, B.; Yin, J.; Chang, J.; Wu, C. Mesoporous bioactive glass nanolayer-functionalized 3D-printed scaffolds for accelerating osteogenesis and angiogenesis. *Nanoscale* **2015**, *7*, 19207–19221. [CrossRef]

109. Fang, J.-H.; Liu, C.-H.; Hsu, R.-S.; Chen, Y.-Y.; Chiang, W.-H.; Wang, H.-M.D.; Hu, S.-H. Transdermal Composite Microneedle Composed of Mesoporous Iron Oxide Nanoraspberry and PVA for Androgenetic Alopecia Treatment. *Polymers* **2020**, *12*, 1392. [CrossRef]
110. Do, A.-V.; Worthington, K.S.; Tucker, B.A.; Salem, A.K. Controlled drug delivery from 3D printed two-photon polymerized poly(ethylene glycol) dimethacrylate devices. *Int. J. Pharm.* **2018**, *552*, 217–224. [CrossRef]

MDPI
St. Alban-Anlage 66
4052 Basel
Switzerland
Tel. +41 61 683 77 34
Fax +41 61 302 89 18
www.mdpi.com

Pharmaceutics Editorial Office
E-mail: pharmaceutics@mdpi.com
www.mdpi.com/journal/pharmaceutics

www.ingramcontent.com/pod-product-compliance
Lightning Source LLC
LaVergne TN
LVHW071006170726
843515LV00006B/1090
* 9 7 8 3 0 3 9 4 3 9 3 9 3 *